THE PULITZER
AIR RACES

The Pulitzer Air Races

American Aviation and Speed Supremacy, 1920–1925

Michael Gough

McFarland & Company, Inc., Publishers
Jefferson, North Carolina, and London

LIBRARY OF CONGRESS CATALOGUING-IN-PUBLICATION DATA

Gough, Michael, 1929 –
The Pulitzer Air Races : American aviation
and speed supremacy, 1920 –1925 / Michael Gough.
p.   cm.
Includes bibliographical references and index.

ISBN 978-0-7864-7100-3

1. National Championship Air Races.
2. Pulitzer Trophy.   3. Airplane racing — United States — History.
4. Aeronautics — Awards — United States.   I. Title.
GV759.2.P85G68 2013   797.5'20973 — dc23   2013011162

BRITISH LIBRARY CATALOGUING DATA ARE AVAILABLE

© 2013 Michael Gough. All rights reserved

*No part of this book may be reproduced or transmitted in any form
or by any means, electronic or mechanical, including photocopying
or recording, or by any information storage and retrieval system,
without permission in writing from the publisher.*

On the cover: the Pulitzer Air Race Trophy (courtesy National
Air and Space Museum); background © 2013 Shutterstock

Manufactured in the United States of America

*McFarland & Company, Inc., Publishers
Box 611, Jefferson, North Carolina 28640
www.mcfarlandpub.com*

Thanks to the taxpayers who paid for the Pulitzer Trophy Race Planes, the designers and engineers who developed them, the manufacturers and craftsmen who built them, the pilots who flew them, the many — mostly unsung and unknown soldiers and sailors — who maintained and repaired them, and the citizens and organizations of the cities that hosted the Pulitzer Trophy Air Races.

Table of Contents

Acknowledgments ix
Preface 1
Introduction 3

ONE   The Pulitzer Trophy, "a perpetual prize for annual closed circuit air races" 5

TWO   1920 — With a Bang: Three Dozen Airplanes Fly the First Pulitzer Race, Mitchel Field, Long Island, November 25 18

THREE   1921 — A Close-Run Event and Narrow Victory: A Curtiss Racer Built for the Navy Wins the Second Pulitzer, North Field, Omaha, November 3 39

FOUR   1922 — Army Curtiss Racers Come Out on Top in the Pulitzer and a General Sets a World Speed Record, Selfridge Field, Michigan, October 14 62

FIVE   1923 — The Greatest Show of All, Lambert Field, St. Louis, Missouri, October 6 103

SIX   1924 — Dayton, Disappointment, and Death, Wilbur Wright Field, Dayton, Ohio, October 4 136

SEVEN   1925 — Back to the Origin, Mitchel Field, Long Island, New York, October 12 167

EIGHT   1926 — "Perpetual" Dies with a Whimper 191

NINE   Pulitzer Legacies 192

Appendix: Pulitzer Racer Specifications 209
Chapter Notes 213
Bibliography 229
Index 231

Acknowledgments

Maybe it's not the case that everyone who writes a book considers it impossible to remember all those who made some contribution to it, but it is for me. I begin by thanking the shopkeeper mentioned in the first sentence of the introduction. For more recent help and encouragement, I thank Dr. Jeremy Kinney, curator, National Air and Space Museum, who hosted me when I volunteered at the museum from 2008 through 2010. Elizabeth Borja and Brian Nicklas of the museum's archives division unfailingly and good-naturedly responded to what must have seemed my unending requests, and I thank all the volunteers and staff at the museum who aided and entertained me.

I thank the staff of the Still Photography Collection at the National Archives and Records Administration in Suitland, Maryland, who guided me through their treasure trove of photographs. Brett E. Stolle, archivist of the National Museum of the United States Air Force, Wright-Patterson Air Force Base, Dayton, Ohio, hosted me twice and generously provided access to the museum's collection of air racing papers and photographs. Thanks to curator Joshua Stoff and Julia Blum at the Cradle of Aviation Museum, Garden City, New York, located on the site of Mitchel Field, where the first and last Pulitzers were flown, and John E. Kordes, Garden City village historian. Rick Leisenring, curator at the Glenn H. Curtiss Museum, Hammondsport, New York, provided me with information and photographs of Curtiss and his company's airplanes. Dr. Anne Chase, a friend of long standing, and my daughter Dr. Laura Gough provided me with graphic analyses.

Some of my most pleasant memories and fruitful hours of research and chatting came at museums, libraries, and historical societies in the cities that hosted the Pulitzer Air Races. I apologize for failing to note the names of all who aided me and for those names not appearing here. In Omaha, site of the second Pulitzer, Gary R. Rosenberg, then of the Douglas County Historical Society, was graciously helpful. Deborah Larsen, of the Mount Clemens Public Library, and Louis Nigro told me about Selfridge Field and its role in the third (1922) Pulitzer and the John L. Mitchell Races, and Ms. Larsen and Cynthia S. Donahue, Macomb County historian, entertained me (and perhaps the reader) with the history of Mount Clemens. The St. Louis Mercantile Library and Charles E. Brown and Deborah E. Cribbs of its staff provided me a wealth of information about St. Louis. The Missouri Historical Library, also in St. Louis, was an excellent source of documents, and Dr. Frederick W. Roos of the Boeing Company in St. Louis discussed the history of St. Louis air racing, especially the fourth (1923) Pulitzer. Dayton, Ohio, is home to Wright State University and its excellent library of aviation history, where Dawne Dewey, head of special collections and archives, and the library's director of public history, was especially helpful. Jeff Opt, NCR Archivist at Dayton History, and Squire Brown, author of a book about McCook Field, shared information with me. Nancy R. Horlacher of Dayton Metro Library was especially helpful with the Fred Marshall Collection of Dayton Aviation History.

I am especially grateful to Dr. Judith Campisi. Her interest in "old airplanes" is limited, but her encouragement was important to my writing this book. Drs. Brent Kerger, Dennis Paustenbach, Marian Pavuk, and Paul Scott's willingness to carry my part of another project allowed me the time complete this one.

Lastly, and I expect "at long last" in their minds, I thank my family for their patience through this book's long gestation.

Preface

Air racing always provides excitement, colorful action, and lots of noise but it currently entertains a limited audience. Ninety years ago, it was far more than that. From at least 1909, the date of the first Gordon Bennett Trophy Air Race, and through the mid–1920s, it was the major testing and proving ground for high-speed aircraft. During six of those years, from 1920 through 1925, the Pulitzer Trophy Air Races were the highlight of American high-speed aviation, and competition for the trophy drove the development of American racers until they were the fastest in the world.

Although the Pulitzers and the pilots and airplanes that flew in them are featured in histories of air racing—first and foremost, Thomas G. Foxworth's *The Speed Seekers*, and also in Terry Gwynn-Jones' *The Air Racers: Aviation's Golden Era, 1909–1936*, Reed Kinert's *American Racing Planes and Historic Air Races*, Birch Matthews' *Race with the Wind: How Air Racing Advanced Aviation*, Don Berliner's encyclopedic *Airplane Racing: A History, 1909–2008* and others—this is the first book devoted to the Pulitzers. Because of the exclusivity of its focus, it provides more contemporary references to the races as they were planned and flown, more information about the cities that hosted them, and comments about the racers' designers and pilots.

The Pulitzers played a pivotal role in pulling American aviation from its dependence on European designs, as had been the case during World War I, to preeminence in world aviation. Two Pulitzer racers, mounted on floats and flown as seaplanes, won two prestigious Schneider Trophy Races for seaplanes, causing changes in European airplane design and introducing American aero engines into European markets. From being nowhere in high-speed flight to best in the world in six years: that is the story of the Pulitzers.

For those few years, the Pulitzers thrilled and entertained millions, but public enthusiasm waned at the same time that aviation experts decided races were not necessary for testing and proving designs. In their six years, Pulitzer racers ruled the skies and changed the development of aviation in the United States and elsewhere.

Introduction

This book about the most important events in American aviation between the end of World War I and Lindberg's 1927 flight owes everything to a shopkeeper's kindness in 1945. Airplanes were seldom seen in the sky over Springfield, Missouri, in those days, and every boy and girl peered upward to see, and possibly identify, an airplane heard flying over. Because of the shopkeeper, my fascination fastened on one particular airplane, which I never actually saw and first knew from a picture on a cardboard box containing a solid wood model plane. I was six years old, the oldest of four kids growing up on welfare, with no money for toys. I used to walk fifteen or so blocks from my house to visit a small shop where I had never bought anything, and the shopkeeper must have figured that I never would. He would hand me the model airplane boxes that I wanted to see and let me examine the pictures and schematic diagrams of airplanes on them. Sometimes, he opened a box to show me the model parts and plans.

The wings and fuselage of the airplane that I liked best looked like those of fighter planes I'd seen in magazines and newspapers, but it had no wheels. Instead, it sat on long struts attached to two floats (or "pontoons," he called them, and what a funny-sounding word it seemed) nearly as long as the airplane. It was the first picture of a seaplane that I had seen. Being six and influenced by the war going on, I looked at the pictures for guns. There were none. I probably asked the shop owner, "What kind of airplane is that?" or maybe I could read "racer" by then. I am sure I was more puzzled than anything else by the idea of racing airplanes with huge floats hanging below them.

When I could read, I learned that the British Supermarine Company had made airplanes like it to fly in the Schneider Trophy Races for seaplanes, the most prestigious international air races, in 1925, '27, '29, and '31. Even hampered by the huge, heavy floats, the 1931 Supermarine seaplane racer was the first airplane to exceed 400 mph in level flight. More importantly, the Supermarine Company developed its legendary World War II Spitfire fighters from the racers.

Aviation played no part in my education or professional life. I earned advanced degrees in biology, did biological research and taught in medical schools, managed a policy research program at the Office of Technology Assessment (OTA), United States Congress, worked in think tanks and at a consulting firm in Washington, and testified as an expert witness in trials about health effects of asbestos, Agent Orange, and dioxin. I wrote 30 scientific papers while a researcher and three dozen articles and newspaper op-eds about health risk assessment, occupational and environmental health, and science policy while at OTA and afterwards. My book *Dioxin, Agent Orange* (Plenum Press, 1986) was favorably reviewed in the *Washington Post, Wall Street Journal*, and other newspapers and magazines. I retired from full-time work in 2000 and from consulting in 2005.

I like research and writing, and rather than continue writing about health risks and associated subjects in retirement, I decided to return to the Schneider Races. I did research in the library and archives at the Smithsonian Institution's National Air and Space Museum (NASM) for the article "Doolittle Wins in Baltimore" (*Airpower*, November 2005, pp. 44–57) about the 1925 Schneider Race. In 2008, I became a behind-the-scenes volunteer at the NASM and continued there through 2010.

I soon gave up any idea of writing a book about the Schneiders. Amazon.com and Google Books list more than a dozen books about them, and another one would add little.

I first became aware of two other airplane races, the Gordon Bennett Trophy Races and the Pulitzer Trophy Races, when researching the 1925 Schneider. My ignorance of those races chagrined me at first, but I soon excused myself. The Gordon Bennetts and the Pulitzers reside in obscurity compared to the Schneider Races. There is one book about the Gordon Bennetts, Henry S. Villard's *Blue Ribbon of the Air*, published by Smithsonian Press in 1987. There were none, until this one, about the Pulitzers.

The Pulitzers, overlooked as they are, lifted American high-speed aviation from humiliation in 1920 to world championship in 1922. In September 1920, three United States organizations shipped racers, amid high hopes and hoopla, to France to fly in the Gordon Bennett. None of the three completed a single 100-km lap of the 300-km race. Three years later, the United States owned every land-plane speed record, and the records belonged to airplanes developed for the Pulitzers. In 1923, the victor in that year's Pulitzer was mounted on floats and flown to first place and a world speed record for seaplanes in that year's Schneider Trophy Race. Two years after that, in 1925, the winning racer in the last Pulitzer Race set new closed-course speed ("around-the-pylons") records for land planes. Less than two weeks later, mounted on floats and piloted by "Jimmy" Doolittle, the Pulitzer winner set world closed-course and absolute speed records for seaplanes.

Every writer wants readers to share in the excitement of discovery. My finding the Pulitzers opened the door to airplanes and people and places that made the first half of the 1920s an exciting time in American aviation, and my book places the Pulitzer Races into the context of the times and the places they were flown. I try to capture the spirit of the times in the four cities—Omaha, Detroit, St. Louis, and Dayton, Ohio—and two small towns with nearby large army airports—Garden City, Long Island, and Mt. Clemens, Michigan—that hosted the races.

The racket from unmuffled engines was deafening, the air smelled of exhaust and hot metal, gray smoke swirled and streamed behind small airplanes flying flat-out at less than 300 feet above the ground, and the crowds at times numbered in the hundreds of thousands. The races were among the great moments in aviation. I try to recapture those events and tell the stories of the people who built, flew, and maintained the racers and the individuals and organizations that hosted the races.

NASM displays the seaplane version of the racer that won both the 1925 Pulitzer and Schneider races. The decision to display the seaplane and not the land plane surely reflects the greater importance attached to the Schneider. That is probably the right decision, but I would like to increase the attention paid, and the importance attached, to the Pulitzer Trophy Races. After all, racers built for the Pulitzers won the two United States victories in the Schneiders.

CHAPTER ONE

The Pulitzer Trophy, "a perpetual prize for annual closed circuit air races"

He flies away from us, flies over the plain that extends before him, to the forests in the distance that seem to emerge just now. For a long time, he cruises over the forests, he disappears, we see the forest, not him. Behind buildings, from God knows where, he appears in unchanged altitude, races against us; while he ascends, the lower wings of his machine darkly incline; when he descends, the upper wings shine in the sun.[1]

Franz Kafka's comments about Glenn Curtiss, capturing the mystery and thrill of seeing an airplane fly for the first time when few people had seen an airplane at all, appeared in a Prague newspaper in 1909. Kafka and two friends had traveled 500 miles from Prague to northern Italy to vacation and to see an air meet at Brescia. In those early days of heavier-than-air flight, people made such journeys to air meets because of the long odds against their seeing an airplane in flight anywhere else. Only eleven years later, in 1920, long trips were unnecessary to see airplanes, but opportunities to see the latest airplanes close up and to watch and hear fast, noisy airplane racers attracted thousands. Late that year, spectators at the first Pulitzer Trophy Race would see the first of six air races that propelled the United States from essentially no standing in high-speed flight to first place.

On Thanksgiving Day, November 25, 1920, there seemed barely room for airplanes to fly between the flat ground and the flat gray clouds on an ordinary late November Long Island day. The chilly wind and clouds threatened snow. Forty thousand or so people sitting on a grandstand and standing behind a fence on the edge of the U.S. Army Air Service's Mitchel Field near Garden City snuggled into their coats, tugged at their gloves, hats or caps, and tucked their scarves around their necks and faces to keep away the wintry wind.

A few minutes after 11:00 A.M., the ordinary day became extraordinary. The 40,000 people would witness something seen by no one before and no one since.

On the starter's signal, the first racer, an army pursuit (the army called its fighter planes "pursuits" until after World War II) plane took off, to be followed by three dozen more airplanes at staggered intervals. They raced away, fading into the overcast to the southeast, headed for a tethered balloon that marked a turning point seven miles distant. There, they veered to the northeast, flying 11 miles to the second turning point. The expected balloon had not arrived at the second turning point, and race officials had draped a huge windmill, reportedly visible from Mitchel Field 15 miles away, in white bedsheets to mark the turn. A sharp turn to the west brought the fliers on course to fly back to Mitchel Field. They sped back into the spectators' view, and completing one 33-mile-long triangular lap of the aerial racecourse, turned about 130 degrees to the southeast to begin their second lap.

The airplanes first off flashed past the grandstand before the last airplanes to take off joined the race, and racers roared by the grandstand for the next hour. Two thirds, 24 of the 36 starters, completed the four-lap Pulitzer Trophy Air Race.

More racers flew in the inaugural Pulitzer Trophy Race than in any other closed-course air race (a race flown with turns around markers—balloons, pylons, windmills, churches, or other landmarks—that identified the corners of the courses) before or after. When the winner of the first Pulitzer landed, the United States could claim that one of its pilots flying one of its airplanes had set the world record for closed-course speed. More lastingly, events set in motion at the 1920 Pulitzer brought world-speed records to the United States over the next five years and propelled United States airplane and engine design from second-rate, at best, to foremost in the world.

No one had envisioned such a race three months earlier. Most everyone associated with aviation, and many in the general public, had known that the heirs to the Pulitzer publishing empire had commissioned a sculptor to make a silver trophy to honor their father, Joseph Pulitzer, who had died in 1911. His three sons had planned to begin presenting the trophy and a cash prize annually for the fastest transcontinental flight in 1916 but delayed initiation of the Pulitzer Trophy competition until after World War I.

There was never a transcontinental Pulitzer Trophy Race. Several fliers died in long-distance contests in 1919, making transcontinental races seem very dangerous affairs. A year later, after a nasty reorganization battle in the Aero Club of America, which governed civil aviation and sanctioned aerial competitions, the club changed the Pulitzer Trophy Contest from distance flights to closed-course speed races. Ralph Pulitzer, the Pulitzer son most involved in aviation, endorsed the change and "agreed to the suggestion that the Pulitzer Trophy be made a perpetual prize for annual closed circuit air races."[2]

The story of the Pulitzer Trophy begins with the birth of Joseph Pulitzer Sr., now best known for being the donor for the Pulitzer Prizes for excellence in journalism and the arts. He was one of nine children in a middle-class merchant's death-haunted family in a small town in Hungary. Seven of Joseph's siblings died before he was 18, and his father's death when Joseph was eleven in 1858 tumbled the family from economic comfort into near poverty.

In 1864, 17 years old and unable to find work in his home country, Pulitzer journeyed to Hamburg, Germany, where he accepted passage on a ship to the United States in return for his agreement to serve in the Union Army, which was losing 13,000 soldiers a month in the Civil War. The recruiting officer in Hamburg, eager to earn the commission for signing up enlistees, ignored Pulitzer's nearsightedness, accepted his statement that he was 20, and put him aboard a ship headed for Boston. The often repeated story that Pulitzer traveled across Europe and tried without success to enlist in the Austrian, French, and British armies before going to Hamburg appears to have been manufactured by Pulitzer. For one thing, he had no money for such travels.[3]

At Boston, he jumped ship and swam and waded ashore to evade the recruiter who expected to receive a part of his enlistment bonus; then he took a train to New York City. There he enlisted for one year and received a $200 bonus, keeping it all for himself. Having learned to ride in his childhood, he became a cavalryman, served in an all–German unit from New York, fought in some skirmishes n Maryland, and rode in the victory parade on Pennsylvania Avenue in Washington, DC.

After the war, looking for work, he followed temporary jobs and advice about where work was available and made his way to St. Louis, where he perfected his English and obtained a job on a German-language newspaper. He studied law, was elected to the Missouri State Senate, became a force in the Missouri Democratic Party, and made a comfortable nest egg by investing in the construction of James Eads' bridge, which was the first to span the Mississippi River at St. Louis.

In 1878, at age 36, he bought the *St. Louis Post*, soon merged it with the *Dispatch*, and established the *St. Louis Post-Dispatch*, which he managed, edited, and published, making it a commercial and journalistic success. Five years later, in 1883, he purchased the money-losing *New York World* from Jay Gould to realize his ambition of owning a New York newspaper. Ten years later, Pulitzer had built it into the best-selling New York paper and made himself one of the richest and most influential men in the United States.

In the late 1890s and early 1900s, fiercely competitive New York newspapers sponsored attention-getting events to keep their names in front of the public. Partly for publicity and partly to spur aviation achievement, the *New York World*, then edited by Pulitzer's oldest son, Ralph, posted a $10,000 prize for the first Albany-to-New York City flight. Glenn Curtiss won the prize in 1910, a major event in his rise to prominence in United States aviation. The *World*'s sponsorship was not so successful in bringing publicity and sales to the newspaper. The *New York Times* garnered more by chartering a train to carry Mrs. Curtiss along the route of the flight.[4] Soon after the Albany-to-New York flight, Ralph's *New York World* and the *St. Louis Post-Dispatch*, edited by Pulitzer's second son, Joseph Jr., announced a $30,000 prize for the first flight between New York City and St. Louis in either direction.[5] Henry N. Atwood, who had learned to fly at the Wright brothers' school at Huffman Prairie, Dayton, Ohio, won the prize on a trip that lasted eleven days, from August 14 to August 24, 1911, flying 1265 miles in 28½ hours, making 15 stops.[6]

William Randolph Hearst and his *New York Journal* (later the *Journal-American*), Pulitzer's most successful and bitter competitor, followed Ralph Pulitzer's offer of $10,000 for the first Albany-to-New York flight by offering $50,000 to the first pilot to fly across the United States. After more than 70 stops, including 15 crash landings, pilot Cal Rodgers completed an east-to-west flight in November 1911 in the Wright biplane, the *Vin Fizz*, named for the soft drink made by the company that sponsored the flight. The crashes and other delays caused him to complete the trip in California after Hearst's deadline. Rodgers won no prize.

Joseph Pulitzer Senior's health had seriously deteriorated in 1889 when the formerly vigorous man went blind. He became hypersensitive to noise, isolated himself in special quiet rooms in buildings, homes, and yachts, and spent long periods abroad seeking cures for assorted illnesses—some undoubtedly psychosomatic. Even so, entirely dependent on secretaries to read and transcribe his editorial work and correspondence, he actively managed, edited, and participated in publishing the *World* until his death in 1911. His isolation caused him to neglect his family, leading to their alienation.

Whatever their personal feelings and reservations about their father, five years after his death, the three Pulitzer sons, Ralph, Joseph Jr., and Herbert, announced they would award a trophy and cash prize for the fastest transcontinental flight each year. The noted sculptor Mario Korbel designed the silver Beaux Arts trophy featuring a winged female figure holding aloft laurel wreaths with a stylized monoplane supported atop the figure's wings. On display at the Smithsonian National Air and Space Museum, it stands four feet four inches high.

Before the Pulitzers and the Aero Club of America had decided on the organization of their transcontinental race in 1919, the year after the war's end, the United States Army Air Service (USAS), precursor to the Army Air Force (AAF) and the United States Air Force (USAF), announced its own plans. Instead of competing with the army, the Pulitzers offered a $5,000 prize and a trophy for the greatest distance flown to or from Atlantic City between May 20 and 30, 1919, during the Second Pan-American Aeronautic Convention in that city. Newspapers in other cities offered prizes for the fastest flights to or from their home cities and Atlantic City. Few stipulations were made for the flights; pilots could start at any time between May 20 and 30, make repeated attempts, and land and take off to continue as many times as they wished.

On May 27, Mansell R. James, a popular Royal Flying Corps veteran, flew into Atlantic City in a war surplus Sopwith Camel. The next day, he took off and flew four hours through bad weather and heavy winds to land at a racecourse west of Boston. As it turned out, James had flown far enough to win the prize posted by the Pulitzers and additional prizes offered by the *Boston Globe* and the *New York Herald* worth, in aggregate, about $10,000. He ate lunch and started his return flight to Atlantic City to claim his prize and, perhaps, better his time. With inadequate directions, he mistakenly flew off to the west, landed at a small town in Massachusetts, and stayed overnight. Next day, he took off and disappeared. His body and airplane were never found. Inauspiciously, the first Pulitzer-sponsored contest, a cross-country flight, ended in a pilot's disappearance and death.

The Air Service fooled no one with the ponderous title "Transcontinental Reliability and Endurance Test," bestowed on its transcontinental contest. The "test" was a transcontinental race, pure and simple. The Air Service measured success three different ways, all related to speed: "shortest

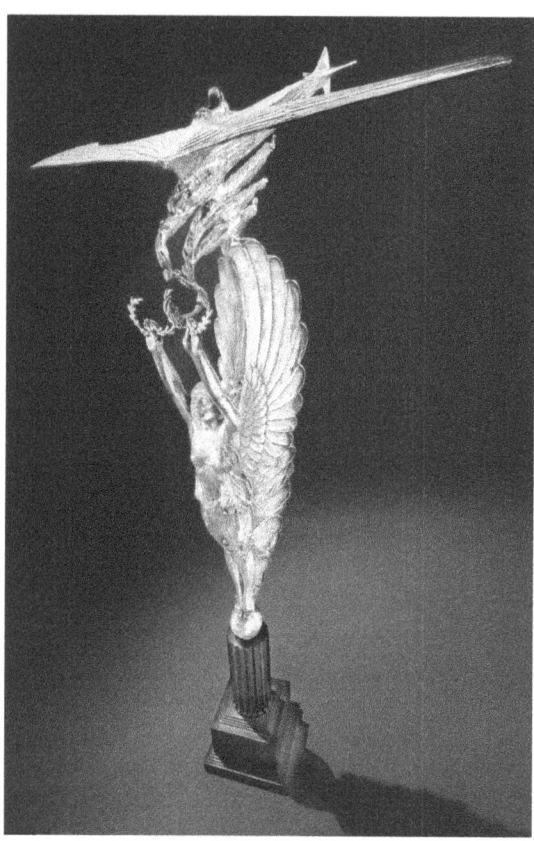

The Czech-American sculptor Mario Josef Korbel created the Pulitzer Trophy featuring a winged female holding a stylized monoplane aloft on the apex of her wings and laurel wreaths, symbolizing victory, in each of her upraised hands. The wings and fuselage are sculpted with feathers and the elongated forms beneath the fuselage have long necks and beak-like structures emphasizing the airplane-bird hybrid nature of the sculpture. The elongated beaks join together and a projection from them reaches down to one of the laurel wreaths. The National Aeronautic Association, which had administered and sanctioned the Pulitzer Races, donated the sterling silver, four-foot, four-inch high statue to the Smithsonian Institution, and it is on display at the National Air and Space Museum (courtesy National Air and Space Museum, negative 2009-30178).

elapsed time, the shortest flying time, and the fastest flying time based upon the handicaps given the various kinds of machines."[7] The *New York Times*[8] called it "the greatest air race ever attempted"; then-assistant secretary of war Benedict Crowell, speaking at the takeoff for the flight, said it was "the greatest aerial contest in the world"[9]; the *Aircraft Year Book* described it simply as "the greatest aerial contest in history."[10]

The army's transcontinental race was part of efforts by General William "Billy" Mitchell and others to establish a United States Air Force independent of the army and navy, along the lines of Britain's Royal Air Force, which had come into being on April 1, 1918. Mitchell and other advocates had managed to have legislation introduced into Congress to establish an independent United States Air Force in 1919, but the legislation met tremendous opposition from President Woodrow Wilson and many senior army and navy officers. Mitchell and others "estimated that it [the transcontinental race] would awaken a new interest in aviation in this country, reveal ... the possibilities ... of this new and powerful arm of national defense, which promises to equal in importance the land and sea forces."[11]

To make the race appear to include a test of the Air Service's capacity to respond quickly in case of a national emergency, General C.T. Menoher, chief of the Air Service, delayed announcing the race until September 20, allowing only eighteen days before the race's start on October 8 for preparing airplanes, landing fields, and other logistics. (Anyone who has served in one of the armed forces would expect that preparations for the race started the minute that a rumor leaked from somewhere that it was planned.) Initial plans called for one-way flights in either direction between Roosevelt Field, Long Island, and the Presidio in San Francisco. On October 6, two days before the start of the race, the announcement was made that the race would be a round-trip in order to reduce the influence of the prevailing westerly wind on speeds. The Air Service laid out a 2701-mile-long course with 20 control points—landing and refueling spots—dotting the course at intervals of about 190 miles. All pilots were required to land at each control point and, in the interests of rest and safety, remain on the ground for at least 30 minutes. Flying was restricted to daylight hours and was not allowed on Sundays.

Some communities selected to host control points worked miracles in preparing excellent facilities; but limited by the time available for preparation, many control points consisted of tents and little else. The army dispatched units of ten men, equipped with a few tools, to each control point, and the soldiers, aided by volunteers, made repairs and carried out maintenance as the planes landed. At some control points, Boy Scouts, surely excited by opportunities to see and touch airplanes and to talk to and shake hands with pilots, carried five-gallon cans of fuel to the aircraft. Local well-wishers visited the control points to see the airplanes and their valiant crews and, in some cases, generate publicity for themselves. For instance, the famous evangelist Billy Sunday, his wife, and the mayor of Rock Island, Illinois, visited Lt. Belvin Maynard, the "preacher-aviator" or "flying parson," who had studied for the ministry before joining the Air Service, when he landed there.

Fifty-eight airplanes were scheduled to start from Roosevelt Field on Long Island headed west and 16 from the Presidio in San Francisco headed east. Not all managed to get airborne, but 46 westbound aircraft and 15 eastbound airplanes took off on October 8, 1919. Twenty-six of the 46 westbound aircraft completed the westbound flight and landed at the Presidio, and seven of the 15 eastbound aircraft landed at Roosevelt Field. Lt. Belvin Maynard, accompanied by his mechanic, Sgt. William Kline, and Maynard's dog Trixie, had the best elapsed time on the first leg, making the westward crossing in 3 days and 7 hours, and the best speed, averaging about 107 mph in 25 hours and 17 minutes of flight.

Top: Rock Island, Illinois, Mayor Schriver (left) greets pilot Lt. Belvin Maynard while "Billy" Sunday and Mrs. Sunday look on. At Rock Island Control Point, 1919 Transcontinental Race (courtesy National Museum of the United States Air Force). *Bottom:* Lineup of airplanes at Roosevelt Field, Long Island, New York, ready to take off on westbound flights in 1919 Transcontinental Race (courtesy National Archives and Records Administration).

Seventeen of the 33 planes that completed one-way crossings of the continent began return flights. Some pilots did not start the return trip because of mechanical difficulties, but at least one simply refused to make the second crossing. On the night of his arrival in New York from the Presidio, Lt. Emil Kiel told the American Flying Club about the terrible snowstorms he'd encountered over the mountains in the West and said he had no wish to make a return flight.[12] "No one can make me race back to California.... The train will be

Crew of De Havilland DH-4 *Hello Frisco*, first-place finisher in 1919 Transcontinental Race: Dog "Trixie," pilot Lt. Belvin Maynard, and mechanic Sgt. William Klein (courtesy Forsyth County [North Carolina] Public Library Photograph Collection/photograph 6550; originally published in the Raleigh *News and Observer*, November 2, 1919).

good enough for me."[13] When his superior officers refused him a train ticket, he delayed his takeoff, citing mechanical problems, until after the deadline for starting the return flight. Kiel's reluctance to fly the return trip did not cripple his career. He remained in the army and was a general in World War II.

Lt. Maynard's wife and two little daughters greeted Maynard, Sgt. Kline, and Trixie when they landed at Roosevelt Field. Maynard and Kline posted the best elapsed time for the round trip, and Maynard was widely heralded and honored for his victory. In December 1919, he flew his by-then weather-beaten DH-4 *Hello Frisco* to Winston-Salem, North Carolina, performed aerobatics that thrilled the crowd, and made the first landing at the new airport, named "Maynard Field" in his honor.

In the same month, he wrote a statement for the Anti-Saloon League that was interpreted by some people "to imply that alcohol had caused some of the deaths in the transcontinental contest."[14] Maynard protested that his words had been juggled, but "the ill-advised statement marred an otherwise brilliant Army career. Maynard left the service the next spring."[15] He continued to fly and would play a surprising role in the first Pulitzer Trophy Race in November 1920. He also remained active in his church and had the off-beat and perhaps dubious honor of officiating at the first wedding to take place in an airplane, flying over Times Square, in 1922. Five months later, he died in an exhibition flight at Rutland, Vermont. The airfield named for him, Maynard Field, was soon too small for Winston-Salem's aviation needs, and it was phased out in 1927:

Today Maynard Field is covered with homes, and only someone with a knowledgeable eye can detect the flattened runway area around Corbin Street and Maynard Drive. A highway marker was dedicated there on May 18, 2008, to commemorate the place where early pilots once took off and landed, advancing the cause of aviation, taking chances with a fledgling industry, and leaving behind only memories of their daring exploits in a city's history book.[16]

In many people's minds, the fatalities and crashes in the Air Service's transcontinental race overshadowed the triumphs of Maynard and other finishers. In all, nine airmen lost their lives, two on flights to the start of the race and seven during the race. Fifty-four accidents damaged or destroyed airplanes.

The waves of negative news reports because of the fatalities did not favor passage of legislation to establish an air force, but, more basically, the United States faced a multitude of postwar economic problems and setting up a new branch of the military was out of the question. The 1920 appropriation for the Air Service imposed great cuts; no independent air force was established; the army sponsored no more transcontinental races.

The death of Mansell R. James in the Pulitzer-sponsored cross-country flight from Atlantic City and the nine army aviators in the transcontinental race underlined the dangers of long-distance cross-country contests. The combination of Pulitzer family misgivings about lending its name to dangerous undertakings that might involve deaths and a change in the leadership of the Aero Club of America led to fundamental revisions in the Pulitzer Trophy contest. They came quickly after August 1920.

The amalgamation of two organizations that competed for leadership of United States aviation — the long-established Aero Club of America (ACA) and the upstart American Flying Club (AFC) — was a critical factor in the transformation of the Pulitzer Races. The

The De Havilland DH-4 flown by United States Air Service 2nd Lieutenant Charles M. Cummings in the 1920 Pulitzer Race. Lt. Belvin Maynard and Sgt. William Klein flew an airplane of this type to first place in the 1919 Transcontinental Race (courtesy National Archives and Records Administration).

Aero Club of America, established in 1905, was the United States' representative to the Fédération Aéronautique Internationale (FAI)[17] (generally referred to as the "International Aviation Federation" in English). The FAI was, and remains, the international nongovernmental, nonprofit organization that sanctions and certifies aerial contests and records.

The Aero Club of American, as the United States' representative to the FAI, had similar functions in United States aviation, sanctioning aerial contests and certifying results and records. The club had authenticated the Wright brothers as the first to fly a powered, controllable airplane, sponsored prizes for public flights, sanctioned air meets in the United States, and entered American aviators in the international Gordon Bennett Races of 1909 through 1912 and in 1920.[18] Until World War I, it was the only pilot-licensing organization in the United States, and it, rather than a government agency, certified civilian pilots until the late 1920s.

From its beginning, the Aero Club's membership included some of the most esteemed and wealthiest people in America, with many members prominent in the "sporting world" of the time — yacht, horse, and automobile racing. Speaking of the Aero Club's early days, a president of the National Aeronautics Association (NAA), the club's successor organization, stated, "We started out as a 'gentlemen's club' for the wealthy balloonist."[19]

The Aero Club's influence and prestige slipped during World War I as the frenzied development of wartime military aircraft made civilian machines seem quaint, old-fashioned and little-changed. Engineers and scientists who had relied on the Aero Club for information and like-minded colleagues found those ideas and people in abundance in aviation companies, the armed services, and universities. Army and navy aviators, trained to military standards, saw no reason to request sanction of their skills by the Aero Club, and the services no longer required it.

During World War I, Henry Woodhouse, prominent in the club and the editor/publisher of *Flying*, *Aerial Age*, and *Air Power* magazines, and Alan Hawley, president of the club, had offered their advice and services to Howard E. Coffin, the chairman of the Aircraft Production Board.[20] In particular, Woodhouse and Hawley proposed that the club take over responsibility for recruitment and initial training of military fliers. Coffin rejected their suggestions. More gallingly, he and the board rejected all of Woodhouse and Hawley's proposals and did not accord them standing as aviation authorities.

Subsequent to those rebuffs, Woodhouse brought charges of mismanagement and scandal against government procurement programs. Woodhouse's intention may have been to highlight waste in aviation programs or it may have been to discredit those who had turned down his suggestions and snubbed him. In either case, investigations did not substantiate his claims. Woodhouse's activities did, however, open a split in the aero club leadership, some of whom wanted the club to have no part of public quarrels. In 1918, disgruntled members tried unsuccessfully to defeat Hawley's bid for reelection as the club's president.

In contrast to the long-standing Aero Club and its aging membership, the American Flying Club, formed at the time of the Armistice (November 11, 1918) was a new organization led by young and energetic military fliers and veterans of the air war. Impressively, more than 500 of the 650 American airmen who had flown over the front lines in World War I had signed the constitution of the club. In August 1919, the club had organized a well-publicized and successful New York-to-Toronto race, which had contests for both military and civilian fliers,[21] and that served as a template for the army's organization and conduct of its 1919 Transcontinental Race.

As fatalities had mounted on the first leg of the army's Transcontinental Race, the

Flying Club registered objections to flying the second leg. The Aero Club of America's silence on the issue made it appear indifferent to the fates of the pilots. At the time, the Aero Club was touting its planned Aerial Derby Around the World, which would be paid for by 10 passengers in a Handley-Page airplane especially prepared for the trip.[22] Depending on one's point of view, the plan for the around-the-world flight was terribly optimistic or simply preposterous. Three years later, when the U.S. Army, with all its material, logistic, and financial resources, made such a flight, only two of the four starting airplanes completed the trip. Enough people considered the Aero Club's proposal preposterous that it further weakened the club's credibility.

As early as April 1919, some officials of the Aero Club recognized the threat posed by the Flying Club to their prominence and importance in aviation and seized an opportunity to besmirch their rival. A group of World War I veterans, calling themselves "The Ancient and Dishonorable Members of the Montmatre and Rue Brey Club," sent invitations to 150 fliers to attend the "greatest souse party of the century," featuring an "orgy" and "wild persons." The Aero Club said the veterans organization represented the leadership of the American Flying Club and wrote letters to call "the attention of the Secretary of War and the Secretary of the Navy to the proposed celebration, which 'disgraces the air service.'" The effort to discredit the Flying Club backfired when it denied any responsibility for the invitations. In the end, the Aero Club looked "rather foolish for concerning itself with a private celebration held by war heroes."[23]

In 1920, the Flying Club had 1000 members, about 200 more than the Aero Club, and it had momentum. Although skepticism is appropriate for both membership numbers—each club made higher estimates for itself and lower ones for its opponent—they provide a measure of the Aero Club's waning.

The Aero Club leaders were right to be concerned about the intentions of the Flying Club. In January 1920, Courtlandt Field Bishop, who had been the Aero Club's first president and was displeased with its current leadership and direction, urged the FAI to recognize the American Flying Club as its official representative in the United States.[24] Aero Club officials pooh-poohed Bishop's actions as "traceable to ill-feeling engendered when he headed a rival officer's ticket which was beaten badly at the election two years ago,"[25] and the FAI continued to recognize the Aero Club as its U.S. representative.

Five months later, in June, allegations about financial misdeeds in its war-related activities descended on the Aero Club—or at least on its management. Following 1915 maneuvers of the National Guard in which not a single airplane was employed, the Aero Club had established the National Aeroplane Fund to seek donations of airplanes to National Guard units and money for purchasing aircraft, training courses, and travel of officers and men for training. The Aero Club reported that it had raised about $390,000 by 1916, when it closed the fund after Congress appropriated $18,000,000 for military aviation.[26]

On June 2, 1920, the *New York Evening Post* asserted that the Aero Club had siphoned off half the money for management expenses.[27] Henry Woodhouse, at the center of the charges about financial misdeeds, filed a $250,000 countersuit against the *Evening Post*. Nothing of legal consequence came from the charge and countercharge; charges were not brought against Woodhouse or other members of the Aero Club, and the suit against the paper was unsuccessful.

Nevertheless, the bad publicity and animosity among Aero Club members proved too much for a majority of the club. By vote of the membership, Woodhouse and other members of the club's board were removed and replaced in June 1920. The new acting officers were agreeable to a merger of the Aero Club and the Flying Club, and the merger was accomplished two months later on August 16, 1920. The new organization retained the "Aero Club of America" name.

General "Billy" Mitchell, emphasizing as he almost always did, the military implications of aviation, later hyperbolically rhapsodized that "the combined clubs bring together all the best elements in aviation that this country possesses. There is nothing which has happened in the development of our whole national defense system which will have a greater effect than this consolidation."[28] At its meeting on August 24, the reorganized Aero Club's contest committee changed the Pulitzer from a cross-country contest to an annual closed-course race. No discussion is recorded in the minutes of the meeting[29] about the reasons for the change, but club leaders had clearly planned for the new races. Ralph Pulitzer, the Pulitzer son with the most interest in aviation, said the change from transcontinental to closed-circuit race had "merit from every standpoint."[30] The Aero Club leadership declared that the club would host the first two races, and that the first would be flown at the Air Service's Mitchel Field on Long Island on Thanksgiving Day, November 25, 1920.[31] No admission would be charged, parking would be free, and the Long Island Railroad would run special trains to bring spectators from New York City.

Ralph Pulitzer would present gold, silver, and bronze plaques to the pilots who finished one, two, and three, and the Aero Club would award the Pulitzer Trophy to the city or state aero club to which the winning pilot belonged. The winning club would hold the trophy for one year and host the 1921 event. The date for the second race was specified as September 4, 1921, but its location could not be spelled out until the 1920 Pulitzer was flown and the winner known. The 1920 Pulitzer would be open to both civilian and military pilots, so long as their airplanes could reach at least 100 mph, but it was clear that military aviation would be dominant. "Although there will be civilian flyers, the contest primarily will be a race between the United States Army and Navy, and the keenness of the professional rivalry that has been engendered ... ought to make it a classic event in American aeronautics."[32] Clearer still, the military would be even more prominent in the years to come: "The requirements of the contestants for next year's [1921] race will be determined by a joint conference board consisting of army, navy and Aero Club officials. The object will be to encourage the construction of sportsman type of air planes that can also be used for training purposes in case of necessity, so as to assure to the government a considerable reserve force of this type of airplane."[33]

The goal of encouraging the designing and flying of sporting/training airplanes was never realistic. Both the army and the navy wanted to win the Pulitzers, and to do that, they contracted for and entered specially built racers in 1921 through 1925. In 1920, the army entered one racer, the navy entered a fighter especially modified for racing, and the army also entered some of its fastest pursuits. There was no chance that sporting/training types, with far less power, would be competitive.

"The Reasons for the Pulitzer Trophy Race," which appeared in the 1920 race program,[34] tied the race to national defense:

1. To stimulate interest in Aeronautics in order that we may by adequate appropriations maintain this most important branch of our national defense.
2. To give the American Public an opportunity of seeing American Aeroplanes developed and produced during the War in competition with some of the airplanes built in England, France, Italy, and even Germany....

3. To compare the Verville Packard Airplane No. 63, ["63" was the race number painted on the wings and fuselage of this airplane in the Pulitzer] developed by the United States Army after the War, with those airplanes that were on the front, but two years ago. This plane has from 50 to 100 miles an hour *more* [emphasis in original] speed than the other contestants, and anyone may see by comparison that all the other planes used during the war are now obsolete.... [As commonly happened during air races, projected racer speeds proved optimistic. The "Verville Packard" proved only 7 mph faster than the next fastest racer, a standard Air Service pursuit.]
4. To stimulate interest in the active and reserve personnel of the Air Service....

Postrace reporting underlined the army-navy competition. The *New York Times*' postrace headline "Army Pilot Wins Pulitzer Air Race" was followed by a subheading, "Naval Officer is Fifth." Under another sub-heading, "Army Wins Pugilistic Bouts," the *Times* reported a fistfight between a sailor and a soldier before the race began. Hardly worth reporting in the *Times*, it would seem, but the newspaper, noting that the soldier was getting the better of his opponent before the fight was stopped, reported that General John J. Pershing was quite pleased and that secretary of the navy Josephus Daniels watched with a "sad face."[35] (It is surely a forlorn hope to think Daniels' expression registered disapproval of the fight rather than disappointment at its apparent outcome.)

At least two conjectures can be advanced to explain the shift to military justifications for the race. Immediately, the military was more likely to have money to invest in new racers. More basically and more frighteningly, World War I had seen aircraft bombing civilian populations, far behind the front, and many citizens worried about air attacks. The Air Service would bear the brunt of defending the country from any such attack in the future and racing should lead to improved aircraft and better protection. There was little reason to worry about aerial attack on the United States in the 1920s, but the fact that no airplane had flown across the Atlantic from east to west was not sufficient to offset fears. And, it was not in the interests of Air Service advocates to downplay fears. Loyalty to the Air Service also played a role in the increased presence of the military. Wartime members of the Air Service, who had established the American Flying Club, constituted much of the leadership and membership of the reorganized Aero Club. Many club members still flew on active duty or in the reserves, and many who had left the Air Service felt allegiance to it.

The Pulitzer Trophy Races fell short of being "perpetual." The six Pulitzer Trophy Races, 1920 through 1925, were flown at five locations: Garden City, Long Island, NY, 1920 and 25; Omaha, NB, 1921; Selfridge Field, Mt. Clemens, MI, 1922; St. Louis, MO, 1923; Dayton, OH, 1924. Each of those locations had the essential ingredients for an air meet — a large landing field, hangar and other facilities, and adequate parking for spectators' cars.

The Gordon Bennetts, the most important air races of 1910s, had ended when airplanes entered by the Aéro Club de France won their third straight contest in 1920 (France had won in 1912 and 1913 before the war forced postponement of the races). The Aero Club of America's decision to make the Pulitzer Trophy Race into a closed-course contest in August 1920, initiated a series of races that was as important to aviation in the 1920s as the Gordon Bennetts had been in the previous decade; and the Pulitzers were far more important to United States aviation. The decision to make the Pulitzers closed-course races resulted in establishing American prestige for high-speed flight. Amid great hoopla, the United States had entered three racers in the last Gordon Bennett at Étampes, France, in September 1920. None finished even one 100-km lap of the 300-km course. The United States teams returned empty-handed and humiliated. Because no United States airplane finished, it is impossible to make accurate speed comparisons with racers that did, but from available measurements, it is apparent that none of the United States airplanes flew as fast as the winner.

What a boost it was, then, when the initial results of the 1920 Pulitzer indicated the winner had flown faster than any airplane in the world, including the winner of the 1920 Gordon Bennett, over a closed course. From then on, except in 1924, the Pulitzer winners posted the best closed-course speeds in the world, and, in 1922, '23, and '25 they churned out the fastest absolute speeds as well.

Chapter Two

1920

With a Bang: Three Dozen Airplanes Fly the First Pulitzer Race, Mitchel Field, Long Island, November 25

> *The 1920 Pulitzer set the stage in American aviation history for a racing craze that swept military and civilian alike. Nothing like it had ever manifested itself before and, except for the early years of space exploration and the first landing on the moon, nothing like it has occurred since.*[1]

Rooster tails of water, soil, and dry, brown grass spurted behind five pairs of wheels rolling across the damp, grassy airfield, until the racers in Flight 1 of the first Pulitzer Air Race lifted into the sky. Forty thousand people watched the racers fly away to the southeast. The crowd could hear its own cheers and whistles and clapping as the distance lengthened between it and the five 300-hp unmuffled engines and before the airplanes in Flight 2 started their engines.

The crowd was pleased and excited, and the organizers of the first Pulitzer Trophy Air Race were satisfied. From a standing start three months earlier—when the Aero Club of America had decided to stage a closed-course race—organizers had mobilized the club's local branch in New York City, the U.S. Army, to a lesser extent, the U.S. Navy, private airplane owners and pilots, and professional and working people around Garden City, New York, secured an airfield and sponsors for the air meet, and were hosting the largest number of racers ever to fly in a single race. No one in the world, except those at Mitchel Field, Long Island, on Thanksgiving Day 1920, would ever see three dozen racing airplanes in the air at the same time.[2]

Thirty years before, as it had been forever, the flat Hempstead Plains on southern Long Island had been treeless, almost shrubless and devoid of human habitation. That had changed from 1889 when Alexander Turney Stewart began its transformation to a landscape crowded with small cities and suburbs. Stewart had immigrated to New York City from Northern Ireland and opened a dry goods store on Broadway in 1823. He and his business prospered. He built and operated America's first department store and owned hotels in New York City and in Saratoga, New York, home to famous and popular horse races. Along the way, he ventured into real estate. He made his biggest real estate purchase—10,000 acres of the Hempstead Plains—with the vision of building his "Garden City." He imported thousands of trees and built a railroad to haul materials and a brickyard to provide building materials, and his city grew quickly.[3] Connected to New York City by the Long Island Railroad and major highways, Garden City was soon prosperous and bustling.

Until 1919, no one owned land in the city. Instead, residents and shopkeepers leased land from the Garden City Company, established by Stewart's wife after his death.[4] The company also leased land to airfield operators, who liked the flat landscape and absence of trees and could tolerate often windy conditions.

The airfields were important in American aviation for decades. In July 1909, Glenn Curtiss had circled Garden City's Washington Avenue Airfield for a total distance of almost 25 miles to win the *Scientific American* Trophy for the first public flight of more than 25 kilometers.[5] Three years later, in 1912, the Aero Club of America and the Hempstead Plains Aviation Corporation opened the Hempstead Plains Aerodrome, a substantial airport with five cement and thirty wooden hangars, four grandstands that seated 1600, and parking areas.[6] The field's name was changed in the years to come. In 1917, it became Hazelhurst Field. A year later, part of the field was renamed Roosevelt Field, in memory of Quentin Roosevelt, youngest son of President Theodore Roosevelt, who had been killed in aerial combat over France. In 1921, the Curtiss Company, which had established an experimental shop and factory in Garden City in 1917, purchased part of the field and named it Curtiss Field. When Curtiss merged with the Wright Company in 1929, Curtiss Field was incorporated into Roosevelt Field. During World War I, in 1917, the U.S. Army Air Service had opened Hazelhurst Aviation Field #2 to the north and east of Hazelhurst Field and soon renamed it Mitchel Field in honor of former New York City mayor John Purroy Mitchel, who had died in flight training. Mitchel Field continued in Air Force Service until 1960. The land formerly occupied by the Air Force Base now hosts Hofstra University, Nassau County Community College, the Cradle of Aviation Museum, other museums and educational facilities, the Nassau Coliseum, and a large shopping center.

Large, flat, treeless, well-equipped and already familiar to the many pilots who had flown over Long Island, Mitchel Field was an inviting venue for 1920 Pulitzer pilots and ground crews. For the Aero Club, it came with the great advantages that it was an operating Air Service airfield, and the Air Service provided police, fire, and emergency services. It had a grandstand, plenty of parking, and could be easily reached from the Long Island Railroad station. Its availability solved the Aero Club's problem of where to fly the Pulitzer.

The 1920 Pulitzer racecourse was laid out in a triangle flown in a counterclockwise direction. Taking off from Mitchel Field, the racers would fly southeast seven miles to Lufberry Airfield (near Wantagh), make a turn northeast and fly 11 miles to Col. Henry J. Damm Airfield (near Babylon), turn almost due west and fly 15 miles back to Mitchel for a lap length of 33 miles. The Pulitzer required four circuits of the 33-mile laps, making the total 132 miles.

In preparation for the Pulitzer, the Aero Club of America Contest Committee announced that it and army and navy representatives would place restrictions and requirements on entrants to encourage development of sporting airplanes that could be adapted as training planes in time of national emergencies. There is no evidence that the restrictions and requirements, if there were any, or any other stipulation had any effect. Instead, purposefully designed Pulitzer racers were high-speed, high-powered designs more akin to fighters, too specialized and too expensive for civilian purchase or use as trainers. In 1920, given the short lead time between announcement of the race and its being flown, almost all the racers were "stock" military aircraft with few changes made for the race.

Even with few airplanes designed specifically for the race, the 1920 Pulitzer provided the template for the races to come: Military airplanes flown by military officers dominated the race, and all but one of the few civilian-owned airplanes that flew in the first two Pulitzers would be ex-military machines. Militarization was complete in 1922. From then on, all the racers were army or navy aircraft, all flown by military pilots. Few individuals, if any at all, suspected that military dominance of the Pulitzers would lead to their demise. The races would end when neither the army nor the navy would foot the bill for racers in 1926.

The Aero Club recognized that international competition had contributed in an important way to the popularity, prestige, and success of the Gordon Bennett Races, and it and other Pulitzer Race organizers invited entries from foreign aero clubs. In mid–October, the *Brooklyn Daily Eagle* interpreted a cable sent by the Aéro Club de France to inquire about the race as "intimating that French entries will join the contestants which are expected from other countries, including Canada and England."[7] The French interest was especially welcome because French airplanes held most world aviation records and the presence of fast French planes would heighten interest in the Pulitzer. Disappointingly, no foreign aero club entered an airplane in the 1920 Pulitzer. More disappointing still, no racer sponsored by a foreign aero club would fly in any of the six Pulitzers.

Likely motivated by the lack of foreign competitors in 1920, Pulitzer organizers emphasized that airplanes manufactured in five countries—France, Germany, Great Britain, Italy and the United States—would fly in the Pulitzer. They drew a comparison with the last Gordon Bennett: "In the recent James Gordon Bennett race in France only eight machines, representing three countries, were entered, while in the Pulitzer race there will be forty-four contesting planes, representing the aircraft productions of five countries."[8]

Even without entries from foreign aero clubs, there would be no shortage of speedy aircraft. The army and navy each entered a racer that had been especially modified from a fighter. In addition, the navy entered two "fighters" that had never been selected for production, but which were very fast. Even if both those failed to start or finish, the *New York Times* wrote, United States aircraft that had been developed near the end of the World War or in the postwar period made "a new speed record likely."[9]

Aero Club officials divided the 44 aircraft entered in the Pulitzer into seven classes, or "Flights," based on expected performance. Members of each Flight would take off together, with an interval of a minute or two between Flights:

- Flight 1: "Contest Committee Invitation Class": Two Thomas Morse MB-3s (army entries), two Ordnance Ds (army), one Fokker D-VII (army), one Loening M-81 S Special (navy entry, Marine pilot), and one Sopwith Dolphin (army). All were army pursuits except the Fokker, a former German fighter, the Loening, which was a modified navy fighter, and the Sopwith, which was a British fighter.
- Flight 2: Nine American-built De Haviland DH-4s. All were army observation planes/light bombers.
- Flight 3: Eight American-built De Haviland DH-4s. All were army.
- Flight 4: Six Vought V.E. 7s. All were navy trainers.
- Flight 5: Five American-built Royal Aircraft Factory S.E. 5s. All were army, originally flown as fighters in World War I, but used as trainers in 1920.
- Flight 6: Two Loening M-8s (navy fighters), three Italian Ansaldo fighters (privately owned), one SPAD S.13 (French fighter), and one Morane-Saulnier AR (French trainer, entered by the company).

- Flight 7: One Verville VCP-R (army racer), two Curtiss T-18 Triplanes (navy racers). Designed to be fighters, these airplanes were never selected for production, and the navy flew them in various races and other contests during 1920, 1921 and 1922). One American-built DH-4 hospital plane (army).

The Pulitzer Trophy would be awarded to the airplane with the fastest time regardless of the Flight it flew in. All entrants were eligible for it, but the Pulitzer winner was expected to come from Flight 7 and less likely from Flight 1 or 6. Airplanes in Flights 2, 3, 4, and 5, standard army observation/bombing and navy training aircraft, had no chance to win the Pulitzer.

The most casual inspection of the entries reveals that the Pulitzer sons' goal of having sportsmen flying sporting airplanes compete for the Pulitzer Trophy was not in the cards. The army and navy had the fastest planes—fighters and modified fighters—and military officers piloted those airplanes. Three of the three dozen racers—the two Ansaldos and the Morane-Saulnier AR of Flight 6—were owned by civilians. All three were former military aircraft. The Pulitzer was a military affair from its first day.

The army's VCP-R (Flight 7) was the clear favorite to win the race. Powered by a 578 "nominal horsepower" Packard engine, it had one-and-a-half to more than twice as much power as any other airplane at the Pulitzer. Nominal horsepower is a measure of the output of the engine that had actually been achieved. It is less than maximal horsepower, which can represent a single measurement made over a very short time under optimal conditions, or an inflated estimate guided by optimism or hoped-for sales.[10]

The VCP-R had originated as a design for a U.S. Army pursuit in 1917. Because no United States-designed pursuit had been available when the country entered World War I (in 1917), United States Air Service pilots had flown British and French fighters in France. The United States Army realized that it needed fighters of its own design, but it was in a position of having to play "catch up." Alfred "Fred" Verville, an engineer at the Army Air Services' engineering center at McCook Field, Dayton, Ohio, was part of a group of army experts who traveled to France to learn about fighter design and production methods. Upon his return, he designed the Verville Chasse Plane (VCP) (*chasse* is the French word for "chase," which translates to English "pursuit").[11] One VCP airframe was tested to destruction in 1919, and the only flying model, powered by a (Wright-built) Hispano-Suiza water-cooled engine of about 300 horsepower, flew for the first time on June 20, 1920.

Verville had spent time at the SPAD design shops and factories, and the VCP reflected SPAD practices. It was a biplane with *monocoque* fuselage (literally "eggshell," or hollow, fuselage) made of two halves formed of plywood steamed and molded on a form and joined together with no internal support except a few wooden formers Unlike the SPADs, which had three-ply fuselages, the VCP had four. It was considerably larger and far heavier than the SPADs. Tail-heavy, it became heavier still when it was necessary to add ballast to the forward fuselage to establish a favorable center of balance. Because of its weight and size, it was seriously underpowered when initially flown with a 300-hp engine comparable to those that powered the lighter SPAD fighters. As originally flown, it was equipped with an annular radiator that added to it bulky appearance.

World War I ended before the VCP was ready for army service. Had it suffered the fate of most planes developed in the last days of a war, it would have made a few test flights, revealed performance not much better than already-purchased fighters, been tossed into the wastebasket of airplanes that did not reach production, and subsequently been limited

Verville VCP, fitted with Wright-built engine and annular radiator (courtesy National Archives and Records Administration).

to appearances in picture books of little-known airplanes. The VCP's fate was different. It did not go into production, but it had a well-publicized career as a racer.

In one of his many endeavors to focus attention on the Air Service, General William "Billy" Mitchell rescued the VCP from obscurity. He initiated its conversion into a racer to compete in the 1920 Gordon Bennett Race, held in France. At his urging, Verville redesigned the VCP's fuselage to accommodate a newly developed 500-hp Packard engine with a "chin" radiator. Mitchell changed the name of the airplane to "VCP-R," for "racer." The chief test pilot of the army, Major Rudolf Schroeder, called "Shorty" because of his over 6-foot height, was designated as its pilot.

The racer incorporated innovative, advanced features such as tapered wings and a smooth fairing between the lower wing and fuselage. Verville systematically reduced drag-inducing protuberances; all control gears, wires, and fittings were enclosed in the wings; single interplane I-struts connected the two wings; the carbane struts (between the fuselage and lower surface of the upper wing) and landing gear chassis were cleanly designed; the few bracing wires on wings and landing gear were streamlined; and the tail was fully cantilevered with no supporting or bracing wires or struts.

At a time when a single man could lift the tail of most single-engine airplanes and move the airplane from place to place, the VCP-R's weight was a novelty — and not a pleasant one. When on the ground, a leather belt fitted with handholds was strapped around the rear of the fuselage, allowing two or more sturdy men to shift the airplane.

The VCP-R made its first flight on July 15, 1920, two months and two weeks before the Gordon Bennett, scheduled for September 28. It was a short flight. The wheels had barely lifted off the ground when the new and little-tested Packard engine quit. Schroeder maintained control and landed successfully, limiting the damage to a bent axel and a blown tire. The VCP-R was repaired and fitted with broad-span wings for additional lift and more responsive flight, and Schroeder made a number of successful flights before the racer was fitted with smaller-span wings, designed for racing. On August 2, he took off in an attempt to measure the VCP-R's speed.

VCP-R, racer number 63, which won the 1920 Pulitzer Trophy Race (courtesy National Museum of the United States Air Force, Dayton, Ohio).

He sped over and around the field, and a car filled with speed measuring devices chased around the field after the airplane.[12] The car's occupants estimated the VCP-R had reached 190 mph, about 20 mph faster than the world speed record of the time. As the VCP-R landed, it and the car, the only moving objects on the field, somehow collided. The lower left wing crumpled, the landing gear sheared off, and other damage was done. The hurriedly repaired racer could not be tested before it and its new and equally untested 578-hp Packard engine were crated and shipped to France. The new engine was installed in the VCP-R at Étampes, site of the Gordon Bennett.

The new engine proved balky and temperamental. Several soldiers took turns swinging its nine-foot-long propeller again and again. After many swings, the engine started with a bang, belching flame and smoke along the fuselage. Perplexed engineers and mechanics suggested that carburetor and air intake problems and gasoline unsuited for high-compression engines caused the engine's rough running, flaming and smoking. Attempted fix followed attempted fix. None was successful. The possibility that the gasoline was to blame undercut Packard's boast that the engine's "fuelizer ... insures perfect combustion even with very poor grades of gasoline."[13] Major Schroeder wanted a larger radiator installed to cool the big Packard engine, but no change was made. More remarkable, although the balky engine was run-up several times while the VCP-R sat on the ground, no test flight was made.

On Gordon Bennett Race Day, Schroeder sat in the cockpit for 15 minutes[14] as 10 men took turns pulling the propeller. When the plane started, flames and smoke shot from the carburetor air intakes on the bottom of the cowling and scorched along the fuselage under the watchful eyes of soldiers standing ready with fire extinguishers. Schroeder, worried about the engine overheating, hurried through the takeoff procedure. The VCP-R, with about twice the horsepower of the British and French racers, roared away.

The cooling water temperature shot up to almost boiling before Schroeder completed the first 50-km-long outbound leg of the course. He climbed from 300 feet, where he preferred to fly, to 1,000 feet seeking cooler air. It made no difference. The engine misfired and the RPMs dropped from 1950 to 1400. Steam streaming from the radiator, Schroeder returned to the starting line and landed, out of the race.

Two months later, a much improved VCP-R was ready for the first Pulitzer Race. Sensitivity to charges that racing was an unnecessary frill and that tax money should not support it may have accounted for an Air Service pre–Pulitzer press release that stated, "The Chief of the Air Service insists that every machine [entered in the Pulitzer Race] must be of standard army specification."[15] How he could include the one-off VCP-R in any list of airplanes "of standard army specification" is a mystery. A contemporary magazine went into detail: "Parts of planes eligible for the race had to be standard parts and could be replaced by a similar part taken from a similar machine without affecting the control or operation of the machine."[16] No "similar machine" to the VCP-R was to be found.

Too late for Schroeder in the Gordon Bennett but before the Pulitzer, army engineers installed a larger radiator on the VCP-R. With it, the Packard engine "ran very cool at 55° C [about 130° F]."[17] In addition, adjustments were made to the carburetors and other engine components, and gasoline better suited to the high-compression Packard engine was available in the United States. The Packard engine started easier, did not flame and smoke, and, thanks to the appropriately sized radiator, did not overheat.

The VCP-R would have a new pilot. Angry because the army had reduced him — the army's chief test pilot — from major to captain in its postwar cutbacks, Schroeder had resigned from the army while still in France after the Gordon Bennett. Schroeder had not been singled out; many officers were reduced in grade after the war's end, but it was too much for him. First Lieutenant Corliss C. Moseley replaced Schroeder in the Pulitzer.

The other speed planes in Flight 7, the Curtiss 18-T triplanes, had originally been built as two-seat fighters for the navy, and one, loaded with full operational equipment, both crew members and two machine guns and ammunition, had, unofficially, reached 162 mph, good enough for the unoffical world speed record in August 1918.

The 18-T and the 18-B (a biplane stablemate for the army) had been designed to take advantage of a new V-12 engine. Charles Kirkham, in charge of engine design and development at the Curtiss Company, had designed the Curtiss K-12 — a small, light V-12 water-cooled engine with a narrow crosssection — in 1916. The K-12 generated a nominal 396 horsepower, about 100 horsepower more than the Hispano-Suiza V-8s that had powered many World War I fighters and were being license-built in the United States by the Wright Company. Kirkham's design would have a bright future. It was the progenitor of V-12s that set world records in the 1920s and led to the Allison and Rolls-Royce V-12s of World War II.

Unsatisfied that any existing airframe could use the K-12's power to best advantage, Kirkham designed the Curtiss (or Curtiss-Kirkham) 18-T triplane and 18-B biplanes.[18] Both the army and the navy tested the airplanes, but no production order was made. In 1920, the navy modified its two 18-Ts for the Pulitzer by fairing over the rear cockpit, removing equipment unnecessary for racing, and reducing protuberances and smoothing surfaces. Many observers considered them to have the best chance against the VCP-R.

The mysterious, last minute inclusion of the De Havilland DH-4 flying ambulance in Flight 7 provided a "somewhat amusing incident.... Its pure white paint and red insignia caused a great deal of comment from the spectators."[19] No one expected it to be competitive. It wasn't. It completed only two laps of the four-lap race before dropping out for unknown reasons. Reasons for its entry in Flight 7 are equally unknown.

Although the VCP-R and 18-Ts were favored to win the Pulitzer, the pursuits in Flight

Pilot, Lt. Cdr. W.B. Havilland, USN, stands beside the Curtiss 18-T (racer number 61) that he flew in the 1920 Pulitzer Race. Havilland was forced down by fuel or carburetor problems after the first lap (courtesy National Archives and Records Administration).

1 and, especially, a modified navy fighter in that Flight were expected to be competitive and, given favorable circumstances, emerge on top. The Air Service had drawn a sharp distinction between it and the navy's preparation for the Pulitzer. The army press release proclaiming that the VCP-R was a standard service airplane asserted, "It is rumored that the Navy Department has constructed a specially built monoplane with clipped wings capable of 165 miles per hour."[20] That airplane was surely not much of a mystery to the army.

In 1918, just as World War I ended, the Air Service placed a contract for 5,000 slab-sided, two-seat, shoulder-wing monoplane fighters designed by Grover Loening. The army designated it "M-8," because it was the eighth monoplane design purchased by the army. In the unusual-looking airplane, the pilot's cockpit was in a cutout of the shoulder-mounted wing, and the pilot looked forward over the wing and through the vee of the V-8 engine. The gunner/observer sat behind in a separate cockpit armed with two flexibly mounted machine guns. The M-8's radiator was "chin-mounted" beneath the engine, in a location that would become common in the years to come. To reduce complexity and production difficulties, Grover Loening had simplified the M-8 and claimed it had only 10 percent of the number of parts found in a comparable European fighter. The design and its construction won acclaim from an RAF official, who called it one of the "most advanced and original designs"[21] of the war.

The Loening Company's future, apparently assured by the army's order of 5,000 M-8s, crumpled when war's end caused cancellation of the order. The army orders ended with two prototypes, but the navy ordered one for evaluation, and subsequently ordered 36 land planes and six seaplanes, all designated "M-80." Two of the M-80 land planes were scheduled to fly in the Pulitzer's Flight 6.

For the 1920 Pulitzer, the navy modified one M-80—clipped its wings, rounded its wingtips, and converted it to a single-seater. The modified M-80, designated the "Loening M-81 S" or "Loening Special"—was the special navy racer mentioned in the army press release. Marine Corps lieutenant Benjamin Bradley was selected to fly the M-81 S, which

Models of Loening M 8 two-place fighter and Loening M 81 S Special racer (from http://modelingmadness.com/reviews/preww2/hammm8.htm, and courtesy of Joel Hamm, who built the models).

started in Flight 1. Entered by the army, two World War I fighters, a German Fokker D-VII and a British Sopwith Dolphin, were scheduled to fly in Flight 1. Neither did. The Fokker turned over in snow at Buffalo on its way to the race, and the Dolphin did not report to the race site. Two of the four Air Service pursuits in Flight 1, the Thomas Morse MB-3s, were given a good chance to win if the real speedsters in Flights 6 and 7 — the VCP-R, the 18-Ts, and, maybe, the M-81 S — had mechanical problems that slowed them or forced them out of the race. The other pursuits, the Orenco Ds, an older design, were not expected to do well.

The Air Service had adopted the Thomas Morse MB-3 as its standard pursuit in 1919. It was a conventional looking biplane except for having two-bay wings (two sets of struts between the upper and lower wings on each side), an unusual arrangement in a small airplane with upper wingspan of 26 feet and lower of 24 ft. 6 in. (see Appendix for racers' specifications). Thomas Morse delivered a preproduction batch of four fighters in 1919, and the first one reached 164 mph, an unofficial record for service airplanes.

The Air Service choice of the MB-3 as its standard fighter did not lead the Thomas Morse Company to commercial success. In the 1910s and 1920s, the government paid companies for designing and building prototype airplanes and took possession of the plans and specifications. With a design in hand, the army or navy invited potential manufacturers, including the designing firm, to bid on producing airplanes. The Boeing Company underbid Thomas Morse for production of the MB-3 and received a contract for 200 airplanes. Thomas Morse built 65 MB-3s, but the loss of the larger contract marked the beginning of a downward spiral in Thomas Morse's manufacturing fortunes. Conversely, it was an early step in Boeing's becoming a major manufacturer.

Top: Loening M-81 S Special Racer (racer number 46) and its pilot, Lieutenant Benjamin Bradley, USMC. Bradley was flying fast enough to finish third in the 1920 Pulitzer, but a water connection to the engine failed on the fourth lap, and he was forced to land one mile short of the finish (courtesy National Archives and Records Administration). *Bottom:* Thomas Morse MB-3. Major Harold Harney, USAS, flew a plane of this type to second place in the 1920 Pulitzer. Note the side-mounted radiator, the rectangular object on the side of the fuselage (courtesy National Archives and Records Administration).

Two Curtiss-built Orenco D biplane pursuits designed by the Orenco Company rounded out Flight 1. The designing company's original name, Ordnance Engineering Company, confusingly suggested the company was a government entity related to the army's Ordnance Corps, and the name was changed to "Orenco." The Air Service had adopted the Orenco D as a standard pursuit at the end of 1918, but the MB-3's superior performance soon displaced it.

Despite their successful designs, neither Thomas Morse nor Orenco became major airplane manufacturers. Although the Air Service selected their pursuit designs for production, neither company was selected for a production contract. Instead, Boeing built the Thomas Morses, and Curtiss built the Orencos. Emerging from a Depression-caused 1929 merger with the Wright Company, the combined Curtiss-Wright Company built airplanes until the end of World War II, but sometime in the late 1930s and early forties, Curtiss, which had prospered in the design and manufacture of pursuit planes for the army, lost its touch. It ceased building aircraft soon after World War II. The successful designers of post–World War I pursuits, Thomas Morse and Orenco, fared poorly. Losing out on production contracts, they soon went out of business as independent companies.

Flight 6 in 1920 was a mixed bunch of military and privately owned airplanes. Two navy M-80s (derived from the army Loening M-8), a French-built SPAD S.13 to be flown by an Air Service pilot, and four privately owned airplanes were grouped together. The M-80s and SPAD did not figure in the results. One M-80 suffered damage to its rear fuselage in an embarrassing prerace collision with an Orenco D and did not take off. The other M-80 overheated and returned to the starting line before completing its second lap. The SPAD 13 was withdrawn before the race began. The four privately owned airplanes in the race flew in Flight 6. Three Ansaldo aircraft, former Italian Air Force fighters, were expected to do well. With its low-powered 120-hp engine, the French Morane-Saulnier AR, a trainer, had no chance.

Umberto Savoia, a colonel in the technical division of the Italian Air Force[22] who was to go on to design successful seaplane racers, R. Verduzio, an engineering professor, and Celestino Resatelli, an engineer, designed the Ansaldo SVA-5 fighter. The airplane was initially designated "S.V." for Savoia and Verduzio, and the name was changed to "SVA" when the Societa Geo Ansaldo, an engineering and shipbuilding firm, was selected to manufacture the airplanes. No initial "R" for Rosatelli, the third designer, was ever part of the designation, perhaps because of his former status as a student of Verduzio.

Both the one-seat SVA-5 and the SVA-9 two-seat fighters were large, graceful airplanes with prodigious range and flying times. Italian fighter pilots liked the single-seat SVA-5 but wanted a smaller airplane. In response, Ansaldo produced the A.1 Balilla ("Hunter"), which retained the smooth lines of the larger airplanes and was powered by the same Italian SPA (Societa Piemontese Automobili) in-line six-cylinder engine of 225 horsepower that powered the SVAs.

Two companies, the Aero Import Company and the Eagle Flying Corporation, which imported Ansaldos for lease and sale, entered Ansaldos in expectation that good performances in the Pulitzer would boost their sales. Both companies entered a SVA-5, and Aero Import also entered a Balilla. All did well in the Pulitzer. The French Morane-Saulier Company's AR trainer finished where it was expected to: dead last. To its credit, it finished, while one-third of the starters did not.

The other four Flights, numbers 2, 3, 4, and 5, were composed of standard army or navy machines with speeds well below those of the airplanes in Flights 1, 6, and 7. To sharpen competition among the contestants in the slow Flights, the winner of each would receive a trophy, and the first- and second-place finishers would receive cash prizes. The pilot and mechanics of winning airplanes would be awarded Liberty (war) bonds worth $350 and $50 at maturity, and the runners-up would receive $150 and $50 bonds.

A "Note" at the bottom of the prize money schedule may have dismayed the forty army and navy pilots and the hundred or more service mechanics: "Army and Navy Officers

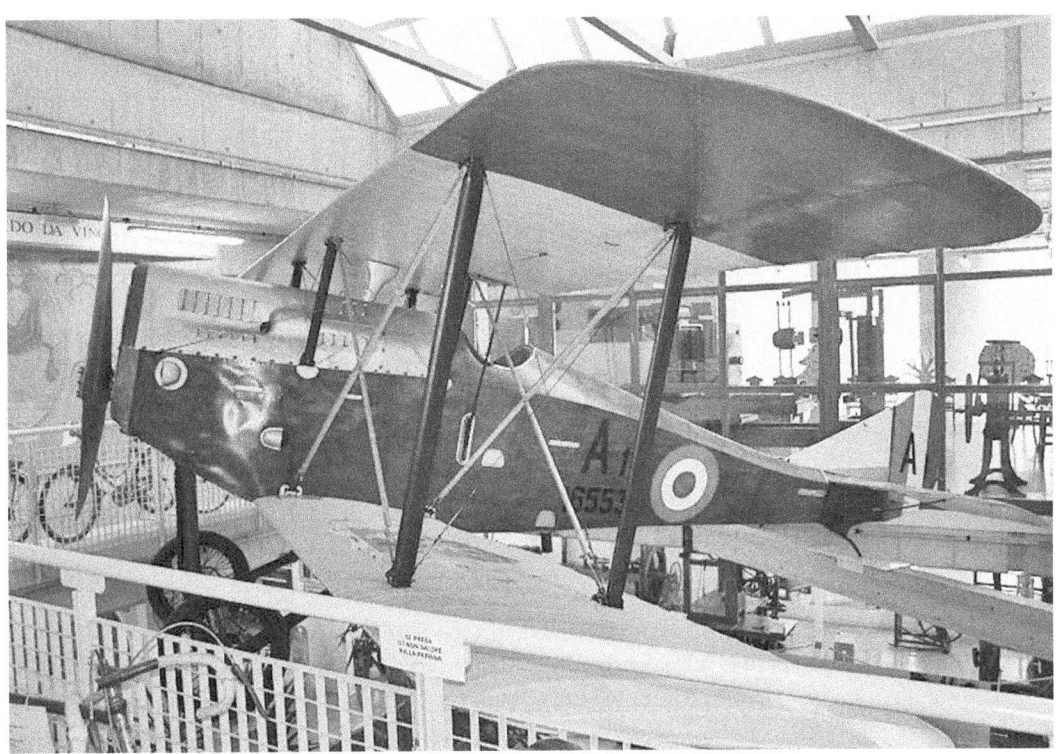

A restored Ansaldo Balilla located at Bergamo Museum of History, Bergamo, Italy. Bert Acosta, a civilian pilot, placed third in the 1920 Gordon Bennett in an airplane of this type (courtesy "Giorces, author, with the kind concession of the Historical Museum of Bergamo," http://en.wikipedia.org/wiki/File:Aereo locatelli.JPG).

competing in the Air Race will not receive cash prizes." The money for prizes earned by service pilots was donated to army and navy relief funds. The author has been unable to determine if the mechanics, who were enlisted men, could accept the war bonds given as prizes. Also unknown is how the prizes for mechanics would have been divided among the members of the ground crews that serviced each racer.

De Havilland DH-4s, the army's standard day bomber/observation airplanes, flew in Flights 2 and 3. British designs, the DH-4s were extensively used during World War I, and hundreds were manufactured under license in the United States, where they remained in service through the late 1920s, until the last one was retired in 1932. Large biplanes built to carry two crewmen and bombs, they were given no chance to equal the speeds of single-seat fighters even if their U.S.–designed and –built Liberty engines, rated at 400 hp, were more powerful than the 300-hp engines common to the single-seat fighters in the contest.

The Vought VE 7s (Flight 4), designed and manufactured in the United States, were advanced trainers. The British-designed Sopwith SE 5s (Flight 5), which had been successful fighters early in World War I, were, by 1920, relegated to use as trainers. Wright engines of about 200 hp were used in both the VE 7s and the SE 5s, making the trainers noncompetitive with the more powerful racers.

A big crowd was expected for the Thursday, November 25, Thanksgiving Day race. Most people had a holiday from work, the Long Island Railroad sped spectators from New York City to the race site in less than an hour, admission and parking were free, and the weather, chilly and blustery, was no worse than many early winter days when large crowds attended football games. It proved important that New York newspapers and national wire services had received written invitations to attend and press passes for reporters and photographers. The newspapers and wire services supplied copy to newspapers everywhere, guaranteeing that the whole country would learn about the race. Ralph Pulitzer's presence at the race to award gold, silver, and bronze plaques to pilots finishing with the first, second, and third best times in the Pulitzer brought additional attention to the race.

Between 25,000 and 40,000 spectators jammed Mitchel Field. The lucky ones sat in the grandstand specially built for the race; others stood. Thousands more saw parts of the race from their cars or other vantage points along the route. Spectators wearing long, heavy coats against the cold and blustery wind may have complained to each other about the weather, but "the day was an ideal one for such an event. With the sun for most of the time concealed behind snow clouds, it was possible to watch the contest without suffering eye strain."[23]

Army captain Harold E. Hartney in an MB-3 made the first takeoff in the Pulitzer Race. Never overtaken, he was the first to land, 47 minutes later. Hartney had had a distinguished career as pilot and leader during World War I. After serving in the trenches with the Canadian Army, he transferred to the Royal Flying Corps in 1915 and flew combat in Italy and France. Transferring to the United States Air Service in 1917, he was commissioned a major and made a squadron commander. A year later, he was a lieutenant colonel and commanded the 1st Pursuit Group. He shot down six German airplanes and was decorated by the British, French, Italian, and United States governments. Postwar, he flew a Fokker D-VII to make a solo two-way crossing of the United States during the 1919 Transcontinental Race. In 1920, he was chief of the Air Service Office of Civil Affairs, which was, in major ways, the precursor to the National Aeronautic Association (NAA).

Hartney had arrived at Mitchel Field early in Thanksgiving week to find his MB-3, the airplane that had set the unofficial speed of 164 mph for service airplanes, in need of maintenance and repair. Working long hours, he, the mechanics, and other ground crew had the MB-3 ready on Wednesday, the day before the race, and he took it up.

The first flight went well until Hartney opened the throttle. When the RPMs of the MB-3's Wright H (a modification of the Hispano-Suiza engine manufactured under license) climbed to over 1500, the plane's docile handling disappeared; the MB-3 fishtailed and wobbled. Hartney decided that the rudder was over-balanced, and he and Benjamin Douglas Thomas,[24] the designer of the MB-3, agreed on removing part of the rudder. The work was completed by 4:00. Following three more flights, Hartney decided that the directional stability was fine. On race day, before the race's start, he made flights around the course in the MB-3 and in an S.E. 5 to familiarize himself with the course.

At 11:00 A.M., Thanksgiving Day, the chief starter raised a red flag to signal the pilots in Flight 1 that they had five minutes to start their engines. As each engine started, the assistant starter assigned to that airplane raised a red flag. When, "but not later than 11:05,"[25] red flags (or blue flags for racers that were not ready) were displayed for all the aircraft, the chief starter raised a white warning flag to indicate that the takeoff signal would follow in ten seconds. While the mechanics rushed to pull the chocks from under the racers' wheels, the spectators watched the red flag being lowered at one-second intervals to count off the ten seconds to takeoff. Simultaneous lowering of the chief starter's red and white flags announced the takeoff. Unlucky pilots whose engines didn't start or balked at running smoothly were not eliminated from the race. The raising of a blue flag signaled a request to make a deferred start, which could be made at any time before the first racer crossed the finish line.

Spectators focused on the start line, deafened by the roar of unmuffled engines and watching ground crew holding the racers' tails to keep them from bolting across the line. At the drop of the starter's red and white flags, the racers lurched ahead. Crossing the line, the racers moved slowly, appearing to waddle across the damp airfield. Accelerating, splashing through puddles, trailing streams of black exhaust smoke and tire tracks filling with water, they gathered speed, raised their tails, ran along for a few yards on the main wheels, and climbed away. They flew in straight lines to a designated point where they began jockeying for position.

Few people ever saw air races with such "race horse starts," with all the airplanes in a flight taking off at once. Undeniably exciting, they were also dangerous. Racers could be skittish on the ground and pilots, focused on getting away as quickly as possible, might veer from straight paths. Fortunately, no accident marred the 1920 takeoffs.

From 1921 on, Pulitzer Races had staggered starts. Each pilot was assigned a time to take off—sometimes a half minute or one or two—after the preceding racer and sometimes inside a time window, say, between 2:00 and 2:05. After takeoff, the pilot climbed to his desired altitude behind the start line, then dived down and across the start. The timing began when the racer crossed the start line.

With all airplanes in Flight 1 in the air, Flight 2 started two minutes later. All three dozen racers were in the air for a time after the VCP-R, the last to take off, was airborne and before the first one to drop out fluttered down with a mechanical problem. Never again would anyone see so many fast and noisy racing airplanes in the air at once, in the small space between the ground and the pilots' chosen altitudes of a few hundred feet. Long Island's low-lying clouds, seemingly hovering only feet above the ground, held the roar of racing engines, the whine of propellers, and the smells of gasoline, hot engines, and gray exhaust smoke close to the crowd.

Hartney, in the MB-3, flew the first (southeast) and second (northeast) legs of the Pulitzer course at 1,000 feet to take advantage of a stout tailwind. On the westbound last leg, he dropped lower. He zoomed around the first circuit in about eleven minutes with the engine "wide open" and overheating. The cooling water boiled, water spurted over the top wing, the oil temperature climbed, and Hartney had to adjust the throttle to keep the engine running. On subsequent laps, he controlled the water temperature by opening and closing the throttle. The oil temperature continued to rise, but he finished the race with a hot engine and no other problems.

Hartney was happy with the Thomas Morse MB-3's performance when he landed, but he did not know how fast he had flown. Standard army pursuits, such as the MB-3, had no speed indicator. Along with all other pilots, Hartney had to wait for the judges to calculate his speed based on the time of his flight, and he would have to wait until the judges made that calculation for all the racers, including Moseley in the VCP-R, to find out what place he had taken in the race. But he felt good: "All machines seemed to be slow compared to the Thomas-Morse.... After the flight I zoomed to 2,000 feet and landed as soon as possible, because of lack of gasoline.... I ascertained that no other ships had finished, that several had dropped out, including the two Curtiss triplanes, and that Moseley [in the VCP-R, which was last to take off] was still in the running."[26]

Lieutenant Benjamin G. Bradley, USMC, in the Loening M-81 S Special, was second off. He was unable to overtake Hartney and gradually lost ground to him. Bradley's race ended badly. As he made the last turn onto the homestretch of the fourth and final lap, a water line cracked, water spurted out, the coolant level dropped, and the engine temperature rose. Bradley kept the airplane in the air as long as possible but the engine stopped and he came down on a golf course one mile short of the finish line.

Two accounts of the end of Bradley's race differ. A writer for *Aviation* magazine, in a nondramatic contemporary account, praised Bradley: "It is greatly to his credit that he landed before the engine froze instead of trying to cover the last mile."[27] In contrast, Foxworth wrote, years later, that Bradley was in sight of the finish line when the Loening's "red-hot engine wrenched violently and almost jumped from its mount. It had seized solid and within seconds Bradley found himself crunching to a safe if splintery stop on a golf fairway, 1 mile short."[28]

Had Bradley been able to fly the last mile, his speed would have been good enough for third place, but "adherents of the Loening monoplane declare that this ship would have won the race but for a disabled radiator."[29] Records offer no support for that contention. Bradley flew the three laps he completed at about 12½ minutes per lap, or between a half-minute and a minute slower than Hartney in the MB-3.

Four of the racers in Flight 1 took off — two MB-3s, one Orenco D, and the Loening M-81 S. Two finished. Undramatically, one of the MB-3s dropped out because of a broken bracing wire, and Bradley's M-81 S dramatically crashed after it lost its cooling water. Both finishers were standard army pursuits — Captain Hartney's MB-3 finishing second and the Orenco D flown by Lieutenant St. Clair Streett fourth.

Hartney would be seriously injured when his airplane crashed in the 1921 Pulitzer, and he subsequently held important positions with the National Aeronautic Association. Streett was already famous for leading the Air Service's first New York to Nome, Alaska, flight and return earlier in 1920. He remained in the army and retired from the U.S. Air Force as a major general in 1952.

As expected, none of the airplanes in Flights 2, 3, 4, or 5 were competitive with the faster aircraft in Flights 1, 6, and 7. A navy VE 7, flown by Lieutenant (junior grade) A. Laverents, United States Navy, posted the best speed of flights 2 through 5 — 125 mph[30] — and finished fifth in the Pulitzer. Army Lieutenant John Roullot flew his DH-4 to sixth place, one mile per hour behind Laverents' VE 7, at 124 mph.

The three Italian-built Ansaldos in Flight 6 performed more reliably than any other manufacturers' aircraft. One of the two that finished the race was disqualified because of the pilot's error in following the course, and one dropped out because of the pilot's problems with his goggles, but none landed because of mechanical difficulties. At least one airplane entered from all other companies had to be withdrawn before takeoff or land early because of mechanical problems.

Bert Acosta, a civilian, flew the Ansaldo Balilla to third place, a credit to the Balilla's good design and Acosta's skill. The Balilla's engine generated about 225 horsepower, more than 100 less than the MB-3's 358 horsepower and more than 300 horsepower less than the VCP-R's 576 horsepower. It finished the race at 134 mph, ahead of twenty service airplanes with more horsepower, and only 14 mph slower than the MB-3 and 23 mph slower than the VCP-R.

The fine qualities of the Balilla and other Ansaldo designs were evident in their speed, reliability, and general good looks. Ansaldo importers had clearly made a good choice when they decided to market the airplanes in the United States, but the weakening U.S. economy and the flood of U.S. and British surplus airplanes — Curtiss "Jennies," for instance, were available for as little as $50 — doomed Ansaldo importers to financial collapse a year later.

Albert "Bert" Acosta, an exuberant, handsome man and excellent pilot, could have been the model for the 1920s' swashbuckling, barnstorming pilot. Born in California, he taught himself to fly in 1910, went to work as an apprentice with Glenn Curtiss in 1911 at age 16, and became an instructor pilot for Curtiss in 1915. He moved to Canada to instruct wartime pilots, returned to the United States in 1917 and was a wartime captain in the Air Service and chief instructor at Hazelhurst Field. Regarded as one of the truly great natural pilots and a master of testing heavy cargo and transport aircraft, his performances in the 1920 and 1921 Pulitzers—third in 1920 and first a year later—showed he could also pilot racers.[31]

Although the French Morane-Saulier Company's low-powered AR finished last in the race, it did better than the service aircraft in Flight 6. The army's SPAD was withdrawn before the race, and the pilots of the navy's Loening M-8s might better have stayed in bed. The engine of one Loening overheated after one lap, certainly a disappointment, but overheating was commonplace that day. The other M-8 "was involved in a tragicomedy collision [with an Orenco D] while taxiing to the starting line; its pilot was unhurt."[32]

Only one airplane in Flight 7, the winning VCP-R, finished. The two Curtiss triplanes, which took off before the VCP-R, came to grief. Fuel line or carburetor trouble forced the triplane flown by navy lieutenant Willis Haviland down on its second lap. Marine lieutenant W.D. Culbertson flew the other triplane to the fastest first lap in the race, 11 minutes, 3½ seconds, compared to the winner's best lap of 11 minutes, 7 seconds. Culbertson's second and third laps were far slower, more than a minute longer than the winner's time, and a broken connecting rod in the engine forced his 18-T down, without further damage, on its fourth lap.

Lieutenant Corliss Moseley, the VCP-R's pilot, had joined the Army Air Service in 1917 and flown with the 27th Aero Squadron, 1st Pursuit Group, in France in World War I.[33] He had been Major "Shorty" Schroeder's back-up pilot at the 1920 Gordon Bennett, and at the time of the Pulitzer Race he was assigned to the office of the chief of the Air Service in Washington.

Moseley had taken the VCP-R up for test flights on Tuesday, November 23, two days before the Pulitzer. The engine functioned well at low and medium RPMs with no flame spitting from the exhausts and none of the overheating problems that had plagued the VCP-R in France. One problem persisted; as the throttle was opened and engine speed increased, the engine "missed badly and continually above 1700 rpm."[34] Despite the engine's rough running, Moseley was satisfied with the airplane and, concerned about mechanical problems possibly arising in additional test flights, did not fly on Wednesday. The VCP-R looked sturdy and powerful in flight, and it proved to be in the Pulitzer.

Moseley's Pulitzer flight lacked the pyrotechnics and overheating of Schroeder's flight in the Gordon Bennett, but Moseley had problems of his own. Scheduled to be the last off, he had to delay his start an additional fifteen minutes while mechanics freed sticky exhaust valves. Airborne, the engine continued to misfire, and Moseley flew circles behind the start line waiting for the engine to steady. The problems were not erased, but his airspeed indicator registered nearly 180 mph, a winning speed in his opinion, and he crossed the start line at about 300 feet. On the first lap, the engine missed badly, the tachometer jumping wildly between 1300 and 1800 RPM. On the second lap, he found the engine ran smoothly at 1700 RPM but misfired if he opened the throttle. Nevertheless, on the last three laps, he opened the throttle wide "for about 30 seconds ... and the airspeed indicator immediately ran up over 200 mph, but the violent missing would commence again and I would have to throttle down."[35]

Lt. Corliss Moseley, United States Air Service pilot, and the army Verville VCP-R that won the 1920 Pulitzer Trophy Race. This picture was taken two years later, at the time of the 1922 Pulitzer. In the 1922 Pulitzer, Moseley flew the same racer (renamed "R-1") 23 mph faster than in the 1920 race and finished sixth (courtesy National Archives and Records).

Completing the four laps of the Pulitzer, Moseley decided to fly one more, "hoping that I could get the throttle entirely open and get a real good record for one lap of the course," but the engine missed badly at 1800 RPM, and the "good record" eluded him. Two suggestions have been made to explain the misfiring engine. In Moseley's quotes that appear in *Skyways*,[36] he blames sticky exhaust valves. Foxworth[37] quotes Moseley (source not specified) as saying the cause was vacuums forming over the carburetor air intakes at high speeds. In support of the carburetor problem explanation, Foxworth reports that Moseley said the engine emitted puffs of dark exhaust, indicating a too-rich fuel-air mixture. Whatever the culprit, the misfiring deprived the engine of its full power.

Moseley had no time to climb from the VCP-R's cockpit before General Mitchell, face beaming, ran over, reached up, and shook his hand. Congratulations were not the only thing on Moseley's mind. "The dust had hardly settled when Moseley was observed leaving for the Big Apple, happily settled between a pair of flappers, the great-grandmothers of today's rock band groupies."[38]

Moseley's landing ended the Pulitzer Race. He and the VCP-R, as expected by many, had won, flying the course in 44 minutes and 30 seconds, two minutes and 30 seconds faster than Hartney in the MB-3. Even so, Hartney's MB-3, with half the power of the VCP-R, had made an excellent showing. Acosta flew the Ansaldo Balilla, with about two-thirds

The VCP-R in flight. The very high water tower in the photograph looks like the one at the Curtiss factory at Garden City, New York, and this photograph was probably taken over or near near Mitchel Field, which bordered the Curtiss property (courtesy National Archives and Records Administration).

the horsepower of the MB-3 and one-third of the VCP-R, to third, about 14 minutes behind the MB-3 and 23 minutes behind the winner. The Balilla would be the only foreign-designed and built airplane to finish "in the money" in any Pulitzer Race. Army Lieutenant Streett, in an Orenco D, finished 4th; navy lieutenant A. Laverents was 5th in a Vought VE 7.

The attractions of New York City and flappers did not keep Moseley long. On Saturday, two days after the Pulitzer, he was back at Mitchel Field and flew six passes—three with the wind and three into the wind—to attempt to set a new world's speed record over a one-kilometer course. The *New York Times* had reported "air service officers confidently expect that he will be able to drive his machine at more than 200 miles an hour."[39] His average speed over the six tries, 186 mph, fell well short of the 200-mph goal and about seven mph slower than the then-current world record of 193 mph set two weeks before by France's Bernard-Henry Barny de Romanet flying a SPAD S.20bis_6, powered by a specially built 320-hp Hispano-Suiza.[40]

Moseley's closest competitor in the Pulitzer, Harold Hartney, also flew in the Saturday speed trials. In six passes over the one-kilometer course, he averaged just over 170 mph in his MB-3, good enough for a new world record for service airplanes.[41] Under the apt subtitle "Tribute to Moseley," Hartney later wrote, "Moseley's work in the race and on subsequent speed trials bears out the confidence General Mitchell placed in him.... The same spirit would overcome other problems and obstacles in the Air Service, and because of that alone, if for no other reason, I take my hat off to Moseley."[42]

Two years later, the army cleaned up and improved the VCP-R, renamed it "R-1," for "Racer 1," and entered it in the 1922 Pulitzer. Moseley flew it again and finished 23 mph faster than he and the VCP-R finished in 1920. It is a measure of progress in high-speed flight that Moseley's speed of 23 mph faster than the winning speed in 1920 was good enough only for 6th place, well out of the money, in the 1922 race.

Soon after the 1922 Pulitzer, Moseley retired from the army and embarked on a career in civilian aviation. In the mid–1920s, he opened and operated an overhaul and maintenance facility at Grand Central Airport in Glendale, California. He became a distributor of aircraft and related items for the newly formed Curtiss-Wright Company in 1929, was one of the founders of Western Air Express, later Western Air Lines, and operated schools for mechanics and other ground crew and provided basic training for pilots during World War II. Following the war, he founded Grand Central Rocket Company, which made and tested solid rockets. He died in June 1974 at age 70. Moseley's aviation career spanned more than 50 years, from the era of wood and cloth biplanes to the space age.

In November 1920, the United States' aviation aspirations and ego were still bruised by the humiliating disappointments at the 1920 Gordon Bennett, where none of the three highly touted United States racers had managed to fly a single 100-km lap of the 300-km race. Spirits soared when the speeds of the Pulitzer racers were announced. Based on his time around the course and the course distance of 132 miles, Moseley had flown the Pulitzer at 178 mph, 12 mph faster than the winner's speed in the Gordon Bennett and faster than anyone had ever flown around a closed course. Equally gratifying, Hartney, in the MB-3, a standard Air Service pursuit, was credited with 168.5 mph, comparable to the Gordon Bennett's winner's speed. Jubilation was to be short-lived. On Saturday, November 27, the day of the time trials and two days after the Pulitzer, an article in the *New York Times* began the rupture of the exhilaration that ballooned with the announced speeds in the race.

The winner of the 1919 Transcontinental Race (see Chapter One), army lieutenant Belvin Maynard, the "Flying Parson," was driven to resign from the service because an article he wrote was interpreted to say that alcohol had been responsible for some of the fatalities in the race. A year later, in November 1920, he was engaged in welfare work for the Brooklyn YMCA when he read that the fastest DH-4 in the Pulitzer had flown a reported 140 mph.[43]

He contacted the *New York Times* to tell a reporter this: "While I was in the army Air Service I tested out hundreds of De Havilland airplanes and know that the average maximum speed of this type is about 130 miles an hour." He concluded that the Pulitzer course had been mismeasured and that each lap had not been the planned 33 miles long. Maynard assumed that the winning DH-4 might have flown 134 mph and, from that speed, calculated the length flown in the Pulitzer at 124.2 miles, not the planned 132 miles. Based on the race length he had estimated, he calculated that the VCP-R had flown "167.574 miles an hour ... by no means a record."[44]

Responding to Maynard's analysis, General Mitchell ordered a survey of the course actually flown, and the Coast and Geodetic Survey determined that the distance around the course was nearly four miles less than the planned length. The reason for the discrepancy was quickly found. The two turning points for the race were to have been marked by captive balloons, but only one balloon arrived on time. The single balloon was flown in the correct position at the first turning point over Lufberry Field near Wantagh. When no balloon arrived at the second turning point at Colonel Henry J. Damm Field in Babylon, race officials looked for a large, high landmark around which the racers would turn. They selected a wooden windmill — over 100 feet high. From the standpoint of visibility, it was an excellent choice. Looking east from Mitchel Field, the windmill dominated the horizon, and it is reported that the windmill was even wrapped with white cloth for the Pulitzer Race![45] (The distance between Mitchel Field and the windmill was about 16 miles.)

The windmill was not a good choice from the standpoint of maintaining the course length. As measured, the length of a lap flown around the windmill was 29.02 miles, not 33 miles as planned. The length of the Pulitzer Race, four laps around the course, had been 4 × 29.02, or 116.08 miles, almost 16 miles shorter than the planned 132 miles. Based on his estimated speed of 134 mph for the winning DH-4, Maynard had calculated the distance to be 124.285 miles, halfway between the nominal 132 miles and the measured 116 miles.

When the measured distance was used to calculate speeds, Moseley's speed in the VCP-R fell from 176 to 156.5 (Maynard had calculated 167). Hartney's speed in the MB-3 was recalculated at 148, fast, but far from a world record for a service airplane. The speed attributed to the winning DH-4 dropped from 140 to 124 mph.

Maynard's quickness of mind and analysis of the discrepancy he saw in the DH-4's speeds make him a model for modern test pilots that we picture with one hand on the controls of the airplane and the other hand on a computer. In Maynard's case, the other hand would have been on a slide rule. Maynard's generous spirit contributed to his overestimating the DH-4's speed, which, in turn, caused his calculation of the VCP-R's speed to be higher than measured.

A writer for *Aviation* wrote "that 1920 will go into aeronautical history as a year of

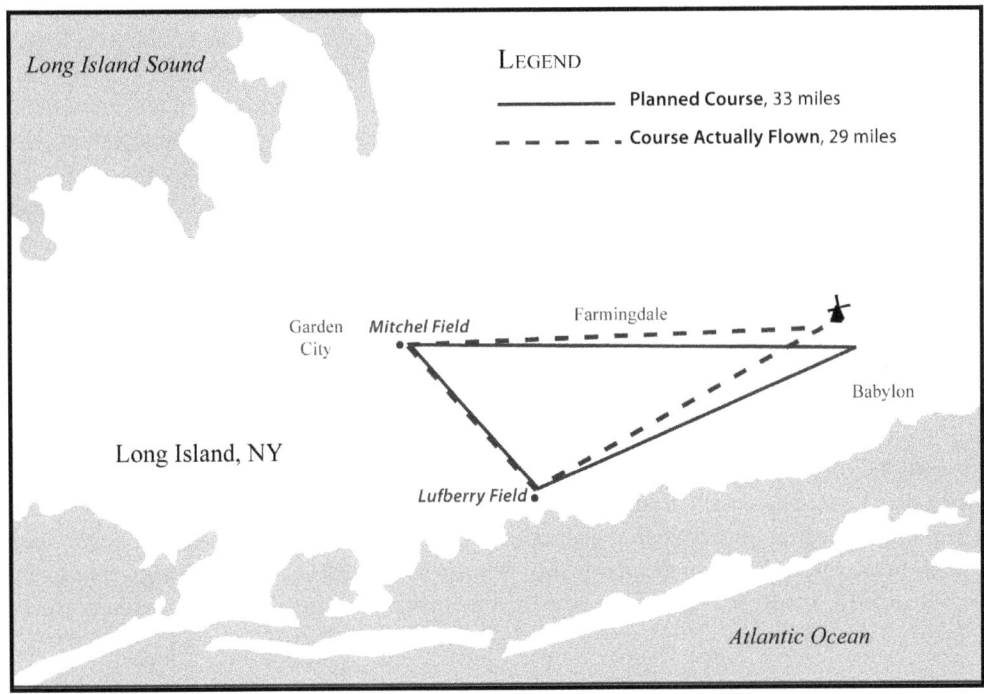

Figure 1. Map of the first Pulitzer Race, Garden City, Long Island, New York, 1920. The course indicated by the solid line was the intended course. When a balloon to be tethered did not arrive to mark the turning point near Babylon, a large windmill (as indicated on the map) over 100 feet high was selected to mark the turning point. The changed turning point shortened the course by about four miles. The Pulitzer Race required completion of five laps of the course (map by Lynn Springer).

great advance in airplane design"[46] and that the Pulitzer results were "further evidence that American pilots are as good as any in the world."[47] Published between the Thanksgiving Day Pulitzer Race and the Saturday time trials, the magazine had high hopes, expecting that the VCP-R would soon best the current record of 193 mph for straightaway speed "by a safe margin."[48] That expectation was not realized, but the army's effort to field a racer had paid off. The Pulitzer victory, reported in major newspapers and magazines, brought attention to the Air Service just as General Mitchell had wanted. The *New York Times*' reporting of the race results underlined the competition between the army and the navy. Its headline "Army Pilot Wins Pulitzer Air Race" was followed by a subheading, "Naval Officer Is Fifth."

Had the $223,800[49] spent by the Air Service on building, repairing, modifying, and preparing and shipping the VCP-R to the Gordon Bennett and the Pulitzer been worth it? Had it been worth $223,800 to win a single race? There is no satisfactory answer to those questions, but the money had been spent and the Pulitzer had been won.

The most obvious lesson to be drawn from the VCP-R's and other racers' performance was that power alone did not guarantee victory or outstanding performance. Weight, size, and streamlining of the airframe as well as careful attention to adequate cooling were also necessary. Had the money spent on the VCP-R bought any special knowledge that could not have been learned from other airplanes flying at the same time? Probably not. The European judgment of the VCP-R at the time of the Gordon Bennett seems appropriate. It was an ordinary design with a big engine.

On the other hand, expecting breakthrough design and performance was expecting too much. The VCP was Alfred Verville's first attempt to design a pursuit; converting the pursuit to a racer was his first try at a racer. He had made a study of SPAD fighter design and production methods, but knew less about other designers' and companies' approaches. More important, the VCP was designed as a service pursuit; its modification for racing consisted almost entirely of installing Packard engines with impressive power but less than reliable operation.

Whatever the VCP-R's shortcomings, an editorial in *Aviation and Aircraft Journal* stated that its "victory is very gratifying to American engineers as it is a vindication of their ability to turn out the world's best designs."[50] A second editorial in the same issue opened with this emphatic sentence: "The Pulitzer Trophy race has given a new incentive to designers and aeronautical engineers."[51] The race did more than provide an incentive for designers and engineers. It provided a template for government-sponsored development of high speed aircraft, built by private companies and flown by military officers in the Pulitzer Races.

And what of average citizens? What had they made of the first Pulitzer? Glenn Curtiss thought he knew: "This is the most representative gathering from the viewpoint of popularity that I ever have seen at an aviation meet. The 1921 race will be a stupendous event. The public, at last, is interested in the airplane."[52] "Stupendous" is a high standard. Whether the 1921 race reached it is questionable, but it was a successful event after it appeared likely to be canceled. It succeeded because of aviation enthusiasm in a somewhat unlikely place: Omaha, Nebraska.

Chapter Three

1921

A Close-Run Event and Narrow Victory: A Curtiss Racer Built for the Navy Wins the Second Pulitzer, North Field, Omaha, November 3

> *The second annual competition for the Pulitzer Trophy has been the occasion of a magnificent demonstration of the intrinsic worth of American pilots, airplanes and aircraft engines. The speed of 176.7 m.p.h. made by the winner, Bert Acosta, on the Curtiss-Navy racer, beats any previous similar performance made here or abroad in competition.*[1]

The Aero Club of America, in cooperation with the army and the navy, had apparently achieved the near-impossible in organizing and hosting the 1920 Pulitzer Race only three months after the race was announced. It seems a small achievement over small difficulties in comparison to the problems faced and overcome to make the 1921 Pulitzer a success and keep the Pulitzers alive. The problems began when Detroit, scheduled to host the race, canceled it two months before it was to be flown. Out of nowhere, Omaha resurrected the event and staged a record setting race. More important, had Omaha not stepped forward and the Pulitzer been allowed to lapse for a year, it might not have been held again.

Following Lt. Moseley's victory in the 1920 Pulitzer, Caleb Bragg, chairman of the Aero Club of America's Contest Committee, announced the Pulitzer Trophy would be awarded to the Aero Club of California, to which Lt. Moseley belonged.[2] Someone made a mistake about Lt. Moseley's aero club membership, or he belonged to two different clubs half a continent apart. Or Bragg and his associates might have decided that holding a race in California, so far away from the Aero Club's headquarters in New York, presented too many difficulties. For whatever reason, two days after the 1920 race, the Aero Club changed its mind, and the *New York World* reported that "Detroit was to have the honor [of hosting the Pulitzer Race] ... because Capt. Mosley [sic] was representing a club from that city when he made his triumphant Thanksgiving Day flight."[3] The *World* got it right when it said Detroit had been awarded the honor of hosting the race, but events took the race elsewhere.

Detroit, rich, booming and enthusiastic, its prosperity rooted in automobiles and auto engines, had the money to sponsor a successful air meet and the population to assure plenty of spectators. The Detroit Aviation Society, Inc., moved quickly after the city was awarded the race. It arranged financing for a week-long event, with five other races to complement the Pulitzer, set September 8, 1921, as "Pulitzer Race Day," and published and distributed entry blanks in March. On July 20, Joseph Sadi Lecointe of France, famous and popular winner of the 1920 Gordon Bennett Race and holder, on-and-off, of the world's speed record, cabled his entry for the Pulitzer.[4]

Nothing was lacking. Funding was assured, Detroit civic and aviation organizations were solidly behind the air meet, the army and navy had contracted with manufacturers for new racers, a famous foreign race pilot had entered the race, and early September promised delightful weather without the blustery wind and cold of the 1920 Pulitzer.

On July 18, two days before Lecointe's cable arrived and a month and a half before the scheduled races, everything came crashing down. Sidney D. Waldon, president of the Detroit Aviation Society, announced that the air meet would not be held. The army and navy, citing shortage of funds, had informed Mr. Waldon that they could not afford to send airplanes to the race. Mr. Waldon and the Detroit Aviation Society, convinced that the absence of military racers meant that no suitably competitive airplanes would be entered and that the Pulitzer and other events would not be sufficiently attractive without them, canceled the week-long event. The *New York World*, a Pulitzer newspaper, emphasized that "no fault attaches to the Detroit Aviation Society for not carrying" through with the race.[5] Evidently counting on government funding being available the next year, Mr. Waldon and his colleagues asked to host the 1922 race.

A 1921 race appeared out of the question. At best, late in July, three or three and a half months of good flying weather remained before winter. Where could a new sponsor be found and how could it organize a race before winter?

A month later, on August 13, with, maybe, two and a half months of good weather left, an unexpected offer from an unexpected source reached Aero Club of America officers. Earl W. Porter, president of the Aero Club of Omaha, on behalf of the club and the City of Omaha, offered to add the Pulitzer Race to an air congress (or "aero congress") already planned for early November. Aero Club officials, surely hoping that no heavy early–November snowstorm would bury Omaha, accepted Porter's offer and committed $1,000 to the air congress' coffers.

Omaha, "the wideawake Nebraska city,"[6] was a prosperous, growing railway and river transportation hub and a center of meatpacking and light industry, including a handful of small airplane manufacturers. Still, it was a surprising location for an important air meet. St. Louis, considered a western city in 1921, is 250 miles east of Omaha, and Omaha was far from the population, manufacturing, and advertising centers of the East. But Omaha had history, as well as enthusiasm, on its side. Omaha had hosted one of the country's first air meets over a decade earlier, and a major army aviation training center was located there.

In 1910, the Omaha Aero Club had offered $10,000[7] to Glenn H. Curtiss to make a flight at its "Mid-West Air Meet" to be held July 23 through 27. Despite stiflingly high temperatures and a "bad gale," more than 4,000 spectators attended the air meet's last day, July 27, to see Curtiss fly. Throughout the 1910s, Omaha crowds turned out to see the "war balloons" from the U.S. Army Balloon School at Fort Omaha. The balloon school, a central element on Omaha's aviation stage for a dozen years, trained many of the air and ground crewmen for the tethered United States observation balloons used in France in World War I. As late as November 1920, the army was recruiting 250 men, with at least high school educations, to attend a ten-month-long course to learn "how to handle his craft [a balloon] in all its phases and to make observations, and he eventually qualifies as a free balloon pilot."[8] Advancing technologies made "war balloons" obsolete, and in 1921, the year of the Omaha Pulitzer, the balloon school moved to Bellville, Illinois, where it soon went out of existence.

Earl Porter, the 1921 president of the Aero Club of Omaha, had been an architect before he joined the Air Service following the United States' entrance into World War I. Eager to go to France and fly in combat, he decided against going to pilots' school and trained to be a "bomber," the backseat man in one-engine reconnaissance, light-bomber aircraft such as De Havilland DH-4s. Credited with shooting down three German aircraft and surviving a serious wound, Porter was awarded the Distinguished Service Cross and "The Drop of Blood," equivalent to the Purple Heart.[9]

In February 1921, he and Aero Club of Omaha officers—all veterans of the World War I air war—initiated plans to piggyback an International Air Congress on the American Legion's third annual convention in nearby Kansas City, Missouri. The American Legion convention, which included a large air show, ended on November 1, and the air congress would begin on November 3. Porter and the Aero Club of Omaha boasted that Omaha would host "the greatest aerial congress ever held in the United States" and that all was going well. At an Aero Club of Omaha-hosted dinner on July 18, "twenty-one businesses and organizations endorsed the Congress." A month later, after Omaha added the Pulitzer to the air meet, "the Chamber of Commerce Journal declared that every business and civic organization was working to bring the Congress to Omaha and that the project had the unanimous support of its own Executive Committee."[10] By mid–August, public subscriptions and fund-raising were deemed sufficient "to finance the transportation of machines and men" to the congress site.[11]

Raymond Farquhar, president of the Nebraska State Aero Association, said that the addition of the Pulitzer Race to the International Air Congress would increase the number of airmen attending the Omaha meeting from 10,000 to 20,000. The mayor of Omaha, who had opened the 1910 aerial meet eleven years before, said the congress would be the "biggest thing Omaha and the West has ever handled." In addition to $6,000 in prize money that had been posted for the Pulitzer, Omaha businessmen had reportedly raised an additional $30,000 for prizes.[12] Omaha looked to be a hero; it had stepped in at a late hour to host the second Pulitzer. The city's leaders and civic organizations were confident.

Apparently the Aero Club of America was not so certain, praising Detroit, which had refused to host the event, and barely mentioning Omaha, which had taken up the challenge. The Aero Club made public a letter from the Detroit Aviation Society "giving to the Aero Club of Omaha permission to conduct the contest." How and why was Detroit's "permission" needed or mentioned? It had already refused to host the air meet. "Detroit in sportsmanlike spirit, agreed to let the other city have the honor this year. At the same time Detroit insisted upon conducting it in 1921." Praise for Omaha was stilted. With all that Omaha had to do and deserved credit for doing, the Aero Club singled out the War ("Defense" in today's parlance) Department's contribution, which cost little and required little if any effort: "Word from Omaha is to the effect that the aerial meet of November looks more promising day by day. It is hoped that 10,000 former airmen of army and navy service will be present.... The War Department has given permission of these fighting flyers to use the dormitories at Fort Omaha during the meet...."[13]

Concerned about the success of the air meet in Omaha with little time to prepare and focusing on the potentially lucrative and publicity-rich meet in Detroit in 1922, the Aero

Club of America may have decided to keep its distance from Omaha's effort. Despite Omaha's willingness, its efforts might not shine, and the Aero Club might have been reluctant to risk its reputation on the abilities of little-known men in a faraway Nebraska city.

Those men moved quickly. Before September 1, the Omaha committees had mailed entry blanks to perspective contestants and exhibitors, and almost 100 were returned by October 1, when the program for the three-day congress was released. Boosters sold subscriptions and tickets to the congress, asked for contributions, and organized a contest for schoolboys to sell "Omaha Aero congress stamps,"[14] with the prize a round-trip by air with the mayor of Omaha to take an invitation to the congress to the mayor of Chicago. One enterprising 10-year-old boy had sold 30,000 stamps by October 4. In a move toward sexual equality, perhaps, or more likely to sell more stamps, the congress considered opening a contest for girls. (The author has not found out if the contest was actually opened to girls.)

Quite out of the control of anyone in Omaha, a potential nationwide railroad strike in late October or early November threatened to doom the congress. Without trains, fewer spectators would travel to Omaha, and fliers and businesses would be unable to ship their airplanes. A far more general effect of a railroad strike would be the stoppage of mail service. Earl Porter, to keep Omaha and aviation in the public eye, offered the assistance of the Aero Club of Omaha to recruit 500 volunteer pilots to fly the mails.[15]

Apart from the impending strike, all looked good for the grand opening on Wednesday, November 3. Reportedly, the coffers were full, and the October 25 *Omaha World-Herald* reported that 73 airplanes had been entered in various events, with ten entered in the Pulitzer.[16]

From nowhere, the news changed. Late on the same day, October 25, the aero congress was belly-up. There was not enough money. Perhaps overly optimistic reports of fund-raising had led the organizers to count on money that had not been collected. Strapped for funds, Earl Porter and the Aero Club of Omaha approached the chamber of commerce with open hands. The chamber's executive committee, which had earlier pledged its support, turned them down. Dodging behind the looming railroad crisis to hide its change of commitment, the chamber announced, "It is imperative that the International Aero Congress be called off because of the impending railroad strike."[17]

Porter resigned the presidency of the Aero Congress on October 25, but he was named to a committee of five members with the task of raising the additional $20,000 necessary to meet expenses and hold the congress. The committee of five did it. Three days later, the October 28 *Omaha World-Herald* headlined another turn-around: "Big Aerial Congress Saved for Omaha."[18] Few details were printed, but "All financial obstacles and other tough spots, both on the field and in business management, have been ironed out."[19]

Omaha had a new airfield—North Field—for the aero congress, built by both paid and volunteer labor. Off-duty airmail pilots, stationed at Omaha, had been among the volunteers. The airmail pilots, with their colleagues across the country, had established an enviable record in 1921. Flying through good weather and bad, day and night, with few navigation aids, they had completed 98 percent of their flights on schedule.[20] The pilots had a mixture of skills unique to the early 20th century. Not only could they fly and navigate the airmail routes, feats not done by generations before, they could also harness teams of horses and walk beside them to pull tree stumps from the ground as their fathers and grandfathers had done before the days of manned flight. The old skills would disappear as tractors and other engine-powered machines replaced horses, but they were vital to the completion of North Field.

Near-steady rains in early October and heavy rains on the 25th had caused terrible conditions on the field. Newly cleared, and bordering the Missouri River, the field was flooded in places. An expert, a local racetrack builder, directed filling of low spots with sand and cinders and drying out wet areas with oil fires. By the 28th, the builder convinced the chairman of the new five-man Aero Congress committee that another rolling of the field would make it ready. Listening to their expert, the committee issued a statement that the field would be ready. The builder and the committee were right.

Porter, with his optimism and faith that people were interested in aviation, aerial demonstrations, and racing, had brought the Aero Congress to Omaha. Despite his resigning from the presidency of the Omaha Aero Club, he must be credited for initiating the congress and assuring that the second Pulitzer was flown in 1921. The *Omaha World-Herald* gave him no credit. In its editorial "The Omaha Spirit," it regretted the months of faintheartedness and confusion and neglectfulness and the plain, or garden variety, of stupidity" that had gone into "a movement worthwhile, even if its promotion and organization has been execrably managed." Porter was not named, but who else could have "execrably managed"? The *World-Herald* lauded the "inspiring spectacle" of seeing "Omaha's leaders in public spirit and civic service come to the front, throw off their coats, spit on their hands and take hold in resolute earnest." Porter's name is missing from the editorial's list of ten men who "devoted themselves to this enterprise, and their names and reputations are the guarantee of success."[21]

Robert Adwers, a historian of Nebraska aviation, interprets the planning and execution of the congress differently. He gives much of the credit to Porter and the "young Turks" of the Aero Committee and blames the chamber of commerce for dragging its feet in the early going.[22] Thomas G. Foxworth, the historian of 1920s air racing, focuses on Porter, who "finally wheedled close to $90,000 — about half coming from the Chamber of Commerce at the last minute to avoid disaster."[23] Wherever praise belongs, funding was wrapped up at the last minute.

After the Aero Congress was over, the chairman of its general committee stated that the total cost had been $55,000 (excluding taxpayers' contributions of $40,000 through city expenditures to improve the airfield and access to it). The chairman said that all debts had been or would be paid, that admission receipts had totaled $16,000 (at $1.00/admission) and that businesses, businessmen, and several prominent individuals provided the $39,000 balance through pledges and underwriting.

In keeping with the $16,000 collected in admissions, about 20,000 people are estimated to have attended the Aero Congress. An equal number watched for free from outside the grounds and from along the racecourse. The number of spectators was woefully less than had been "expected" in the heady days of August, when the *New York World* (a Pulitzer paper) reported in a headline "20,000 Flyers and 100,000 Outsiders Expected by City."[24] In the light of actual attendance, the estimate was wildly inflated. Disheartening, but not deadly. Attendance was less than expected — not enough to cover expenses — but big enough to supply an appreciative audience for world record-breaking flying.

To avoid the embarrassment of a mismeasured race, which had cast a shadow over the 1920 Pulitzer, the course from Omaha was carefully measured at 30.7191 miles (50 kilometers) per lap, with five laps necessary to complete the Pulitzer. The first leg was flown northeast from North Field in Omaha to Loveland, Iowa, where a white wooden windmill marked the first turning point.[25] At Loveland, the course turned nearly due west to Calhoun, Nebraska, where the turning point was to have been an oil fire inside a white circle. When the wind proved too strong for a fire on race day, a 50-foot high and 30-foot wide[26] cross was built and covered with cheesecloth. At the second turning point, pilots turned southeast to head back to North Field.[27] Spectators saw racers fly back to the field, turn sharply, to the northeast and fly away.

The Missouri River wandered through the course. On taking off, the racers crossed the river, which marked the northeastern side of the airfield, and they skirted its east bank

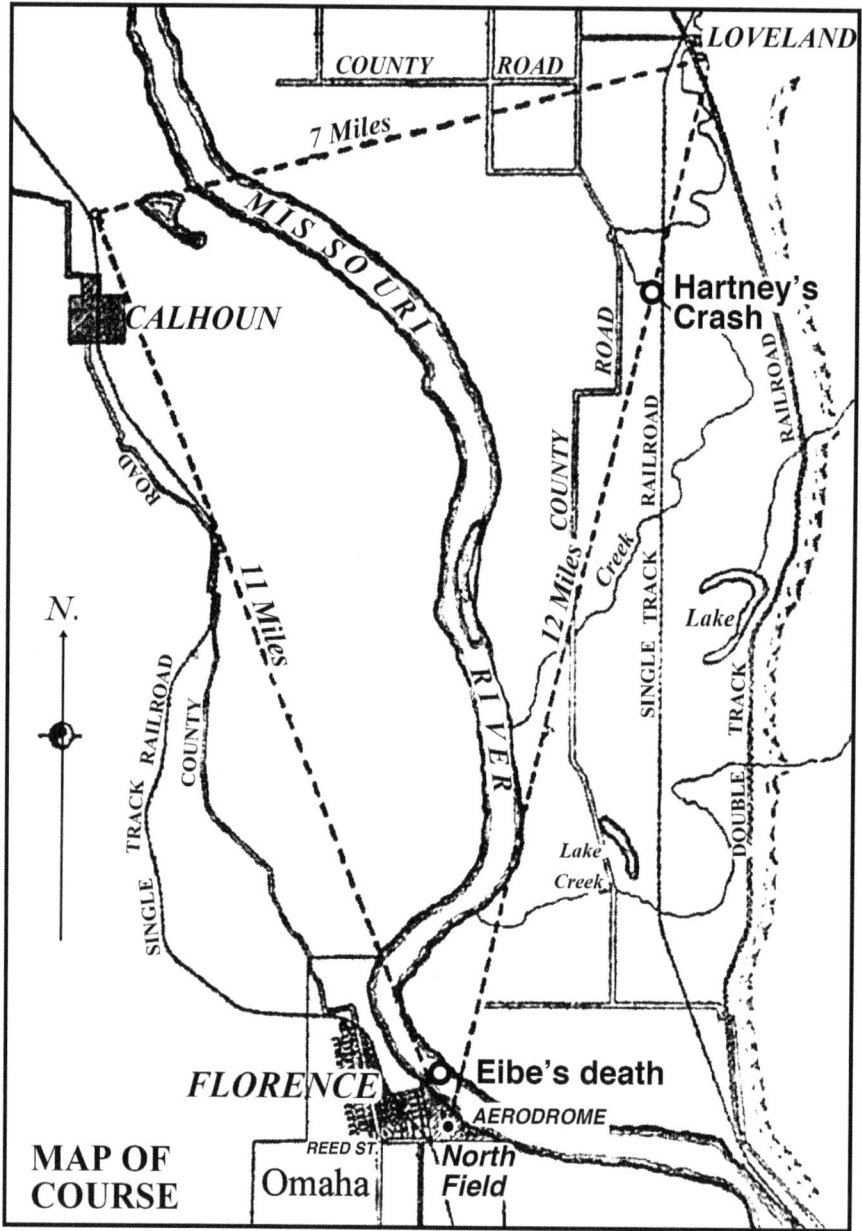

Figure 2. Map of the second Pulitzer Race, Omaha, Nebraska, 1921. The race was flown counterclockwise: North Field to Loveland to northwest of Calhoun and back to North Field. The Pulitzer Race was five laps of the 50-km course (map by Lynn Springer).

for a short distance on their way to Loveland. They crossed the river flying between Loveland and the second turning point north of Calhoun, and flew over it for about a mile on the third leg of the course before completing it at North Field, where they turned northeast toward Loveland (or landed after completing five laps).

With funding assured, the field shaping up, and the course decided, all was ready for the Pulitzer Race, the headline event, to lead off the Aero Congress on Thursday, November 3. There would be no 1920-like frenzy of three dozen racers in the air at one time. Only six racers, the products of three manufacturers, would fly in the 1921 race.[28]

The rules for 1921 stipulated that racers entered in the Pulitzer had to be capable of at least 140 mph. The imposition of the 140 mph minimum reflected the Aero Club of America's confidence about improvements in airplane performance over a year's time. In 1920, only two airplanes had completed the Pulitzer at over 140 mph — the Verville VCP-R and the Thomas Morse MB-3, both military airplanes. A year later, three of the six starters were privately owned airplanes (two war-surplus Ansaldos and a Curtiss-Cox racer originally built for the 1920 Gordon Bennett). The other three, two Thomas Morses and one Curtiss, had been built for the military but were being flown by the companies that built them.

Perhaps as an unexpected consequence, requiring racers to reach a minimum speed was to convert the Pulitzer from a military and civilian competition to a purely military affair. As the minimum required speed was increased from year to year, only military airplanes powered by the latest engines could meet it.

The Aero Import Company of New York, which imported, sold, and leased Italian Ansaldos, entered the same A-1 Balilla that Bert Acosta had flown to third place the year before, but it was far more powerful. For 1921, the company had installed a 415-hp Curtiss K-12 engine, giving the re-engined Balilla about twice the horsepower of the 225-hp engine that had powered it in 1920. On the day before the Pulitzer, the Curtiss K-12-powered Balilla had won the 150-mile-long American Legion Derby, the featured aerial contest at the American Legion Convention in Kansas City, at 149.4 mph.

The Ralph C. Diggins Company of Chicago entered another Ansaldo product, a SVA-9, powered by a SPA inline 6-cylinder 225-hp engine, similar to the engine that had powered Acosta's airplane the year before. Plainly underpowered compared to the other entrants, its only chance was that misfortune would dog the more powerful racers, leaving it and its reliable inline 6-cylinder engine to finish.

The fortunes of the Thomas Morse Company, builder of the MB-3, which had placed second the year before, were sliding downward. Boeing had underbid it and won the production contact to build 200 MB-3s (designated "MB-3As" when Boeing-built) for army pursuit squadrons. The loss of the production contract combined with the failure of other Thomas Morse designs to interest military and civilian clients left the company with few prospects.

Racers provided a bright spot for Thomas Morse. The Air Service, impressed that an MB-3 with a 300-hp engine had gone head-to-head with the VCP-R with twice the horsepower in 1920, had contracted with Thomas Morse for three racers on May 24, 1921. The new racers, called MB-6s, were essentially MB-3s with clipped, single-bay wings. The reductions in size were substantial: MB-3s' wingspans were 26 feet upper and 24 feet, 6 inches lower; both wings of the MB-6s were 19 feet. At 18 feet 6 inches, the MB-6s were a foot and half shorter than the MB-3s (see appendix).

The MB-6 contract was worth $48,000 for three airframes, with the Air Service providing engines, propellers, and instruments. The redesign and construction of the MB-6s

A Thomas Morse MB-6 (courtesy National Archives and Records Administration).

based on the MB-3 would appear to have been a straightforward job, but the racers were a month behind schedule when delivered to McCook Field on September 20.

One of the MB-6s was static tested to failure (loaded with weights until things broke). Army Air Service test pilot John A. Macready took off in the second MB-6 for the first time on October 21, less than two weeks before the Pulitzer. It was fast, timed at 182 mph, but as expected from its small wingspan, it glided poorly and its narrow track landing gear made it skittish on the ground. Corliss Moseley, the winning pilot in the 1920 Pulitzer, made the second flight in the MB-6. He landed hard; the racer somersaulted 1½ times and landed upside down. The racer was written off; Moseley walked away almost unscratched. The remaining MB-6, still in its crate from the factory, was shipped to Omaha.

The navy also contracted with Thomas Morse, ordering two racers (along with ten standard MB-3s) on May 16, 1921. The Thomas Morse designer, Benjamin Douglas Thomas (no relation to the Thomas in the company's name), who had designed the conventional-looking MB-3s and 6s, had become fascinated by the "Alula wing," which is designed to mimic, in some respects, the wing of a gliding bird. Thomas incorporated it in the MB-7. As originally designed and built, the Alula wing had a constant chord (distance from leading to trailing edge) from the fuselage to about half span. At half span, a sharp sweepback began in the leading edge and continued until it reached the trailing edge in a very sharp tip. The part of the wing with maximum chord was thicker than the outer portions, which were made thinner toward the tips, and the outer portions had a slight anhedral, downward angle.

Originally touted as a good choice for heavy, cargo-carrying airplanes, the Alula wing was subsequently promoted as a racing wing, and it was fitted to the British Martinsyde Semiquaver racer which, with conventional wings, had flown in the 1920 Gordon Bennett. The Alula-wing, with its unusually tapered wing and humped center section, bestowed a distinctly odd look to the Semiquaver when prepared to race in the 1921 Aerial Derby (two circuits around London).[29] The writer and editor for *Flight* magazine expressed no surprise when the Semiquaver's many untried experimental features caused its pilot to decide against taking it up on a test flight the night before the derby. It did not start the race.[30] Later, it was flown in competition with a Bristol Fighter — a veteran two-seat World War I fighter — but it performed no better than the Semiquaver with conventional biplane wings would have done.[31] British interest in the wing essentially disappeared in February 1922, when the Air Ministry's director of research reported "wind-tunnel tests have not demonstrated this form of wing to be exceptionally good."[32]

The MB-7 was completed while Alula wing experiments were going on in England, but many of the wing's distinctive characteristics were missing from the MB-7's. It retained a curious, humped-looking, high-mounted monoplane wing but the sharp taper and pointed wingtips were gone. Even so, despite the MB-7 sharing a nearly identical fuselage with its MB-3 and -6 stablemates, it looked very different from them. Whatever anyone else thought about the MB-7, its pilot, Harold Hartney, second-place finisher in the 1920 Pulitzer, "knew the MB-7 could be a winner — intrinsically it was the fastest plane at the meet."[33]

During and after World War I, the Wright and Curtiss companies continued the bitter rivalry that had begun between their founders, the Wright brothers and Glenn H. Curtiss. Both did well during the war. The Wright Company secured a major source of revenue when it obtained a license to build Hispano-Suiza water-cooled engines in the United States. It built thousands of the engines, introducing changes in design, and produced a series of Wright "H" engines that bore strong resemblances to the Hispano-Suizas but differed from them. Wright successfully increased the horsepower from the 180 hp of the standard Hispano-Suiza to about 400. On the downside, increased vibration accompanied horsepower increases above about 300 hp, and vibration would be troublesome in Wright-powered racers.

The Wright Company had no racer at Omaha, but its engines powered the Thomas

Thomas Morse MB-7. This MB-7 flew in the 1922 Pulitzer, and is painted for that race. It is not the MB-7 that crashed and was destroyed in the 1921 race (courtesy National Archives and Records Administration).

Morse aircraft. The MB-6's engine could produce 358 horsepower and the MB-7's 400. Because of the Wright Company's great experience in building engines, their engines were considered more reliable than the Curtiss V-12s, which, manufactured only in small numbers, powered two Curtiss racers and the Balilla in the Pulitzer. Indeed, the Wright engines were "the favorites of the bettors 'in the know.'"[34]

The Curtiss Company's entry into postwar racing began with Charles B. Kirkham designing a revolutionary V-12 engine, the "K-12," in 1916. At 700 pounds, it was heavier than 550-pound Hispano-Suiza and Wright engines, but it generated in excess of 400 horsepower and was almost vibration free. Unsatisfied that any existing airplane design could get the best out of his new engine, Kirkham designed the Curtiss 18-T triplane and 18-B biplane two-seat fighter planes to demonstrate their reliability and power, and they were fast. An 18-T triplane fully outfitted with military equipment and carrying a crew of two flew 163 mph on August 19, 1919, to set an unofficial speed record.[35]

Finlay R. Porter, who took over the Curtiss Company's engine department when Kirkham left the company, modified the K-12 with reduction gears to decrease the propeller's RPMs to a speed less than the engine's RPMs. The reduction had two beneficial effects: A propeller loses efficiency as its speed approaches the speed of sound, and reducing the RPMs increases efficiency. The reduced RPMs also decreased the risk of breaking or splitting the wooden propellers used in the 1920s. Porter's redesigned engine, called the "C-12," was brilliantly designed and generated 338 nominal horsepower, but there was little market for new airplanes or engines in 1920. A small, but ultimately critical, market originated from an unexpected source and set in motion the events that would see Curtiss design and build world record-setting airplanes and engines and cement Curtiss' spot atop American aviation industry.

In Houston, S.E.J. ("Alphabet") Cox, the "Napoleon of Promoters" and "the most seductive and unreliable promoter in America,"[36] made his living selling shares in worthless or nearly worthless "oil land" that, he claimed, would gush oil when drilled. For a modest sum, an investor could buy shares in the oil companies that Cox formed quick enough to dazzle Houstonians, who saw many a fast-talking promoter.

Both Cox, "tall with wavy dark hair and an engaging smile," and his wife, Nelda, or Nelly, "small, shapely, and with the face of a star of the silver screen,"[37] exploited their interests in aviation for publicity. He managed to have his frequent flights reported as news; brought the first airline service to Houston, and used airplanes in his business. She, a pilot and the first American woman to buy an airplane, purchased an Oriole biplane from Curtiss in New York and flew it home to Texas.

Cox leaped at an opportunity for publicity when the Aero Club of Texas asked for a contribution to partially finance building a racer to compete in the 1920 Gordon Bennett. Not settling for any halfway contribution, Cox announced that he would contract for and pay for racers all on his own. After soliciting designs from manufacturers and would-be manufacturers around the country, he selected the Curtiss Aeroplane and Motor Company to build two (or more) racers. (There is some uncertainty about the number of Curtiss-Cox racers that were built. Two were built and shipped to France for the Gordon Bennett; another one was at least started; and there may have been a fourth.)

To satisfy the contract with Cox, Curtiss put together a team of engineers and designers—William L. Gilmore, chief engineer, Arthur "Mike" Thurston, assistant engineer, Charles D. Case, and Bill Irwin, designers—along with draftsman William Wait to design, build, and test the racers in a little over two months. The racers are variously called "Cox,"

"Curtiss," and "Curtiss-Cox" airplanes. The Curtiss Company itself referred to them as "Curtiss-Cox."[38]

The Curtiss designers, worried that cloth-covered wings on the racers could not survive the planned 200-mph speed, covered the wings as well as the fuselage with wood ply. The racers, named by Cox and his wife the *Texas Wildcat* and *Cactus Kitten*, were powered by Curtiss C-12s, geared V-12 engines, generating 338 nominal horsepower and cooled by noisy, whistling, rectangular radiators mounted below and in front of the cockpit on the fuselage sides. The whistling noise was common in aircraft with side-mounted radiators, and the *Wildcat*, the Curtiss-Cox racer that flew in time to be entered in the 1920 Gordon Bennett, was sometimes called "Whistling Billy." Other aircraft with side-mounted radiators were called "Whistling Rufus" and so forth.

The *Wildcat* and *Kitten* could be flown as biplanes or monoplanes with the expectation that they would race as monoplanes. One of two sets of monoplane wings for test and proving flights was thick and spanned 32 feet. The other, thin, wing for racing was 20 feet across. The monoplane wings were shoulder-mounted well in front of the cockpit and braced with solid Y-shaped struts from the landing gear. The cockpit, nestled near the tail, could be completely enclosed — a novelty — with a transparent covering, which slid forward to allow the pilot entrance and exit; it was pushed back into place after the pilot was seated. In the biplane version, the Y-shaped strut from the landing gear passed through an opening in the lower wing to the upper wing, and the biplane wings were joined together by interplane struts and no wires.

Curtiss test pilot Roland Rohlfs made the first flight in the *Wildcat* on August 2 at Roosevelt Field, Long Island. Although fitted with the high-lift monoplane wing, the airplane gobbled up the entire length of the field before leaving the ground. In the air, it porpoised and bobbed, and Rohlfs had his hands full to keep it under control. Aileron flutter jerked the control stick from his hands, and the stick battered the insides of his legs before he could grab it and coax the airplane around and down to land. Engineers adjusted the ailerons and controls, and Rohlfs made a few more flights with the high-lift wing, reportedly reaching 183 mph, about 20 mph better than the world record. No flight was attempted with the

Top: **Pilot Roland Rohlfs with the Curtiss-Cox *Texas Wildcat* monoplane, fitted with high lift wings** (Long Island, summer 1920) (courtesy National Museum of the United States Air Force).

Top: The Curtiss-Cox *Texas Wildcat* biplane. Note the side-mounted radiator, the Y-shaped struts from the landing gear to the upper wing, which pass through the lower wing, the interplane struts, and the absence of bracing wires (courtesy National Museum of the United States Air Force). *Bottom:* The *Texas Wildcat* monoplane and three men important to the Curtiss-Cox racers: Clarence Coombs, who was to fly the *Cactus Kitten* triplane at Omaha, Roland Rohlfs, pilot of the *Texas Wildcat* at the 1920 Gordon Bennett, and S.E.J. Cox, owner of the airplane. The *Wildcat* is fitted with the thick, long-span monoplane wings, and the picture was taken on Long Island during the summer of 1920 before the Gordon Bennett (courtesy Library of Congress).

racing wing because the Roosevelt runway was not long enough for the *Wildcat* to take off with the lower-lift wing. That seemed of little consequence.

French airfields were, supposedly, long and smooth, good news because of the *Wildcat*'s long takeoff run and its landing gear, with solid tires and no springs. Surely there were men in the Curtiss organization or at Roosevelt who had served in France during the war and knew the supposition to be completely wrong. If there were such knowledgeable men and if they said anything, they were ignored.

Before the Curtiss-Cox party left New York, S.E.J. Cox announced that the *Wildcat*—

the fastest airplane in the world by a wide margin, if he was to be believed — had flown 223 mph,[39] which was about 50 mph faster than the FAI-recognized world record. The public announcement of the speed was necessary, Cox said, to block efforts by "foreign interests" to disqualify his racers because of their great speed. Cox left unexplained how high speed could disqualify a racer.

Not one of the 20 members of the Curtiss-Cox team checked to be certain that the larger, high-lift monoplane wings were included in the 100 tons of airplanes, equipment, tools, and baggage shipped to France. The expedition would have been saved a world of trouble if someone in the entourage had interpreted forgetting the high-lift wing as a harbinger of mishaps, missteps, and bad luck and convinced Cox and the others to turn back to the United States.

In France, only one flight was made with the racing monoplane wings. Engine roaring and the propeller thrashing the air, the *Wildcat* sat dead still on the runway. The *Wildcat's* propeller, engineered for high-speed flight, did not grab enough air to move the plane from a standing start. Mechanics and engineers swarmed behind the *Wildcat* as if it were a stalled car, and pushed it down the runway to start it moving. After a long take-off run using all the runway's length, the racer climbed into the air.

Rohlfs called the flight a "nightmare." The monoplane *Wildcat* pitched violently up and down, tearing off all the inspection doors, which had been held in place by strong springs. It was obvious that biplane wings, which had not been tested in flight, would be necessary for takeoffs from the short airfield at the Morane-Saulnier Company where Cox based his team to facilitate his frequent trips to Paris. The biplane wings brought from the United States were judged unsatisfactory, and Curtiss engineer Arthur "Mike" Thurston stayed awake all night to design new wings and tail surfaces. Four days later, the Curtiss team, aided by Morane-Saulnier workers, rolled out the biplane *Wildcat*.

Worried that travel over rough roads would damage the *Wildcat*, Cox refused to have it trucked to the race site. With a final coat of bright red paint still drying and fuel sufficient to fly 50 miles, Rohlfs took off to fly the 30 miles to Ètampes, the Gordon Bennett's race site, in the late afternoon of the 26th. The small red biplane accelerated down the field, hit a rut near the end of the takeoff run, and bounded into the air. The *Wildcat*, Rohlfs reported later, flew better than ever.[40]

Perhaps fuel consumption was greater than expected; perhaps Rohlfs did not fly a direct course to Ètampes. For whatever reason, the engine sputtered from an empty tank when Rohlfs touched down. The wheels, damaged when hitting the rut on takeoff, collapsed, the landing gear struts dug into the ground, the *Wildcat* flipped on its back, and the tail section broke off. With no fuel, there was no fire, and Rohlfs survived but suffered a bloody head wound and a dislocated shoulder. The other Curtiss-Cox racer, the *Cactus Kitten*, remained in its shipping crate in France; there was no time to work on it. S.E.J. Cox's and the Curtiss Company's foray into the 1920 Gordon Bennett ended when the *Texas Wildcat* crashed.

One year later, two Curtiss-Cox racers were present at Omaha. The Curtiss staff had rebuilt the *Cactus Kitten*, shipped to France but never uncrated, as a triplane. Its three wings, identical in shape and size, were originally built for the S-3 triplane Scout, four examples of which Curtiss had built near the end of World War I. They had narrow chords and thin, high-speed airfoils, making them natural choices for the Kitten. They had the advantage, too, of being available at minimal expense. Bert Acosta, the Curtiss test pilot, flew the *Kitten* triplane to 193 mph on a test flight at Curtiss Field on October 27, but it was tricky and difficult to fly. Landing, Acosta said, "Never again."[41] Cox employed Clarence Coombs, who had been scheduled to fly the *Kitten* in the 1920 Gordon Bennett, to fly it in Omaha. Coombs made no test flight before he flew the *Kitten* in the Pulitzer.

The *Kitten* was painted handsomely: the fuselage was bright red, the wings silver, the interwing struts black. It was powered by a 350-hp Curtiss C-12 engine. The side-mounted radiators are visible as open-wire structures in the nose-on photo of the *Kitten* and as "packing case"—appearing rectangles on the side view. The other Curtiss-Cox racer, the "bright red-orange" *Texas Wildcat*, remained on display in "racing monoplane form"[42] throughout the Aero Congress. Its origin is not clear. The *Wildcat* that crashed at Étampes may have been rebuilt. Alternatively, Curtiss had started at least one additional racer when it built the *Wildcat* and *Kitten*, and that airplane might have been completed for the 1921 Pulitzer.

Someone, probably S.E.J. Cox, told a reporter that the *Cactus Kitten* had flown 220 mph.[43] Although 220 mph was faster than the speed achieved by any airplane at that time, it was 3 mph slower than the 223 mph Cox had publicized for the *Wildcat* in 1920 before the Gordon Bennett. No one was ever to know the *Wildcat*'s top speed. Cox withdrew it from the 1921 Pulitzer, saying that officials had declared it was too fast for the race.[44] A racer

Wrecked Wildcat biplane, Étampes, France, September 27, 1920. The broken wheel, severed fuselage, and intact wings are visible (courtesy National Museum of the United States Air Force).

Front and side views of the *Cactus Kitten*. A handwritten note on the edge of the photograph at the National Archives and Records Administration identifies the man sitting in the cockpit as Clarence Coombs, who piloted the *Kitten* in the 1921 Pulitzer. The open wires of the side-mounted radiators are visible in the nose-on view, and their "packing-case" appearance is apparent on the side view. The *Cactus Kitten* finished second in the 1921 Pulitzer. It was faster on the straightaways but lost ground on the turns (courtesy National Archives and Records Administration).

too fast for a race! Imagine that! Imagine the gall of a man who proclaimed that as the reason his racer would be disqualified from a race. In reality, the *Wildcat*'s landing speed of about 95 mph was judged unsafe. Cox, true to his con-man persona, preferred to misstate the reason.

With the *Wildcat* grounded, its pilot Roland Rohlfs, who would sit out the 1921 Pulitzer, made some comments about flying the *Wildcat* to Étampes on the day before the 1920 Gordon Bennett: "It was not an airplane I flew that day—it was a possessed thing I cannot describe ... nothing more than a dining room table with 1,000 horsepower driving it."[45] If it was "a possessed thing," the *Wildcat* must truly have been a beast to fly, and Rohlfs may have little regretted the *Wildcat*'s being disqualified from the 1921 Pulitzer.

Curtiss engineers and designers had learned from the performance of the *Wildcat* and *Kitten* that focusing solely on high speed produced racers with terrible flight characteristics. The success of the Nieuport 29V, a conventional-looking biplane that won the 1920 Gordon

Bennett, had rested on careful attention to a variety of factors: weight and drag reduction and a smooth finish, a reliable engine — the 300-or-so-hp Hispano-Suiza — and Lamblin "lobster pot" radiators had been blended in the Nieuport to produce a racer with good-handling characteristics as well as speed. The Lamblins, barrel-shaped structures mounted between the landing gear struts, produced less drag than other radiators of the day and they worked sufficiently well to keep the Nieuport's engine within near-normal operating temperatures over a 300-kilometer course on a hot day. The *Wildcat* and *Kitten* looked rakish and fast; the Nieuport 29V was smooth and graceful and fast to boot. (Lamblin radiators were used on the 1921 Curtiss racers.)

The navy's 1921 budget for racers was sufficiently large to allow it to contract with two companies. The navy contracted with the Thomas Morse Company for the Alula-winged MB-7 and, in late May, contracted for two racers with the Curtiss Company. In early June, Curtiss designers and engineers began working quickly to prepare two racers for the Pulitzer, at that time scheduled for September in Detroit.

The first Curtiss CR, for "Curtiss Racer," a small, neat biplane equipped with a Curtiss CD-12, a V-12, 376-hp (nominal horsepower rating) engine, and Lamblin radiators, was rolled out nine weeks later, on August 1. It lacked the rakish lines of the *Wildcat* and *Kitten*, and it lacked their outlandish landing speeds of nearly 100 mph. The CRs landed at 75 mph, as specified by the navy contract. The second CR, identical to the first except for the shape of the cowling and having a smaller vertical tail fin, was ready a week later.[46]

The two CRs "were conventional biplanes with docile landing speeds and gentle flying qualities — airplanes that pilots had confidence in riding to their limits of speed (about 190 mph)."[47] They were also the progenitors of a succession of Curtiss biplane racers that were to dominate air racing and speed records for both land and seaplanes in the next five years.

The CRs' fuselages, built in halves and joined together lengthwise, and wings were wooden framed and covered with plywood, and the engine mounts were of wood. A refinement of the K-12 and C-12 engines, the new Curtiss CD-12, a direct-drive development of the C-12 engine with reduction gears, powered the CRs. The C-12's reduction gears had successfully decreased propeller RPMs, but proved troublesome and reduced the engine's reliability. When Finley Porter left Curtiss, the company hired Arthur Nutt, 26, as its chief motor engineer, and Nutt designed a slower-turning engine and discarded the gears. Curtiss built four (count them) of the new CD-12 engines ("D" for "direct" drive), and installed two of them in the CRs.

Bert Acosta made the first flight in a CR. After a smooth flight and a good landing, he sat, horrified, in his seat as the fuselage split into two halves in front of and behind the cockpit. Acosta thought the plane was falling apart around him. It was not. The split was limited to immediately in front and behind the cockpit, and the problem was quickly identified. Reinforcement was needed around the cockpit opening where the fuselage halves could not be joined, and it had not been installed. It was hastily added.

Both CRs were flown throughout August, September, and the early part of October. The almost leisurely flight schedule of the CRs can be contrasted to the near absence or total absence of flight time for the MB-6 and MB-7 racers and the *Cactus Kitten* before the Pulitzer.

When the federal government decided not to participate in the 1921 Pulitzer, the Curtiss Company managed "to borrow back" one of its racers and enter it on behalf of the company in the Pulitzer. Acosta, confident that he would finish better than third place, as he had in the Balilla the year before, and the well-tested racer were at the race site on November 1.

Curtiss had built three V-12 engines—the K-12, the C-12, and the CD-12—and all three were present at Omaha, the K-12 in the Balilla, the C-12 in the *Cactus Kitten*, and the CD-12 in the CR. Giving the lie to those "in the know" bettors who favored Wright over Curtiss engines, *Aviation* magazine commented after the Pulitzer about the "magnificent showing of the Curtiss engines."[48] Before the race, the *Omaha Daily Bee* had ignored the "in the know" betters to select the *Cactus Kitten* and the CR as the favorites in the Pulitzer: "Either of these machines is capable of speed in excess of 200 miles an hour on the straight-away. Both pilots [Coombs and Acosta] have established reputations for intrepidity and rare ability."[49] A *Daily Bee* reporter wrote that Acosta's "sole ambition is to cop" the Pulitzer.[50]

In the last days of October, Aero Congress officials scurried to tidy up last-minute details. All members of the Aero Club were requested to wear their Air Service uniforms so that they could be readily identified during the congress. The chairman of the Aero Congress' contest committee "scoured" the city to find stopwatches to use along with an electronic timer, and a local jewelry store synchronized the stopwatches.[51]

The decision to fly the race counterclockwise, meaning that racers would make left-hand turns, was made literally at the last minute. Some race-goers speculated that the change in direction favored the Americans because the two Italian SVA airplanes' propellers turned right and the Americans' turned left,[52] which, marginally, made left-hand turns easier.

After the worries, confusion, and accusations about finances, the sodden field, and the weather, race day was near perfect. It was "[a]ll any speed pilot could ask. The sun shone, the atmosphere was clear with splendid visibility and a ground temperature of 48 deg. A mild breeze ... blew throughout the afternoon."[53]

Decided by drawing lots, the pilots started in the following order: Acosta first; Lloyd Bertaud in the Ansaldo Balilla second; Coombs in the *Cactus Kitten* third. Hartney in the MB-7 was to take off fourth, but the MB-7's engine refused to run smoothly, and he yielded his starting position to James Curran in the Ansaldo SVA-9, and Captain John Macready, USAS, in the MB-6 followed him.[54] Macready was able to fly the privately owned Ansaldo because he was on leave from the army. After a number of delays and only minutes before expiration of time for starts to be made, Hartney took off last in the MB-7.

The starter flagged Acosta off at 2:30, and the Curtiss racer roared along the ground for about 400 feet before leaping into the air. Acosta circled once, checking out the racer, climbed a bit, dived down and, aided by the speed gained in the dive, flashed across the start line at an estimated speed of more than 200 mph. On his second turn, at Fort Calhoun, a flying wire[55] broke, and some spectators "noticed that one of the wings appeared slightly unsteady"[56] when he arrived back at North Field. Unsteady wing or not, Acosta kept his speed up, flew across North Field and turned onto his second lap. He flew all of the five laps at a constant pace, averaging 10 minutes 26 seconds per lap, with the fastest and slowest laps five seconds off that time. Acosta finished the course of about 153 miles at 176.7 mph. His speed, a new (unofficial) record for a closed-course flight, bested the speed posted in the 1920 Gordon Bennett and other races in Europe.

Well before Acosta had completed his first lap, Bertaud in the Balilla and Coombs in the *Cactus Kitten* were in the air. The spectators had the rare thrill of seeing three racers thundering toward their grandstand, flying at less than 500 feet, turning, and flying away. Their $1.00 admissions had bought them an exciting race.

Coombs was a small man. His head sometimes disappeared below the cockpit edge, leaving the *Kitten* to look pilot-less. Visibly faster than the CR on the straightaways, the

Kitten lost ground as Coombs flew wide around the pylons. Coombs finished at 170.3 mph, about six mph slower than Acosta.

"Bert Acosta on the Curtiss-Navy won the race in the opinion of observers ... due to the fact that Clarence Coombs had never flown the triplane until he stepped into the cockpit for the race and consequently flew more cautiously than Acosta."[57] Pilot experience and ability are harder to evaluate than are the specifications of racers—size, weight, power, etc.—but they can be the crucial factors in winning races. It is also possible that the *Kitten* was such a dog in the air that no amount of piloting skill could make it behave on turns.

The *Kitten*'s flight ended spectacularly. As Coombs was nearing touchdown, the elevator stuck in the down position, forcing the nose down. The racer slammed into the ground. With the lessons learned from the collapse of the *Texas Wildcat* that lacked shock absorbers in its landing gear, the *Kitten* had been built to survive hard landings. It did. The wheels hit the ground, the well-designed shock absorbers worked, and the racer rebounded and flew for a short distance. It hit again, bounded again, and "bounced like a rubber ball" for over 100 yards. S.E.J. Cox, the *Kitten*'s owner, "was seen to clench his hands during the tense moment when it seemed uncertain whether Pilot Coombs would be able to land the triplane."[58] Exhausted, Coombs struggled slowly from the cockpit. The *Kitten* was not damaged.

The day had not started promisingly for Captain Macready and the MB-6. While being run up at full throttle, the MB-6's engine had stopped. Mechanics traced the cause to a faulty fuel pump. The pump, not the Wright Company model that was standard on Wright engines, had been fabricated at the Air Service Engineering Center at McCook Field. The MB-6 sat on the ground while mechanics and engineers ran trials and made adjustments to the balky fuel pump, which were finally successful. Macready took off.

He "flew perfectly,"[59] in what was "probably the prettiest, most consistently flown race of the day."[60] Turns nearly vertically banked, throttle wide open, and speed constant, the times for his fastest and slowest laps differed by about eight seconds. His perfect flight completed, he zoomed upwards to 1,000 feet on his way to the 2,000 feet required before turning to return to the landing field. That's when the engine quit. Remembering the MB-6's clipped wings and its startling "sink rate," Macready dropped the nose to fend off stalling and guided the airplane to a downwind landing. The engine came back to life. Macready immediately turned the racer and made a normal landing into the wind. He finished at 160.7 mph, good enough for third place.

Thrilling the spectators, beginning "within twenty-five feet of the ground [Lloyd] Bertaud [in the Balilla] would swing into an almost vertical bank and soar aloft as he rounded the pylon."[61] Skillful pilot? Yes. Spectacular flight? Yes. Fast enough to win? Not even close. Bertaud finished fourth at 149.7 mph, about 25 mph slower than Acosta in the CR despite having an engine as powerful as the CR's and the *Kitten*'s. The Balilla was a four-year-old design, and, like airplane after airplane, its design may have reached a speed limit that could not be overcome by more power.

The MB-7 was the lightest racer at Omaha, and its 326 nominal horsepower Wright H engine gave it a power-to-weight ratio ("power loading," the weight of the airplane divided by the horsepower of the engine) comparable to the Balilla and a little higher than the CR and *Cactus Kitten* (see appendix). Daylight was fading when race officials told Harold Hartney, the MB-7's pilot, that he had to take off before 4:40 P.M. or be disqualified. Hartney and his mechanics had wrestled with the MB-7's fuel pump, identical to the troublesome one on the MB-6, for almost two hours before removing the pump, replacing a broken spring, and reinstalling it. The engine ran but refused to run smoothly.

Hartney, who had resigned from the army and taken the position of executive secretary of the Aero Club of America, had no thought of not starting the race. He had beaten Bert Acosta for second place in the 1920 Pulitzer, and he was determined not to default to him by not starting in 1921. Worried by the balky engine, his wife, who had flown from Kansas City to Omaha that morning, pleaded with Hartney and urged him not to race. He took off two minutes before the 4:40 deadline. Ten minutes after Hartney had flown off to the northeast, spectators knew something had gone wrong. Peering to the northwest, they expected to see a dot appear in the sky and materialize into Hartney's MB-7. There was no dot. Five minutes later, search planes were readied to take off. There was no need. A farmer telephoned that a racer had crashed.

Robert Adwers, author of *Rudder, Stick, and Throttle: Research and Reminiscences on Flying in Nebraska*, recounts his father's recollections of Hartney's crash.[62] An army lieutenant was the judge at the first turn, and with him were two physicians, two other men, and Adwers' father. The first turn was on the Henderson farm, and a telephone at the farmhouse provided communication with the airfield.

Adwers writes that Hartney was flying at about 300 feet. The engine sputtered, stopped, restarted, and stopped again. As the racer began to stall, Hartney shoved the stick forward, and the plane hit the ground in a cornfield: "My Dad [said] ... it just nosed down and crashed in the cornfield." The two physicians at the windmill "took a wild ride" across the cornfield in a sidecar-equipped motorcycle. By the time they arrived at the crash site on Jim Gilmore's farm, about a mile and a half away, Hartney, who was conscious and in much pain, had been moved to the porch of Gilmore's house. The doctors diagnosed a broken hip, treated him as best they could, and sent Adwers' father and another man back to the telephone at the turning point with the news and a request that an ambulance be sent for Hartney.[63]

Hartney was lying on a cot in the farmhouse waiting for the ambulance when one of many souvenir hunters who swarmed over the MB-7 wreck dropped a match. The gasoline-soaked ground and wood and fabric airplane went up with a whoosh, and the MB-7 was reduced to ashes. Hartney cried when he heard the news.[64] He told a reporter, it was "the prettiest and most graceful plane I ever handled."[65]

He may have been right about the MB-7's being the fastest airplane at Omaha. The other MB-7 that was built to the navy contract was later flown at 181 mph in April 1921.[66] That recorded speed is less than the 200 plus mph reported for the CR and *Kitten*, but there's no confirmation of those speeds. In any case, the MB-7 was competitive with them.

The day after the race, the *Omaha Daily Bee* provided a more dramatic first-person report of the crash than Adwers' father, who said "it just nosed into the ground": "'It was the gas pump,' he [Hartney] told a reporter for the *Bee*. "The motor started to pop. Then I went into the tailspin at an altitude of 500 feet. I pulled out of that all right and the rest of the fall I cannot describe. I only know the machine came down on its landing wheels and crashed through the plowed field and turned over.'"[67] The paper also reported that Hartney's first words to rescuers were to "telephone my wife and tell her I've only sprained my ankle." The words may have reassured his wife, who had urged him not to fly, but Hartney's injuries were much worse. He was hospitalized for weeks and forever after walked with a limp.

Worse news was to come. At nearly the same time as Hartney's crash, Harry Eibe died when making a parachute jump to demonstrate a new model parachute being marketed by the company he represented. The plane carrying Eibe had climbed to about 2,000 feet, and, reportedly, Eibe shrugged off his pilot's suggested jump place and signaled him to fly farther north. The jump looked good, and Eibe was drifting down when some spectators heard him cry, "Help me! I can't swim!" Few spectators, standing on the west side of the field, could see that the northwest wind was driving Eibe over the Missouri River, which, below a steep bank, was out of their sight.

He plummeted into the middle of the river. No water rescue boat or personnel had been provided. The swift current prevented any swimmer from trying to reach Eibe. The absence of oars made it impossible for a would-be rescuer to reach him in a rowboat. The current swept him as much as a quarter mile downstream; his cries continued for a while, then the weight of his parachute and clothes pulled him under. In keeping with then-current ideas about drowning, Eibe reportedly went under three times. Eibe's death cast a pall over the air meet. A scheduled parachute competition was canceled, and the prize money was sent to Eibe's widow in Chicago.

The next day's *Omaha Bee* headlined: "One Dead. One Hurt. At Air Meet."[68] Some criticism was made of the Aero Congress organizers and Omaha police for failing to provide rescue services, but Eibe's death was treated as an accident in a dangerous business. A partner in the parachute company had drowned a few months earlier in Baltimore. The *New York Times* put its story about the Omaha Aero Congress on page 17, and its headline focused on the Pulitzer results: "Bert Acosta Wins Air Race Trophy."[69] The article went on to describe Hartney's crash and Eibe's death.

Both Coombs and Hartney thought that they could have won the Pulitzer. Coombs reportedly said, "Sure, I could have won the race if I had flown a course of right-hand turns."[70] Hartney said, "If I could have stayed in the air I would have beat Acosta by two minutes."[71] No matter what was said and what second guesses were made, Acosta had won. His airplane—the new Curtiss racing biplane—if not the pilot was a preview of racing for the next four years.

Acosta was the only civilian to win a Pulitzer Race. The Air Service and navy pilots who followed him lacked his flamboyance and, most at least, apparently went through life unaccompanied by the demons that dogged Acosta. In *Time* magazine's judgment, Bertrand Blanchard Acosta, born in San Diego, "probably had a greater talent for flying than any man before or since."[72] The public knew him as the third-place finisher in the 1920 Pulitzer and winner of the 1921 race, but his skill in flying heavy aircraft earned the respect of his peers. Lester Maitland, a far above average pilot, finished second in the 1922 Pulitzer, set an absolute speed record that year, and, with Albert Hegenberger, made the first flight from California to Hawaii. He said, "Acosta, as a pilot, is in a class by himself. He understands ships [airplanes], particularly heavy ships, better than any man I know. He is a marvelous test pilot, and has an uncanny gift for discovering, within a few seconds, the strong and weak points in a plane."[73] Acosta made some notable duration and distance flights after winning the Pulitzer. Along with Clarence D. Chamberlin, he flew repeated circuits over Long Island and New York City in a single-engined Wright-powered Bellanca monoplane, the *Columbia*, to set a duration record of 51 hours, 11 minutes, in April 1927.

On June 29, 1927, the Fokker trimotor *America*, commanded by Richard E. Byrd, with Acosta and Bernt Balchen as relief pilots and George O. Noville as flight engineer and radio operator, took off from Roosevelt Field to attempt a flight to Paris. Accounts of the flight vary in how much credit should go to each of the three pilots, but some reports credit Acosta with flying 36 of the flight's 42 hours. The *America* completed the third crossing of the Atlantic, following Lindberg, who was first on May 20–21, and Chamberlin and Charles

A. Levine, who flew the *Columbia*—the airplane that Chamberlin and Acosta had flown to the duration record—to Berlin on June 6.[74] But the *America* did not land in Paris, its intended destination. Thick fog over Paris prevented landing, and the pilots flew the *America* west to the English Channel, where they ditched it off the beach that would become famous as "Omaha Beach" on D-day. The four-man crew paddled ashore.

Two days later an X-ray revealed that Acosta had a broken collarbone.[75] It's likely that the ditching caused the broken bone, but it also gave rise to the often-repeated story that Acosta had become so unruly from drink on the flight that Byrd had hit him with a fire extinguisher or flashlight to quieten him. True or not, the fire extinguisher story illuminates Acosta's descent from fame into alcoholic obscurity. He lost his job with Curtiss in 1925 because of various antics. He spent time in jail for stealing an airplane, failure to pay alimony, abandonment, and other infractions. He rallied in 1937 to fly for the Loyalist forces in the Spanish Civil War, but none of his efforts was sufficient to reestablish himself. He collapsed from tuberculosis in a New York bar in 1951 and died in a sanatorium in Colorado in 1954.

Acosta had a well-earned reputation as a "ladies' man" in the parlance of the times. On the morning after his victory at Omaha, Curtiss engineer Arthur Nutt went to Acosta's hotel room to talk about the performance of the CD-12 engine. Invited in, he saw Acosta grinning in bed "with two females, one on each side." Before boarding a ship to return to the U.S. after the *America*'s flight in 1927, he lured a frosty Hungarian beauty from the 32-year-old Prince of Wales' table. The Hungarian beauty must have proved not too frosty; the next morning, his copilot, Bernt Balchen, went to visit Acosta and found him and the woman still in bed.[76]

Acosta was "grounded" for various flying infractions in 1928. Seven years later, in 1935, the Department of Commerce lifted its ban on his flying and issued him a "learner's permit," which allowed him to receive a transport pilot's license after five hours of solo flight. Alford J. "Al" Williams, the winner of the 1923 Pulitzer Trophy Race and a famous and acerbic pilot, paid Acosta what may have been his greatest compliment. When he learned that Acosta would fly again in 1935, Williams said, "Aviation needs Acosta badly. Seeing him take a ship off the ground is the best eye tonic I've had in years."

Before Eibe plunged to his death in the river at Omaha, S.E.J. Cox, elated at Coombs's second place finish in the Pulitzer, invited 25 people to dinner. Even if the death had subdued him, Cox predicted more wins for his airplanes in future races. He was wrong. From 1922 on, no civilian airplane or pilot would figure in the Pulitzer. It became an entirely military event—all military airplanes and all military pilots. In any case, Cox's legal problems, including stints in Leavenworth Federal Penitentiary after convictions for mail fraud, ended his career as a racer owner.

Aviation magazine's postrace editorial commented on the *Cactus Kitten*'s performance: "It becomes obvious that the 'Cactus Kitten' was at the time it was built the fastest machine in the world—as the Curtiss engineers intended it to be."[77] Whatever the *Kitten*'s top speed, it was never measured on an FAI-sanctioned straight-away course.

The merger of the Aero Club of America and the American Flying Club in 1920 had not provided the robust leadership of aviation activities desired by some aviation leaders. To address those concerns, a new organization, the National Air Association, was created at meetings held during Omaha's Aero Congress.[78] The new association was to represent all aviation interests, and it established a special committee "to care for affairs of national legislation for the future of aviation."[79]

The new association was not constructed as a rival to the Aero Club of America but as an organization that would work harmoniously with it. The relationship became a merger. In January 1922, National Air Association leaders announced that their organization would amalgamate with the Aero Club at the air meet scheduled for Detroit later in the year.[80] The amalgamation took place, and the combined organization emerged as the National Aeronautic Association (NAA) in 1922, and it continues to this day.

Earl Porter and the Omaha Aero Club overcame lagging funding efforts, an uncompleted airfield, and threats of wintry weather to stage a successful Aero Congress. Their last-minute intervention to host the second Pulitzer may have saved the Pulitzer Races from ending after a single race. Had there been no race at Omaha in 1921, ennui and resignation about a lack of popular support might have reversed Detroit's announced intention to hold the 1922 race. After all, the city had pulled out of the 1921 race when U.S. government financing was unobtainable. It is a good bet that government financing after a year's hiatus would have been difficult or impossible, and without it Detroit would likely have turned its back again.

The 1921 Pulitzer was flown, the Pulitzer was a success in terms of speed, the tragedy of a death before thousands of spectators and the wreck of a new racer with serious injuries to its pilot saddened spectators and led to inevitable questions about continuing air meets in the face of their dangers. It made no difference that parachutist Eibe had died in an exhibition, not a contest; critics saw air meets as inherently dangerous. Hartney had not died in the crash of the MB-7, but he had been seriously injured and might have died. Surely air

Curtiss CR, Bert Acosta, and the Pulitzer Trophy. Note the Lamblin radiators on the landing gear struts. The streamlined, conical projections from the radiators were for display purposes and were removed before flight. The Pulitzer Trophy, in the foreground, is now in the National Air and Space Museum. The Curtiss factory building at Garden City, New York, in the background, is still there, but occupied by another company (courtesy National Archives and Records Administration).

races were dangerous, and they required more rules and regulations. Better yet, for critics of their dangers, suspension of the races would eliminate their risks.

Enthusiasm far exceeded criticism. The second Pulitzer Trophy Race had been "a magnificent demonstration of the intrinsic worth of American pilots, airplanes, and aircraft engines.... The speed of ... the Curtiss-Navy racer beats any previous similar performance made here or abroad in competition."[81] Aviation officials and enthusiasts in the United States perked up with that result. Another Pulitzer would surely bring more rewards.

What a change there had been in one year. In 1920, no United States racer had managed to complete a single lap in the Gordon Bennett Race, and, embarrassingly, the record-breaking speeds initially reported in the 1920 Pulitzer had resulted from mismeasurement of the course. Nineteen twenty-one was different. Three of four United States racers—the CR, the *Cactus Kitten*, and the MB-6—had completed the race, and the CR had posted the fastest speed ever recorded around a closed course.

Still, it had been a close-run thing. Omaha was an unexpected host city, far from the big cities of the East, and the Aero Club of Omaha had required financial rescue. The airfield was barely ready. Stress and uncertainty had added to the difficulty of organizing and hosting an air meet, but both had been borne and the effort ended in success. Likely everyone concerned with the air meet would have been content with less drama and worry in getting it off the ground, but they had made it work.

The winning pilot and the winning airplane and the trophy that went to the winner were proudly posed in front of the Curtiss Building in Garden City sometime after the 1921 Pulitzer. Whatever demons were to pursue Bert Acosta afterwards, they must have been out of sight and mind that day.

After the sturm und drang of the 1921 Pulitzer, the members of the Aero Club of America and the NAA, which replaced it, must have breathed a sigh of relief. Detroit, rich, big, vibrant Detroit, close to the population centers of the East and already preparing for it, would host the 1922 Pulitzer.

CHAPTER FOUR

1922

Army Curtiss Racers Come Out on Top in the Pulitzer and a General Sets a World Speed Record, Selfridge Field, Michigan, October 14

America now, for the first time since 1909, holds supremacy alike in altitude, in duration, and in speed, with all three of the major records in the possession of officers of the Army Air Service.[1]

Earl Porter's determination to hold an "International Aero Congress" and host the 1921 Pulitzer at Omaha had paid off. The congress had attracted a good crowd, some national attention, and, most importantly, the Pulitzer had showcased American racers equal to or faster than their European counterparts. European experts, who had previously largely ignored United States high-speed aviation, paid attention. *Flight* magazine in Britain, which had printed less than one page about the 1920 Pulitzer, devoted three pages to the Curtiss CD-12 that powered the Curtiss CR racer to victory in the 1921 race.[2]

The success of 1921 raised expectations for 1922, and its location, Detroit, raised expectations even higher. Detroit, five times the size of Omaha — with almost a million population as compared to Omaha's 200,000 — was the fastest-growing city in the United States and far closer to the eastern centers of population, influence, wealth, advertising, and industry. Detroit was the center, the heart, and nearly the entirety of the United States automobile industry in the early 1920s. Cadillac, Ford, Oldsmobile, Packard, and other manufacturers, along with companies that built components and replacement parts were flourishing and the city along with them. The city's population would increase more than tenfold between 1910 and 1930 — growing from 146,000 in 1910 to 993,000 in 1920 to 1,569,000 in 1930.[3]

Automobile industry executives, the Detroit Chamber of Commerce, realtors, wholesalers, retailers, and others who benefited from the swelling population and influx of money, confident about the city and its future, wanted to move the city to the forefront of the country's attention and keep it there. In that spirit, the Detroit Aviation Society had decided to ask to host the second Pulitzer Trophy Race in 1921. It had changed its mind about the 1921 contest when the federal government declared it would not pay to fly military airplanes in that year's race, but it asked to host the 1922 race and expected the army and navy to return to the competition. The confidence was richly rewarded. The army and navy returned to racing in a big way, sponsoring 24 entries in the 1922 Pulitzer.

The Pulitzer would be the centerpiece of six closed-course races. Three, the Curtiss Marine Trophy Race, the John L. Mitchell Race, and the Pulitzer, were flat-out speed race featuring the fastest airplanes available. The other three races included airplanes not necessarily built for speed.

TABLE 1
1922: Closed-Course Events of the First National Air Race, Third Pulitzer Race, Detroit and Selfridge Field, Michigan, October 7 through 14

Event	Airplane Type	Course	First Three Finishers (hp) Speed
Curtiss Marine Trophy Race	Seaplanes, flying boats. All entries from U.S. Navy.	Eight laps of 20-mile course, alighting, taxiing, and turns on water required on laps 5, 6, 7	1. Lt. W.A. Gorton, Naval Aircraft Factory TR-1 (240) 113 mph 2. Lt. H.A. Elliott, Vought VE-7H (250) 109 mph
Detroit News Aerial Mail Trophy Race	Planned for two-engine mail planes, but none was ready. Flown by twin-engine army bombers and a twin-engine army transport	257.7 miles (10 laps of 25.77-mile course)	1. Lt. E.H. Nelson, Martin transport (2X400) 105 mph 2. Lt. C.B. Austin, Martin bomber (2X400) 101.4 mph 3. Lt. C.M. Cummings, Martin bomber (2X400) 101 mph
Aviation Country Club of Detroit Trophy Race	Light, commercial airplanes carrying two or more passengers plus pilot at greater than 80 mph	257.7 miles (10 laps of 25.77-mile course)	1. Lt. H.R. Harris, *Honeymoon Express*, a Dayton Wright Co. modified de Havilland DH-4 (400) 135 mph 2. Tie. Lt. R.S. Worthington, Fokker T-2 (400) 110 mph 2. Tie. C.S. Jones, Curtiss Oriole (120) 110 mph
Liberty Engine Builders Trophy Race	Army observation planes[a]	257.7 miles (10 laps of 25.77-mile course)	1. Lt. T.J. Koenig, Packard-Lepere biplane (400 hp) 129 mph 2. Maj. F. Bradley, De Havilland DH-4 (400) 126 mph 3. Lt. W.L. Boyd, De Havilland DH-4 (400) 122 mph
John L. Mitchell Trophy Race	Thomas Morse (Boeing-built) MB-3As, all from 1st Pursuit Group, USAS	200 km (4 laps of 50-km course; ca. 124 miles.)	1. D.F. Stace, MB-3A (340) 148 mph 2. A.M. Guidera, MB-3A (340) 136 mph 3. O.W. Broberg, MB-3A (340) 135 mph
Pulitzer Trophy Race	Racers capable of at least 140 mph and landing speed less than 75 mph	250 km (5 laps of 50-km course; ca. 155 miles)	1. R.L. Maughan, USAS, Curtiss R-6 (450) 206 mph 2. L.J. Maitland, USAS, Curtiss R-6 (450) 199 mph 3. H.J. Brow, USN, Curtiss CR-2 (450) 194 mph

[a]The navy had entered two airplanes in this race but withdrew them after a controversy about the definition of "observation plane."

The fierce 1922 Curtiss Race would prove a landmark not only because of its competitive nature but also because its results would foretell navy decisions about airplane engines. The Mitchell Race, added to the race program at the last minute, seemed at first of little interest. It would be flown by pilots of the 1st Pursuit Group in identical pursuit aircraft, men who were Air Service pilots—not specially selected and trained race pilots—flying standard airplanes. Where was the excitement in that? To the surprise of many, the test of ground crews' abilities to ready and refine service aircraft for racing and of pilots' skills in handling airplanes they flew on an everyday basis proved exciting and engrossing, and the Mitchells continued until 1936, eleven years after the last Pulitzer was flown.

The Pulitzer, the last race of the 1922 air meet, met all expectations. Racers flying 200 mph roared over the heads of crowds at 20- and 50-foot altitudes, disappeared out of sight over water, and, preceded by ear-splitting roars, swept back into sight from a different direction. Each of the first seven finishers flew faster than the 1921 Pulitzer winner only 11 months before. The winner flew the fastest airplane in the world.

The other races were for production model airplanes, which could be competitive and exciting but had no promise of truly fast speeds. On the positive side, they allowed some civilian competitions, and the races of Air Service two-engined transport and bomber aircraft produced sights of the big airplanes being "bent" around pylons at low altitude and high speeds.

The five races for land planes, including the Mitchell and the Pulitzer, would be flown from the Air Service's Selfridge Field, near the town of Mt. Clemens, about 25 miles north of the center of Detroit. The Curtiss Marine Trophy Race, which opened the eight days of the National Air Race, would be flown over the Detroit River, which flows in the middle of the city, and across an arm of Lake St. Clair to the city's east.

The Detroit Aviation Society, Inc., published *Rules and Entry Blank: Detroit's Aerial Contests, 1922*, in the spring of 1922. It closely resembled *Entry Blank and Rules: Second Annual Aerial Contest for the Pulitzer Trophy* booklet, which the society had published the year before in advance of the canceled Pulitzer that was held in Omaha. The 1922 booklet may have been the first publication to call an aviation meet the "National Air Race."

The 1922 National Air Race initiated an event, the "On to..." contests that were to be repeated at other large race meets. Flyers from more than 200 miles away from Detroit could compete for prizes that were awarded on a formula based on elapsed time, miles flown, average speed and motor power."[4] A contemporary *New York Times* account described the 1922 event in surprisingly dismissive words: "One of the unusual features of the meet was the arrival of gypsy fliers from other cities."[5] *The Aircraft Year Book* was more upbeat and accurate, saying that the "On to Detroit Race" was "calculated to be of wide-spread influence in the stimulation of civilian flying."[6] In all, about 200 visiting airplanes were parked or hangared at Selfridge during the race week.

Glenn Curtiss had first offered the Curtiss Marine Flying Trophy (hereafter Curtiss Trophy) in 1911 to encourage higher speeds and better performance from seaplanes and flying boats, but the contest was slow to get started. The first Curtiss Trophy competitions were endurance contests flown in 1915 through 1918. Following three years with no contest, the competition was revived in 1922 as a speed contest. Foxworth writes, "For no-holds-barred rivalry the 1922 Curtiss Marine Trophy has to rank as one of the most aggressively flown air races of the 1920s."[7] It also had unique rules that guaranteed excitement and required piloting skills likely never tested in another race.

The 1922 Curtiss Race was a U.S. Navy affair, and the navy made it easy for the most naïve spectator—one who had never seen an airplane before—to identify the racers. Nine were painted a different color — purple, blue, yellow, green, gray, red, brown, black, or white. Evidently the navy had only nine paint colors on hand; the 10th racer was painted a black-and-white check. The Curtiss Race must have been the most colorful ever flown.

The navy dusted off its two Curtiss 18-T triplanes, which had been in storage since the 1920 Pulitzer, and readied them for the Curtiss Trophy contest. Along with the radical Gallaudet D-4, the 18-Ts were the pre-race favorites. The remaining seven aircraft — standard one- or two-seat advanced trainers, "shipboard combat planes,"[8] and observation airplanes — were given little chance.

The navy replaced the 18-Ts' Curtiss K-12 engines, which had failed in the 1920 Pulitzer, with new and reliable CD-12s, and installed four-bladed propellers. The airframes were reconditioned and refurbished, equipped with a large fuselage-mounted pontoon, two wingtip pontoons, and an auxiliary rudder beneath the tail. On sheer speed, the 18-Ts were the clear favorites in the Curtiss.

From our vantage point of 90 years in the future, it is difficult to understand how anyone held any hope for the Gallaudet D-4 as a racer. The most radical airplane at Detroit and an airplane that would have been radical anywhere, it was a clean, well-finished, two-seat patrol bomber, powered by a 421 (nominal) hp Liberty V-12 engine built to navy standards. But the D-4's long (45+ feet) wingspans and 5,400-pound loaded weight made it at

Curtiss 18-T in the Curtiss Marine Trophy Race at Detroit in 1922. Painted yellow and flown by Lt. Lawson Sanderson, USMC, the number 4 Curtiss 18-T was comfortably in the lead as it entered the last lap of the Curtiss Marine Trophy Race but ran out of fuel and did not finish (courtesy National Archives and Records Administration).

The original Gallaudet D-4, two-engined, two-seater patrol bomber, powered by two Duesenberg engines in the fuselage and cooled by side-mounted radiators. As modified for the 1922 Curtiss Marine Trophy Race, a Liberty engine was installed in place of the Duesenbergs and the forward cockpit was faired over. The racer was designated "D-4." Note the propeller blades amidships and that the seaplane is resting on special supports while out of the water (courtesy National Archives and Records Administration).

least one and a half times larger and heavier than other racers at Detroit, many of which had engines of equal or greater power. Like the Curtiss 18-Ts, the Gallaudet was built in what became the standard configuration for U.S. navy seaplanes, with a large central pontoon and balancing pontoons at the wingtips.

The mounting of the D-4's propeller and engine contributed to its smooth appearance. Edson Gallaudet, the airplane's designer, had decided on the unusual propeller location — midway between the wing and tail, with only the propeller tips protruding from the fuselage — to capitalize on propeller properties. The tips of a propeller generate much of the propeller's thrust, and the more closely spaced blades nearer the hub are the source of much of the propeller's drag. Gallaudet's design was intended to capture the propeller's thrust, with the tips protruding from the fuselage, and reduce drag by enclosing the propeller blades nearer the hub inside the fuselage.

Gallaudet had built a single D-1 as an experimental airplane for the navy, and the navy accepted it as satisfying the terms of the contract but ordered no more. Gallaudet built four D-2s for the army; there was no D-3; and Gallaudet built two D-4s for the navy in 1918. The Gallaudet airplanes may have shared structural or handling problems. Two of the seven Gallaudet aircraft, one of the army's D-2s and one of the navy's D-4s, crashed, killing the pilots in both cases. The navy made the other D-4 into a racer by fairing over its front cockpit. It was painted blue, with race number 3 painted on its side for the Curtiss Race.

The Curtiss Marine Flying Trophy race was eight laps around a 20-mile-long course with its start in the Detroit River near the center of the city. After takeoff, the racers followed the course of the river east to Lake St. Clair and flew along the south shore of the lake to an anchored barge. There, they turned north to cross an arm of the lake to a prominent water intake pipe. Once past the intake pipe, they turned left and followed the north shore of the lake to the west to the mouth of the river. There, they veered left and flew back to the start line.

Glenn Curtiss was interested in the commercial possibilities of seaplanes and flying boats, and such airplanes had to be able to navigate on the water as well as fly. As a test of the racers' capabilities and the pilots' skills, racers were required to alight on the river as they rounded into laps 5, 6, and 7, execute a complicated taxi course around the starters and referees' barge, and then take off to continue.

Bad weather forced postponement of the Curtiss Race from Saturday, October 7, to Sunday, October 8. Sunday was not a great day either; it was raw and gray, with wind and

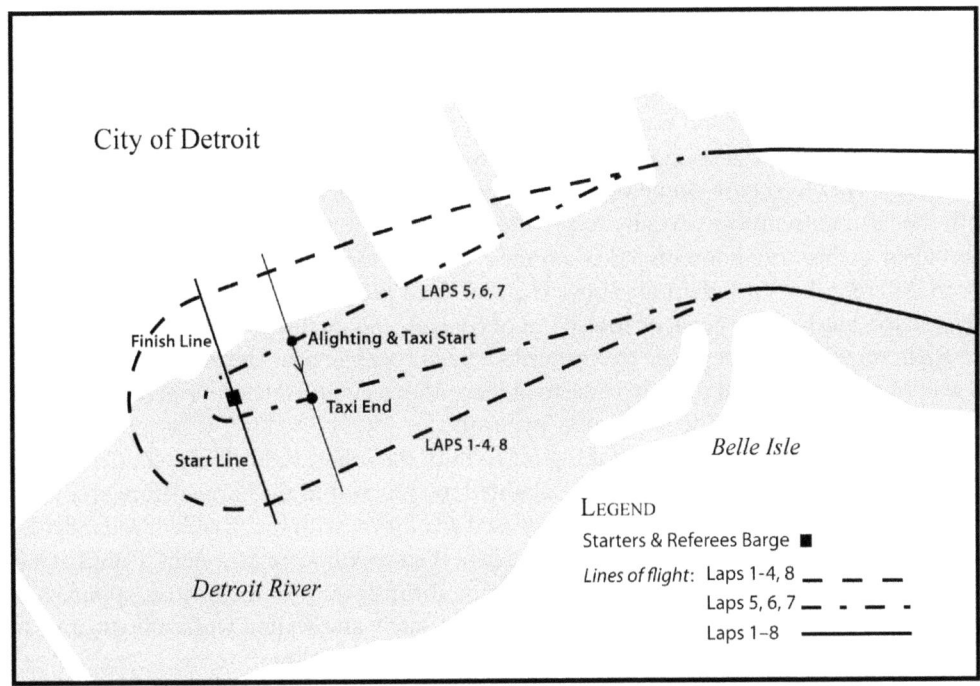

Figure 3. Map of the navigation area of the Curtiss Marine Trophy Race, Detroit, Michigan, 1922. The race started and ended on the Detroit River in downtown Detroit. The racers took off and flew east from the city along the south shore of the river and near the south shore of Lake St. Clair to a barge that marked the first turning point. There, they turned left and flew north over the lake to a large water intake pipe, which marked the second turning point. The racers turned left again to follow the lakeshore to the river and back along the north side of the river to make a turn around the starting barge to begin the next lap. On laps 5, 6, and 7, the racers were required to alight within the "Alighting and Taxi" area, taxi around the starting barge, and take off again before reaching the "Taxi End" area. The course to be followed on laps 5, 6, and 7 is indicated by the line of dashes and dots. The dashed line indictates the courses flown on laps 1–4 and 8. The solid lines indicate the courses flown to and from Lake St. Clair on all laps. Total length of the race: 8 laps of 20-mile-long course, or 160 miles (map by Lynn Springer).

low-hanging clouds, but at least 200,000 people, many from outside Detroit, arrayed themselves along and under the racecourse, with the crowd concentrated near Belle Isle in downtown Detroit where the Detroit River is about three-fourths a mile across.

Still-gusty winds tossed the racers lined up in shallow water, just off the beach, as they were flagged off at one-minute intervals, one after another, in a spectacular, crowd-pleasing race. The racers flew east along the river about two miles to where the river opened up into Lake St. Clair. They followed the north-facing shore of the lake a couple miles to a barge, where they turned north, crossed the lake to a waterworks intake pipe, turned left and flew along the south-facing lakeshore back to the river, turned right, flew along the river to pass a barge that marked the starting line, and started a new lap. The Curtiss Race required eight laps of the 20-mile-long course.

From anywhere along the course, spectators saw faster seaplanes gain on and pass slower ones. Spectators near the start line were treated to the sight of fast-moving seaplanes slowing as quickly as possible to alight on the river as they passed a marker and before turning into the 5th, 6th, and 7th laps. On the water, they taxied along the north shore of the river, made a left 180 degree turn around the barge that marked the starting point, and took off to fly the next lap. Nothing like the water navigation had been seen before. Nothing like it would be seen again.

Both Curtiss T-18-1s, one starting in position 1, the other in 6, came to grief. Rutledge Irvine, flying the green-painted race number five, 18-T — the first 18-T to take off — was leading when he alighted to make the taxi and turn on his 5th lap. Alighting hard, the 18-T skipped and bounced several times. A crosswind nearly blew it into the starting barge, but Irvine snaked it past. Somewhere in alighting, maneuvering on the water, and taking off, the left pontoon was partially ripped from the wing. Back in the air, the whole plane throbbed as the loose pontoon jerked on the racer. The racer's fate in calm air might have been different, but turbulent air above the coal-fired power plant near the starting point proved too much. In an airplane tossed out of control, and unable to glide or fly far enough to alight on water, Irvine chose to crash into a huge coal pile near the power station. After a wait of minutes, which must have seemed longer for spectators, Irvine emerged from the coal, unhurt and filthy. The 18-T was a write-off.

Lawson "Sandy" Sanderson, taking off sixth in the other, yellow-painted, Curtiss 18-T, was comfortably in the lead when he alighted to taxi around the barge before taking off on his eighth and final lap. He narrowly missed a collision with the disabled Gallaudet D-4, which was drifting in the river, and was back in the air with victory in sight. Comfortably in the lead as he approached the finish, his engine abruptly quit for the least acceptable reason possible. It ran out of gas. Sanderson alighted safely and waited while a boat came to his aid and towed the only remaining flyable Curtiss 18-T to shore.

Sanderson would have easily won the race had he not run out of gas. He completed the first five laps at greater than 120 mph; his sixth and seventh were slower, just over 116, but still three mph faster than the winning speed. It may have been of little comfort to Sanderson, but he did take home a special trophy for fastest one-lap speed. Both pilot and 18-T would race again. Sanderson flew in the 1922 Pulitzer later in the week and again in the 1923 Pulitzer. The by-then venerable 18-T would fly one more race, as a land plane in the 1923 National Air Races to be held in St. Louis, where it crashed with its crew of pilot and mechanic/observer in the Liberty Engine Builders' Race. The 18-T was destroyed; the crew were unhurt (see Chapter Five).

The Gallaudet D-4 did not finish the race. When making a required alighting, something fouled and damaged the propeller, and the blue seaplane was unable to take off. The mishap underlined the questionable utility of the propeller's mid-fuselage location, and Sanderson's 18-T narrowly missed hitting the D-4 as it drifted around the alighting area.

The three favorites—the 18-Ts and the D-4—were out of the race, leaving the contest to be decided among the seven airplanes expected to be also-rans. A rather prosaic standard Naval Aircraft Factory (NAF) biplane won the Curtiss Marine Flying Trophy. The NAF entered four airplanes, all designed to be fighters, in the Curtiss Race. The two TRs, the TR-1 (painted white for the race) and the TR-3 (black-and-white check), were equipped with thin "racing" wings. The two TSs, the TS-1 (brown) and the TS-2 (red), had thicker "standard" wings. Of far more significance than the different wing profiles, the airplanes were powered by different engines.

Both the "1" racers, the TR-1 and the TS-1, were powered by Lawrance J-1, 240-hp maximum, air-cooled engines, which were novelties in 1922. Essentially every other service airplane, army and navy, had water-cooled engines. The TR-2 had a new liquid-cooled engine, the Aeromarine U-8-D liquid-cooled engine of 228 maximum horsepower, and the TS-3 had a version of the much-used Wright E-3 liquid-cooled engine of 250 maximum horsepower.

Lt. A.W. Gorton, USN, flew to victory in the TR-1, the only one of the four NAF aircraft to complete the 160-mile-long Curtiss Race. Problems of various sorts forced the other three NAF airplanes out of the race. Dropping oil pressure ended the race of the other air-cooled-engine powered racer, the TS-1; it alighted on the 6th of the required eight laps of the 20-mile-long triangle course. The TS-2 lost all its radiator water and was forced to alight on its 4th lap. The TR-3 also alighted on its 4th lap because of its loosening propeller.

Naval Aircraft Factory (NAF) TS-2. One of four generally similar NAF fighter designs entered in the Curtiss Race, it lost its coolant on the 4th lap of the eight-lap race and was forced to alight. The similar NAF TS-1, powered by a Lawrance air-cooled engine, won the race (courtesy National Archives and Records Administration).

Curtiss 18-T and Naval Aircraft Factory TR-1 in flight, Detroit, 1922 (courtesy National Archives and Records Administration).

Only one airplane other than the winning TR-1 completed the Curtiss Race. Lt. H.A. Elliott, USN, flew a Vought VE-7H, a standard navy training plane (versions of which had flown in the 1920 Pulitzer) around the Curtiss course at 108.7 mph, only four miles per hour slower than the winner. The TR-1's winning speeds of 112.6 mph was 30 mph slower than the 146-mph winning speed in the 1922 Schneider Race for seaplanes. No direct comparison of speeds is possible because of the time-consuming, tricky alighting, taxiing around an obstacle, and taking off again that were required on laps 5, 6, and 7 of the Curtiss race.

The TR-1's engine was to play a significant role in the future of naval aviation. Admiral W.A. Moffett, director of Naval Aviation,[9] had questioned the reliability of water-cooled engines (about one third of all failures of those engines resulted from cooling problems or loss of coolant) and objected to the size and weight of radiators, plumbing, and coolant. The weight, size, and complexity of liquid cooling were special problems aboard ships, where storage space for airplanes and parts and workspace were always limited. The success of the J-1 engine in the Curtiss Race added weight to Moffett's contention about the superiority of air-cooled engines and contributed to the navy's adopting of those engines for almost all its airplanes.

American air-cooled aircraft engines of all descriptions can be traced back to 1919 and Charles Lawrance's manufacture of three-cylinder air-cooled engines. The Air Service's Engineering Division at McCook Field ordered and tested a nine-cylinder radial from Lawrance in 1920 and was sufficiently impressed to order three more, one of which passed the then-standard 50-hour continuous-running endurance test in 1921. Lawrance developed a slightly larger engine, the "J" model, for the navy, which powered the Curtiss Race–winning TR-1. The J-model air-cooled radials were the first ones manufactured in quantity[10] and led to the very successful Wright Whirlwind engines.[11]

The staging of the Curtiss race, with its three required alightings, turns, and takeoffs, was, "quite an original and interesting feature for a seaplane race, and that helped considerably to liven up the proceedings."[12] The entry of the exotic Gallaudet and the speedy Curtiss 18-Ts contributed to spectators' enthusiasm for the Curtiss Race, and the win by a nearly stock production navy airplane highlighted the capability and reliability of the airplanes of the armed services. The Curtiss Race was a great start to the National Air Races.

Selfridge Field, the site of all the remaining races of the 1922 National Air Race, was located near the small town of Mt. Clemens. In contrast to Detroit's recent rapid industry-fed growth, Mt. Clemens, about 25 miles to the north, had initially prospered because of the curative and restorative powers of the local waters. From the 1870s through the 1930s, thousands of patrons came from around the world to the town's bath houses. The baths brought wealth and bestowed on the town a peculiar ambience — the sulfurous smell of rotten eggs, which local historian Don Green describes in his talk entitled "Awful Smell."[13] The smell was a constant presence in Mt. Clemens' history: "[T]he stench of ... the bath era. Who could forget it?"[14]

Local legends differ about the discovery of the health-giving effects of the water,[15] but it is clear that Dr. Henry Taylor transferred the claims into commercial reality when he built the first bathhouse, "The Original," in 1873. "Taking the waters" was one of few "treatments" 19th century medicine could offer to sufferers from many diseases.[16] Advertising and word-of-mouth spread stories of near-magical relief from pain and easing of diseases following immersion in, and the drinking of, Mt. Clemens' water. After Dr. Taylor, other healers and businessmen built more bathhouses and saw them quickly become busy. At one time or another, a total of 78 hotels — many with baths — catered to bath patrons from across the country and Europe.

Some bathhouses had dozens of baths, each in its private room — some of marble — with female attendants for women and male attendants for men. Estimates for the number of visitors during the summer range up to 50,000 per month.[17] The 50,000 figure is surely too high. Even if each visitor stayed only a single day — and those who took a complete "cure" could stay three weeks — visitors would have added about 1,700 people, more than 10 percent of the town's population of 14,000 at the height of the bathhouse era.

Whatever the number of visitors, the population of the town grew. It was 6,600 in 1900, when the bathhouses began to flourish, about 14,000 in the mid–1930s, when the bathhouse era began to fade, and 20,000 in 1973, when the last bathhouse closed. Dorothy Larsen, a local librarian and historian, says the availability of cortisone to treat aches and pains combined with the shortage of clients who could take 21 days for a "treatment" killed the bathhouses.[18]

Mt. Clemens had two other major businesses, seemingly out of step with healing and curing. The erection of the first building in Mt. Clemens — a distillery that began operation around 1797 — heralded the town's subsequent notoriety for its many saloons and its flaunting of Prohibition (1920–1933). Alliteration makes it easy to remember the sources of Mt. Clemens' prosperity. Added to "Baths and Booze" was "Babes," referring to the town's nearly wide-open prostitution industry. Drinking and prostitution may seem a peculiar fit with health-bestowing baths, but the "bathing industry" wanted to provide its patrons with access to entertainment "in the style of other spas."[19] The bathhouse owners and managers surely tolerated and may have abetted both "booze and babes."

Prohibition was the law of the land in 1922, but it mattered little in Mt. Clemens. A brewery and a "blind pig" (a "blind pig" was a "speakeasy" in Detroit parlance); the name perhaps coming from police officers turning a blind eye on such places, operated next door to the police station throughout Prohibition,[20] and whiskey was the highest-value product coming down the Clinton River from Canada. A different kind of business came to town in 1915. The Packard Motor Car Company, denied permission to expand its factory in Warren, Ohio, had moved to Detroit in 1902. Packard prospered, and Henry Bourne Joy, its president, expanded the business to aircraft engines in 1915. He brought aviation to Mount Clemens when he purchased land on the shore of Lake St. Clair near Mt. Clemens to build a test track and the Joy Aviation Field.

The purchased land had advantages. It was flat, and nearby Detroit provided mechanical, electrical, and engineering expertise, manufacturing and repair facilities. The site had its disadvantages; much of the land was wet and soggy, fierce, bitter Michigan winter winds and storms made flying uncomfortable if not impossible for a few months every year.

Wanting to make money from his Lake St. Clair property, Joy and "the Mount Clemens Business Men's Association set about convincing the War Department that the largely unimproved but strategically located Joy Aviation Field was an ideal place to establish a military airfield"[21] in 1917. The effort succeeded, and the Air Service leased the land for three years. Facing up to the problems of the location, the Air Service drained and improved the "sodden marsh" that covered much of the field and completed 12 hangars, several barracks, shops, offices, and other structures during 1917. Nevertheless, "conditions at the site were still miserable. Large pools of standing water"[22] returned with heavy rains. To reflect its change from civilian to military ownership, the Air Service changed the name of the airfield to honor an army aviation pioneer. The new name, Selfridge Field, honored army lieutenant Thomas Selfridge who had been part of Dr. Alexander Graham Bell's Aerial Experiment Association and had designed the association's first powered airplane, the *Red Wing*, which he flew. He was one of the first three army officers trained to fly the army's first dirigible. Despite his accomplishments, his death is more often remembered than his achievements. He was the first army pilot to die in an airplane crash, perishing when flying as Orville Wright's passenger on a demonstration flight at Fort Meyer, Virginia, on September 17, 1908. Six months after Selfridge Field opened, dreadful winter weather made flying impossible, and flight operations were transferred to airfields in Louisiana and Texas until the weather improved. March 1918 brought spring floods, but an aviation gunnery school opened the following month. Short-lived, it closed before war's end in November 1918.

Selfridge Field faced an uncertain future in 1919. No training school remained, parts of the field shone with standing water, winter weather was reliably bad, and spring floods threatened yearly. Despite those disadvantages, an advisory committee highlighted the advantages of the location because of it proximity to the automobile industry.[23] Whatever the field's merit, it is difficult to escape the conclusion that Joy unloaded a not-so-good property on the Air Service when a U.S. Representative from Michigan added an amendment (it would now be called a "legislative earmark" to identify it as legislation likely to benefit a particular individual or organization) to the 1922 army aviation appropriation that directed the War Department to purchase Selfridge Field, about 650 acres, for $190,000.[24]

In summer 1922, soon after the purchase had been assured, the Air Service opened Selfridge Field as a fully operational airfield. Its future remained no sure thing. General Mason Patrick, commander of the Air Service in the mid–1920s, considered Selfridge unsatisfactory and wanted it closed. Lobbying and more appropriations brought the improvements necessary to make the airfield a reliable base. A seawall completed in 1931, finally ended flooding and wet, soggy land. Selfridge served as a United States Army Air Force (USAAF or AAF) and United States Air Force (USAF) field until 1971, when it was transferred to the Michigan Air National Guard.

The 1st Pursuit Group, the Air Service's only pursuit group in the 1920s, was the most famous unit to be based at Selfridge. It moved to Selfridge in 1922 and remained there until 1941. Its brightly painted Curtiss P-6Es were probably the most photographed airplanes of the early 1930s.

Major Carl A. "Tooey" Spaatz,[25] commander of the 1st Pursuit Group when it moved to Selfridge, oversaw preparations for the air meet to be held in October 1922 as he dealt with other problems. Mt. Clemens ignored Prohibition. Selfridge Field, at least the officers club, ignored it too. When questioned by General Patrick about on-field drinking, Spaatz admitted it went on and argued that it was good policy. Otherwise the officers would go elsewhere to indulge — the blind pig next to Mt. Clemens' police station was a couple miles away — where Spaatz could not keep an eye on them. Spaatz apparently prevailed.[26]

Mr. Joy, who had sold Selfridge Field to the government, and whose estate bordered it, raised another problem. He complained about airplane noise from the field. Spaatz ordered his pilots not to fly over Joy's property and altered landing patterns. It wasn't enough. Joy's complaints reached the office of the secretary of war, but, evidently, the powers-that-were in Washington accepted Spaatz's efforts as sufficient.[27] Surely, anyone who knew the Selfridge Field history found it amusing or ironic that the man who began development of the land as an airfield and sold it to the government for use as an airfield complained about airplane noise.

Cold, choppy winds across Lake St. Clair and Selfridge Field tested the enthusiasm of 25,000 to 35,000 spectators on October 12 and 13 when three races were flown. The crowds included the secretary of the navy, Generals Patrick and Mitchell, the chief and assistant chief of the Air Service, respectively, the mayor of Detroit, Henry Ford, and Orville Wright. Edsel Ford, who had guaranteed $10,000 in prize money when a lawsuit brought by Henry

Curtiss P-6E of the 1st Pursuit Group, Selfridge Field, Michigan, early 1930s (courtesy National Museum of the United States Air Force).

Woodhouse against the Aero Club of America threatened suspension of the Pulitzer, was an especially valued spectator. (Ford's money wasn't needed; the Woodhouse suit failed.)

None of the airplanes that flew in races 2, 3, and 4 — two-engine mail planes (Table 1 event 2), light commercial planes (event 3) and observation planes (event 4) — on Thursday and Friday of the Detroit race week was a racer. Because there were no two-engined mail planes to race in event 2, the Detroit *News* Aerial Mail Trophy Race was flown by two-engined army airplanes. The Martin transport plane's victory was expected because it was essentially the same design as the Martin bomber, but far less burdened with military equipment and weighed less.

The *Aircraft Year Book* commented about the mixed bag of aircraft that competed in the Aviation Country Club of Detroit Trophy Race: "A speed contest between a light commercial carrier, such as the Curtiss '*Oriole*'; a heavy weight lifter, like the Fokker transport; a training type, the Vought three-seater [VE-7]; and, finally, the very fast and powerful '*Honeymoon Express*,' was curious, if not bizarre, and the conviction was generally expressed that very unfair handicaps were imposed."[28] The "fast and powerful "*Honeymoon Express*'" won. The Fokker T-2 and the Curtiss Oriole biplane, flown by Curtiss executive C.S. "Casey" Jones, tied for second. Jones and the Oriole were to become fixtures at the National Air Races, winning a number of races for light planes and almost always placing well.

The Packard Lepere biplane that won the Liberty Engine Builders Trophy for observation planes was already famous. It had set the world's altitude record at 37,800 feet a year earlier.

General William Mitchell, who established the most exciting and lasting of the races unveiled at Selfridge to honor his brother who had died in France in World War I, announced it in time to be flown in 1922, but not early enough to have it listed in the program. Neither did it appear in *Aeronautical Digest*[29] and *Aviation*[30] reports written in advance of the races. Even had it been mentioned, it might have escaped attention.

The Mitchell seemed no more than a convenience or courtesy to the Air Service to allow its pilots to race their standard aircraft. Making it seem even more like an expedience, General Mitchell established it only after it became apparent that the 1st Pursuit Group, stationed at Selfridge, could not fly their Boeing MB-3A pursuits in the Pulitzer because they lacked the range to complete the 250-mile-long race.

The Mitchell Race, limited to 1st Pursuit Group pilots flying their everyday standard pursuits, could be seen as having only local or, at most, Air Service interest. There was no advance indication that although it was "a pure 'horse race' it soon came to far exceed in spectator interest the more sophisticated Pulitzer."[31] The first running of the John L. Mitchell Race was set for the morning of Saturday, October 14, and the third Pulitzer for the afternoon. The mayor of Detroit declared "Race Day" a municipal holiday. The weather, warm and muggy with blustery winds blowing across Lake St. Clair, was better than on the previous two days but not ideal for racing, and a light haze, which thickened through the day, hung over the racecourse.

Cars began arriving at Selfridge at 7:00 A.M.; four hours before racing would start, and the Michigan State Police estimated 10,000 cars were parked around the field. More spectators arrived on special Grand Trunk Railway trains carrying passengers from Detroit and back home for $1.80, admission to the races included. In all, the state police estimated a crowd of 75,000 spectators, whose interest was kept at "fever heat" so "that scarcely a person left until the last plane crossed the line, just before dark."[32]

The triangular 1922 course for the Mitchell and Pulitzer races could appear to have been laid out for seaplanes because all but a few miles were flown over Lake St. Clair. Safety considerations apparently dictated the course; racers would seldom be over crowds of people or areas of habitation, and the impact of crashing a faltering racer into the lake's water would be less than hitting land. (This author has found no reference to pilots' being questioned about their swimming ability.)

The race began at Selfridge, located on the north side of a peninsula that extends east into Lake St. Clair. The racers crossed the peninsula, and continued south-southwest to skirt the lake's shore to Gaukler Point, where they turned to the east to fly over the lake to an anchored ship, which marked the second turn. From there, they flew northwest and back to Selfridge. The course was 50 kilometers long, and the Mitchell required four laps, the Pulitzer five. (See Figure 4.)

Reflecting local pride, that day's afternoon edition of the *Detroit News* headlined "Michigan Flier Is Race Winner," when Michigan-born Lt. Robert Stace won the 200 km-long Mitchell at an average speed of 148.1 mph.[33] Stace had completed West Point in two years and qualified as a pilot on June 30, only three and a half months before the Mitchell Race. Assigned as the 1st Pursuit Group's assistant engineering officer, he infused the mechanics and other enlisted men of the Service Squadron with determination to defeat competitors from the group's "operational" squadrons, which, as always, looked down their noses at service and support squadrons. They tuned his airplane's engine, removed all possible protuberances, and varnished the airplane to greater smoothness and luster.

Stace was first to take off in the Mitchell race. He climbed and circled as the other five competitors took off and lined up. When all were ready, the six MB-3As dived across the starting line in a roaring burst of speed, immediately jockeying for better positions while flying the straightest possible courses to keep the distance flown to a minimum. The race,

Lt. Donald F. Stace (USAS) and his MB-3A, winners of the first John L. Mitchell Race. The MB-3A's side-mounted radiator is clearly seen (courtesy National Archives and Records Administration).

flown by identical standard-issue pursuits would go to the pilot whose ground crew best prepared his airplane and who best judged his distances to the pylons, turned closest and fastest, and most quickly regained a good flight line and best speed toward the next turning point.

Spectators watched pilots jockey for position on the straightaways and bank steeply around the pylon on Selfridge Field. They watched the MB-3As fly away headed southwest, disappear from sight, and, heralded by a roar, come back into sight in the southeast, so low over the water that propeller wash churned up whitecaps. Spectators held their breaths as the pilots pulled up to clear the trees lining the lake. Stace was never headed and won. He'd flown only 100 hours as a pilot, but he'd clearly learned his craft.

The excitement of the Mitchell Race whetted the spectators' appetite for the Pulitzer, and lunch between the two races gave everyone an opportunity to look around Selfridge and talk over what they had seen and expected in the afternoon. The Pulitzer started on time at 1:00 and was run in three heats to accommodate the 15 racers that took off.

By 1922, the army had decided to identify its racers with the prefix letter "R" and a number. The VCP-R, the army's first racer, was renamed R-1, and the R-2 was the Thomas Morse MB-6 originally built for the 1921 Pulitzer. The R-3s through R-6s were all built for the 1922 Pulitzer. The R-3s were Verville-designed and Sperry Company-built monoplanes; the R-4s were Loening designed and built monoplanes; the R-5s were Thomas Morse high-wing, all-metal monoplanes; and the R-6s Curtiss biplanes. All the army's "R" for racer airplanes were represented at Selfridge.

Newsman Harold V. Wilcox wrote about the racers and pilots entered in the Pulitzer in the "Home Edition" of the *Detroit News,* printed on "Race Day" after the conclusion of the Mitchell Race but before the Pulitzer. If pilots saw the article, some would have been pleased and encouraged. Others would have had opposite reactions, bordering on being offended:

> All 12 of the new planes frankly are untried experiments, and some of them are certain to prove bitter disappointments to the Government and their designers, who have gambled fortunes on them. Two or three, however, are expected to prove to be the fastest airplanes in the world....
> Some of the swiftest planes will be flown by pilots unable to get the most out of them, and it is likely to develop that some of America's finest airmen, such as Capt. St. Clair Street [Streett's name was misspelled] and Capt. Frank O'D. Hunter, will have mounts unworthy of them but it is believed that today at least three of the finest flyers in the world will handle the controls of planes worthy of their utmost skill and intrepidity. This trio includes Lieut. L.H. Sanderson of the Marine Corps, pilot of the Navy "mystery ship;" Lieut. A.J. Williams, U.S.N., pilot of the Navy Racer No. 1, and Lieut. R.L. Maughn [Maughan's name was misspelled], U.S.A; pilot of the Army Curtiss Racer No. 2....[34]

Reading Wilcox's article, a pilot might have found confirmation of his worries that his airplane was out-classed. Few pilots would have considered themselves "unable to get the most out" of his racer—race pilots seldom lack self-esteem, confidence, and ego—but some might have cast a wary eye at one or another competitor. And, a pilot who had seen himself come up short, somehow, in practice flights, would surely wonder if he was included among those "unable."

Streett, scheduled to fly the Verville-Sperry R-3, and Hunter, who would fly a Thomas-Morse R-5, would have disliked seeing their planes written off as also-rans, but, having seen their competitors fly, they likely had already discounted their own chances. Sanderson, Williams, and Maughan would have seen the reporter's write-up as confirmation of their already high levels of confidence. As it turned out, Wilcox had much of it right.

The 1922 Detroit aviation meet had been "primarily scheduled for the purpose of holding the Third Annual Contest for the famous Pulitzer Trophy,"[35] and everyone at Selfridge Field awaited its start. The Pulitzer's plethora of airplanes—some well-proved veterans and some brand new—did not disappoint. Old machines broke their earlier records. New ones set world records.

Few monoplanes had flown in the 1920 and 1921 Pulitzers, but nine raced in the 1921 contest along with five biplanes and one sesquiplane—an intermediate between monoplane and biplane, with a major and a minor wing. Modern they might have looked and harbingers of the future they were, but the monoplanes (and the sesquiplane, which turned out to be a dead-end) would finish behind new Curtiss biplanes and, more surprisingly, the Curtiss biplane racers of the year before.

Heat 1 of the Pulitzer pitted one high-wing monoplane, the navy's MB-7, an updated version of a Thomas Morse airplane designed for the 1921 Pulitzer, and three new low-wing aircraft, the products of two new aviation organizations. A company formed by former Curtiss Company employees designed and built the navy's Bee-Line BR-1, and a collaboration between Alfred Verville, who had designed the first army racer, the VCP-R (now the R-1), and the Sperry Company, which had not previously built a racer, produced three R-3s for the army. Two R-3s flew in Heat 1 and one in Heat 3.

TABLE 2
1922: Pulitzer Racers, Pilots, and Results[a]
for Race of Five Laps of 50-km Course

Heat 1

Aircraft	T/O[b] Time	Pilot	Finished	Speed mph[c]	Laps Flown
Thomas-Morse MB-7	1:00.0	Capt. Francis P. Mulcahy, USMC	DNF[d]	140	3
Bee-Line BR-1	1:00.5	Lt. Steven W. Calloway, USN	DNF	163	2
Verville-Sperry R-3	1:01.0	Lt. Eugene H. Barksdale, USAS	5th	181	5
Verville-Sperry R-3	1:01.5	Lt. Fonda B. Johnson, USAS	7th	178	5

Heat 2

Aircraft	T/O Time	Pilot	Finished	Speed mph	Laps Flown
Curtiss R-6	2:20.0	Lt. Lester J. Maitland, USAS	2nd	199	5
Curtiss R-6	2:20.5	Lt. Russell L. Maughan, USAS	1st	206	5
Curtiss CR-2	2:21.0	Lt. Harold J. Brow, USN	3rd	194	5
Curtiss CR-2	2:21.5	Ens. Alford J. Williams, USN	4th	188	5
Wright NW-1	2:22.0	Lt. Lawson H. Sanderson, USMC	DNF	186	4

Heat 3

Aircraft	T/O Time	Pilot	Finished	Speed mph	Laps Flown
Verville R-1 (VCP-R)	3:46.0	Lt. Corliss C. Moseley, USAS	6th	179	5
Loening R-4	3:46.5	Lt. Ennis C. Whitehead, USAS	8th	170	5
Loening R-4	3:47.0	Lt. Lester D. Schulze, USAS	9th	161	5
Thomas-Morse R-5	3:47.5	Capt. Frank O'D. Hunter, USAS	11th	149	5
Thomas-Morse R-5	3:48.0	Lt. Clayton L. Bissell, USAS	10th	155	5
Verville-Sperry R-3	3:48.5	Capt. St. Clair Streett, USAS	DNF	164	4

[a]From T.G. Foxworth, *The Speed Seekers* (New York: Doubleday, 1974] 453–455.
[b]Take Off.
[c]Speeds were calculated over the number of laps flown.
[d]Did Not Finish.

In 1920, the navy had contracted with the Thomas Morse Company to build two MB-7 monoplanes with semi-Alula wings (see Chapter Three). When budgetary cuts forced the navy to drop out of the 1920 Pulitzer, Thomas Morse borrowed back one MB-7 and Harold Hartney flew it, representing the company. It crashed and was destroyed at Omaha. The navy claimed that it had not benefited from the MB-7's Omaha appearance and demanded that Thomas Morse replace the lost racer, which would have been a complete monetary loss to the company. Through negotiations, the company and the navy worked out a deal in which the navy accepted and paid for the remaining MB-7, and the wrecked one was forgotten. To soften the financial blow to the company, the navy agree to buy one additional MB-3, which was added to ten already purchased, and those 11 MB-3s equipped the Marine Corps' first fighter organization, F Flight of the Third Marine Air Squadron.

Powered by a 333-hp Wright H-3, the second MB-7 was completed in February 1922 and delivered to Mitchel Field, where it sat for two months, waiting for a propeller and other equipment. First flown in April by Marine first lieutenant Francis "Pat" Mulcahy, who made 17 flights in it between April and October, it had a top speed of 180.8 mph. Mulcahy reported the MB-7 was "stable and behaved well" but noted that the unusual wing interfered with forward vision, especially on landings. Everyone involved, according to Foxworth, "felt reasonably ready for the upcoming Pulitzer Race."[36]

Arthur "Mike" Thurston and Harry Booth, who had participated in the design of the Curtiss CR racers in 1921 (Thurston had also worked on the 1920 *Texas Wildcat* and *Cactus Kitten*), had left Curtiss to follow their own instincts and proclivities in the design of racers. Their Aerial Engineering Company's entries in the 1922 Pulitzer were designated "BR" for "Booth Racer," but the racers were more often called "Bee-Line Racers." Foxworth says flatly "nothing like [the Bee-Lines] ... had ever raced before."[37]

The Bee-Lines were cantilever low-wing monoplanes with graceful, Curtiss-like, propeller spinners and noses, slender, flat-sided fuselages made of reinforced wood veneer with internal bulkheads, and conventional tail assemblies with large fins. Located behind the gas tank, the cockpit sat over the trailing edge of the wing. A high-compression 400-hp Wright H-3, bringing with it the engine's characteristic vibrations, powered the racer.

One Bee-Line (the BR-1) featured two innovations—wing radiators, which Arthur Thurston, while at Curtiss, had patented,[38] and retractable landing gear, a combination not seen on any other racer in the 1922 Pulitzer. The other Bee-Line (BR-2) had retractable gear but relied on the proven Lamblin radiators, mounted on the fuselage sides above the wings, for cooling.

The wing radiators were made by sweating together two 5½-inch wide sheets of corrugated brass—each thinner than a sheet of paper you hold in your hand—with the corrugations aligned to produce little channels running the length of the sheets. The brass sheets, 22 to each wing, were screwed down to spruce ribs and molded to form the airfoil surface of the wings, with the channels running front to back across the wing.

Hot water from cooling the engine was piped to the front of the wing and flowed back through the channels, where the on-rush of air across the wing cooled it. From the trailing edge of the wing, the water flowed forward in channels on the wing's undersurface, still being cooled, to a pipe that ran along the leading edge of the wing and back to the engine.[39] The cooling system was finicky. All air pockets had to be carefully purged or water flow would be interrupted or stopped and steam would burst the thin tubing. Leaks, marked by spurting clouds of steam or streams of water, were common. When all went well, the system provided adequate cooling for the 400 or so horsepower engines of 1922, and it reduced added drag to zero.

Top: Aerial Engineering Company "Bee-Line BR-1" with wing radiators. It was forced down by loss of oil on the second lap of the five-lap 1922 Pulitzer Race (courtesy National Archives and Records Administration). *Bottom:* Aerial Engineering Company "Bee-Line BR-2." The BR-2, its Wright engine cooled by Lamblin radiators on the fuselage sides above the wings, was withdrawn before the start of the 1922 Pulitzer (courtesy National Archives and Records Administration).

The BR-1 retractable landing gear was operated by a hand crank, and the wheels and landing-gear struts retracted toward the midline of the aircraft. Retraction took eight seconds and when complete the wheels were "completely within the wing stub contour, and entirely concealed by a veneer and aluminum housing when drawn up."[40]

Thurston and Booth had designed the Bee-Line to reach 216 mph, and two weeks before the Pulitzer, Bert Acosta had been unofficially clocked at 213 mph, about the same as the then-current official speed record of 212 mph held by Joseph Sadi Lecointe flying a Nieuport sesquiplane. Acosta's forgetting to lower the landing gear marred his flight, but he was uninjured, and the racer was quickly repaired.

The navy entered both Bee-Lines in the Pulitzer, but withdrew the BR-2 (with Lamblin radiators) before the race. Two reasons have been given for the withdrawal: engine trouble[41] and inadequate time for the designated pilot, Lt. David Rittenhouse, USN, to become familiar with the racer.[42] Maybe it was both. The BR-1 equipped with both wing radiators and retractable landing gear was considered to be the faster Bee-Line, and its entry in the race was a fair test of the design.

The Wright H-3 engine, which powered the Bee-Lines and several other racers, reliably delivered 400 or so horsepower (maximum). The H-3 had its origins in a World War I Hispano-Suiza ("Hisso") 150-hp engine. The Wright Company purchased a license to build the Hispano-Suiza from the Simplex Auto Company in New Brunswick, New Jersey, during the war, and over the next few years enlarged the engine and increased its horsepower. By 1920, the Wright-built engines differed so much from the Hispano-Suizas that few parts remained interchangeable. The well-known vibration of the more powerful Wright engines was so extreme[43] that it would cause structural damage to several 1922 racers.

Interservice politics played a role in General William Mitchell's instructing Alfred Verville, the designer of the army's first racer, the VCP-R (see Chapter Two), to design the R-3 for 1922. In 1921, the Washington Conference on the Limitation on Armaments was meeting to consider, among other things, reducing the number of battleships in the world's fleets. Champions of those ships, seeing Mitchell as a vocal and constant critic with easy access to the press, wanted him away from Washington. Lo and behold, he was ordered to Europe for four months to study aviation developments. Verville, who was on contact to the army, accompanied the general. During Mitchell's absence, the Washington Conference scrapped 68 battleships from the world's fleets. What difference would Mitchell have made if he had been in Washington?

On Mitchell's return to the United States, he wrote a report that resulted in American manufacturers and designers being

> invited to compete in the design and construction of new types of craft....
> The Air Service thereupon requested the Curtiss Aeroplane and Motor Corporation of Garden City, L.I.; Loening Aeronautical Engineering Corporation, New York City: Lawrence Sperry Aircraft Co., Inc., Farmingdale, L.I.; and the Thomas-Morse Aircraft Corporation, Ithaca, N.Y., to lay out pursuit planes. The engineers were permitted a free hand ... the only restriction, or guidance being that the planes must be of a design suitable for military work and must have a speed of more than 190 miles an hour, or approximately 30 to 35 miles an hour faster than the most advanced type in use in the air services.
> At the same time, the Navy Department authorized Messrs. Booth and Thurston of Hammondsport, N.Y., the Curtiss Aeroplane and Motor Corporation, the Thomas-Morse Aircraft Corporation and the Wright Aeronautical Corporation of Paterson, N.J., to undertake the development of specialized types or the adaptation of existing types, with the predetermined object of attaining speed and maneuverability applicable to duty on ship or shore.[44]

In an end-run around the process of soliciting bids and plans from airplane builders, which might take a long time, General Mitchell one day called Verville into this office and ordered him to design a "modern airplane" and present him a plan the next morning. Verville spent the night with sketches and calculations and designed a cantilever monoplane to eliminate the drag associated with biplane wings and the strut and wire braces holding them together.

Verville had wanted the direct-drive Curtiss D-12, the best high-speed engine available — small, powerful, and literally vibration free — for the R-3. He did not get one. Curtiss had made it clear to the army that it did not want any Curtiss engines used in army racers that would compete with the company's aircraft. Verville had to settle on the almost equally powerful but vibrating Wright H-3. Nor could he use wing radiators, which were protected by a Curtiss patent. Verville installed two Lamblin radiators under the fuselage, behind the wheel wells. He wanted a metal propeller. None was ready in 1922. Perhaps worse, the army insisted he use a McCook Field-built propeller rather than the commercially available one he had ordered.

One of Verville's own decisions can be questioned. After considering both wood and metal wings, he chose wood. There were good reasons. The airplane manufacturers had limited experience with metal construction, which was unproved, heavy, and costly. Even so, metal construction was increasingly used, and it would have provided a more solid structure, one more resistant to the vibration of the racer's Wright engine. The wooden wing's nearly 18-inch thickness did, however, provide plenty of depth for the storage of the retracted wheels.

The necessary compromises Verville had to make on the R-3 were some of many small setbacks that contributed to a summation of Verville's career: "Alfred Verville, a kindly genius ... had a penchant for just missing the brass ring of commercial success, the R-3 was years ahead of its time when it first appeared in 1922 as a certain winner of the Pulitzer Trophy Race."[45] The R-3 would not have necessarily won if it had been equipped with a Curtiss engine and wing radiators, but its chances would have been far better.

Mitchell asked Verville about which airplane manufacturers should be approached to build the R-3. The three obvious choices because of their work on racers — Curtiss, Loening, and Thomas-Morse — were already at work on their own designs, and the decision was made to contract with the Sperry Aircraft Company on Long Island, which built three R-3s over the summer of 1922.

The R-3 passed a load factor test of 4.8 (sandbags totaling 4.8 times the weight of the wings and fuselage were loaded on the racer without causing structural failure), but the static load test could not reveal the problem dramatically revealed on the racer's first flight. The wings visibly vibrated. To address the wing vibration, Verville had the wings' fabric covering replaced with plywood. He also devised rubber spools to go around the engine mounting bolts to lessen the vibration from the Wright engine.

Verville passed on his solution to the wing vibration problems when he learned that the Bee-Line engineers were encountering the same problem with their Wright engine–powered racers. The stiffer wings corrected the vibrating wing problems but did nothing to reduce the engine vibration from loosening engine fittings and couplings. The R-3 flew better with less vibration on four short test flights on October 4. The next day, it was crated and shipped to Selfridge.[46]

"Traditional" or a similar word might be used to describe the four Curtiss biplanes in Heat 2; the fifth racer in that heat, the Wright NW-1, a sesquiplane, was something of a hybrid (Table 2). Because of their long commercial rivalry, beginning when Orville and Wilbur Wright first sued Glenn Curtiss for patent infringement 1909, the Curtiss versus Wright competition made Heat 2 the "grudge match" on the 1922 Pulitzer program.

According to William "Bill" Wait,[47] who had joined Curtiss as a draftsman in 1917 and was made director of the Curtiss racing program in 1922, General Mitchell played a role in the selection of the R-6 design, which would win the 1922 Pulitzer. Geoffrey Rossano inter-

Lt. Eugene Barksdale (USAS) and the 1922 Verville-Sperry R-3 (one of three R-3s in the race) that he flew to 5th place in the 1922 Pulitzer (courtesy National Archives and Records Administration).

viewed Wait, then 94 years old, in 1982, and provides a verbatim record of Wait's comments about the day that General Mitchell visited the Curtiss factory to discuss racers for the 1922 Pulitzer[48]:

> I had already designed a little racer using the OX5 engine. I'd hoped to fly it myself afterward, but I never did. I had shown it to Bill Gilmore, our chief engineer, however, and the day Mitchell visited the office they told me to come up to Curtiss' office. Bill asked, "Have you got that little racing design?" "No, it's down at the house." "Well, you go down and get it for General Mitchell." I did that, brought it back, and Mitchell liked it just fine....
>
> We soon discovered that my original design with the OX5 motor was about big enough to accommodate the new Curtiss 12-cylinder engine [the D-12].[49]

Wait's comment that the new Curtiss 12-cylinder engine almost fit into the space designed for the V-8 OX5 underlines the D-12's small size and great efficiency. It was about the same size (the length, height and width of the OX5 were 55, 35 and 30 inches, and the D-12's were 57, 28 and 35). The D-12 produced 450 hp (nominal), five times the horsepower of the 90-hp OX-5, and weighed only 1.8 times as much.[50]

The D-12, "one of the few truly great aero engines of all time,"[51] was a development of the Curtiss V-12s that began with Charles Kirkham's K-12 in 1917, which had been followed by Finlay Porter's C-12 of 1920 and Arthur Nutt's CD-12 of 1920 and 21. *Aerial Age* magazine, surely quoting or closely parroting a Curtiss press release, ended an article with this: "The Curtiss D-12 engine represents more nearly the ideal airplane engine than any other motor owing to its light weight and low head resistance [drag] per horse power."[52] Whatever hyperbole accompanied it, there was no better engine than the D-12 in 1922. The original design for the R-6 paired the D-12 engine with a four-bladed wooden propeller, but it always flew with a two-bladed propeller. The R-6s (and other airplanes) could have flown with two-bladed metal propellers in 1922. Dr. Sylvanus Albert Reed, an acoustical engineer, had used rapidly spinning metal blades to produce the sounds he needed for his research, and the idea for a metal propeller had emerged from his work. He convinced Curtiss to support his research on metal propellers, and when they were ready, Curtiss manufactured them.

Great commercial successes, the Curtiss-Reed metal propellers were built in many sizes for small to large airplanes and for low- and high-powered engines. Reed took one of the first to be made to Detroit for the Pulitzer,[53] but rules forbade its installation.

The army's R-6 clearly resembled the navy Curtiss CR of 1921, but it was smaller and lighter, had a new engine—the D-12—and featured new "wing" or "skin" or "surface" radiators and other innovations. To cool and maintain oil temperature, the oil radiator was located in the water cooling system, which "insures not only that the oil will be warmed up quickly, but also that its temperature will be kept uniform once it reaches the same temperature as the water in the radiator."[54]

The R-6 retained the single interplane I-struts of the CRs, and single landing-gear struts replaced the CRs' V-struts. The wings were tested to, and withstood, a load factor of 10, and the fuselage, designed for a load factor of 7, was tested to 4. The fuselage, stabilizer, and long, graceful fin were painted black; the wings and elevators gold.

A month before the Pulitzer, *Flight* magazine reported on Joseph Sadi Lecointe's new world speed record and credited part of its success to "Lamblin radiators [which] ... appear to be an essential accessory to all racing aircraft, and it seems only a matter of time before they will be found on every aircraft the designer of which professes to consider head resistance and efficient cooling."[55] The Curtiss racers and their wing radiators would end the dominance of the "essential accessory."

Comparison of the R-6 to the CR of 1921 highlights the small size of the army racer. The 19-foot spans of the upper and lower wings of the R-6 were three feet eight inches

Lt. Lester Maitland (USAS) and the army Curtiss R-6 Racer that he flew to 2nd place in the 1922 Pulitzer (courtesy National Archives and Records Administration).

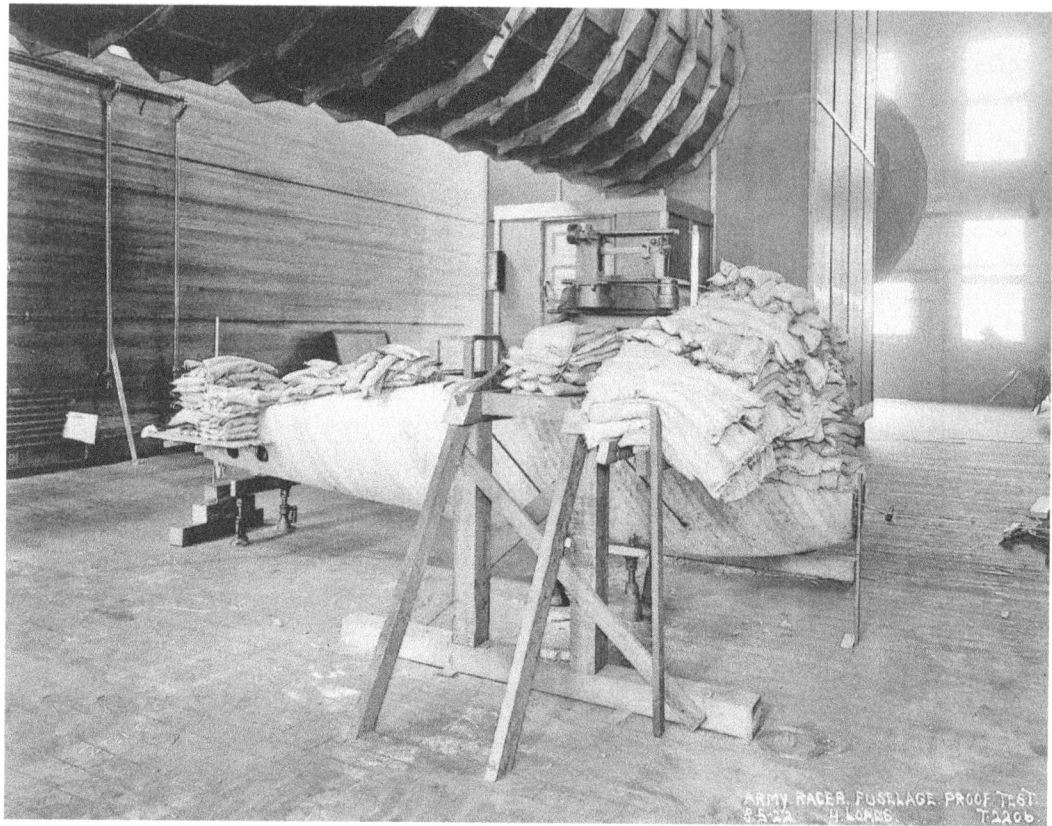

Sandbags loaded on the fuselage of R-6 to test its load-bearing capacity (courtesy National Archives and Records Administration).

shorter than the CR (see Appendix), and the 18-foot, 10-inch fuselage was two feet four inches shorter. Like the CR, the R-6 fuselages were built of spruce bulkheads covered with "Curtiss-ply," a two-layer laminate of $5/32$-inch spruce. The entire R-6 fuselage weighed 127 pounds. Overall, the R-6 weighed 30 pounds less and had an engine of about 100 more horsepower than the CR.

Great things were expected of the R-6. Foxworth writes that the greatness was seen from the start, "When the exuberant Maitland took the R-6 aloft on her very first flight, he poured on full power and blistered the Curtiss Field at 223 mph! This was the more remarkable because the official world speed record then stood at 205; Maitland's average from six sizzling passes (carefully timed but unofficial) beat that by 6 mph."[56] Foxworth's report of the "first flight" may be a report of a later flight, because Bill Wait, the designer of the R-6, remembered a very different first flight:

> The Army sent Lester Maitland[57] and Russell Maughan[58] out to Garden City to test the racers and get accustomed to them. Maitland had a reputation for being a pretty hotshot pilot and he made the first flight of the racer. After he took off, he flew out a ways and then circled back towards us. Suddenly we saw our plane do a half roll over and quickly snap right back again, no more than 50 or 100 feet off the ground. We thought he was crazy and I said to myself, "Well, that's really hot." Maitland soon came around and landed. He looked half scared to death. "Did you see what happened?" he yelled. "I didn't do it, the plane did it. I tried to make a turn but the rudder is so sensitive it damn near went the whole way over." He didn't want to do any more flying that day.[59]

Maughan flew the R-6 the next day in stocking feet, convinced that he would have better control of the rudder bar.[60] He was spared the snap half-roll that had frightened Maitland, but his feet cramped, and he landed poorly. That afternoon, he purchased a pair of sneakers, and from "that day on he always flew with them, which worked okay."[61] Air Service pilots, including Lt. Cyrus Bettis, who would win the 1925 Pulitzer (see Chapter Seven), appear to be wearing tennis shoes in 1920s photographs. Either other pilots leaned the same lesson as Maitland or they followed his example.

Aviation magazine and the *New York Times* reported that Maughan flew his R-6, which proved to be the faster of the two R-6s, to a new electronically timed but unofficial world record of 220.4 mph on October 2, 1922.[62] The test flights at Curtiss Field complete, the two R-6s were crated and shipped to Selfridge. There they joined 13 other aircraft that would race in the Pulitzer and dozens of other airplanes entered in other races.

The navy contracted with Curtiss to upgrade the 1921 CR racers, and the company replaced the CR's Lamblin radiators with wing radiators and replaced the CD-12 engine with the new D-12 engine. The radiators and new engine produced measurable improvements in speed; the racers' top speed went up 16 to 17 mph, and the racers, now called CR-2s, were reportedly capable of more than 200 mph in level flight. Other changes were made. To bear the greater weight imposed by the radiators on the upper wing, a strut replaced the rear flying wire from the fuselage to the lower surface of the upper wing. An oil-cooling duct was installed in front of the central pylon that supported the center of the wing. The already narrow gaps between the stabilizer and elevators, the fin and rudder, and the wings and ailerons were sealed with narrow strips of rubber, and the wheels were streamlined.

The navy selected Lt. Alford "Al" Williams to fly the second CR-2. Variously described

Lt. Harold Brow, USN, flew this Curtiss CR-2 to 3rd place in the 1922 Pulitzer. Note the wing radiators, visible on upper wing. The absence of Lamblin radiators, which were used on the 1921 CR, gave the CR-2 a more streamlined appearance and better speed (courtesy National Archives and Records Administration).

as "acerbic"[63] or "cocky"[64] or "taciturn, with a commanding, almost intimidating bearing,"[65] Williams was a remarkable man. He graduated from Fordham University in 1915, pitched in the minor leagues for the New York Giants baseball organization for two years, volunteered to fly in the navy during World War I and qualified as a navy pilot, earned a law degree from Georgetown University in 1925 while on active duty with the navy, and was a consummate test pilot and stunt flyer. When the United States decided not to enter the Schneider Trophy Races after 1926, he organized and led efforts to raise money for, and to build, Schneider racers. Although two—a monoplane and a biplane—were built, neither raced.

During the 1930s, Williams wrote prolifically about aviation for magazines and newspapers and presented brilliant flying demonstrations in a bright-orange navy fighter, stripped of operational equipment, for the Gulf Oil Company. (Williams' Grumman F3F "Gulfhawk" is on display at the National Air and Space Museum's Udvar-Hazy Facility). During World War II, he instructed Army Air Force pilots in aerobatics and precision flying, and after the war he continued to work in aviation and fly Gulf Oil Company-sponsored Grumman aircraft in exhibition flights. He retired in 1951 and died in 1958. He is buried in Arlington National Cemetery.

Earlier in the summer of 1922, Williams had been designated to fly the Wright NW-1, but Marc "Pete" Mitscher, the chief of the navy's racing pilots, changed his mind and selected Marine lieutenant Lawson Sanderson. At some time during the summer, the navy decided to enter the Curtiss *Cactus Kitten* triplane in the Pulitzer and selected Williams to fly it.

By June 1922, S.E.J. Cox, who had paid for construction of the *Texas Wildcat* and the *Cactus Kitten*, no longer owned the airplanes. Financial problems had descended on Mr. and Mrs. Cox, and they returned the racers, valued at between $200,000 and $250,000, to the Curtiss Company in lieu of far smaller debts of about $22,000.[66] In various places, it is reported that S.E.J. Cox sold the *Kitten* to the U.S. Navy for a dollar or that he had donated it to the navy, and that the navy used it as a trainer for the 1922 races. Given the *Kitten*'s hair-raising flying characteristics, it's improbable in the extreme that anyone would have considered using it as a trainer. Curtiss prepared the *Kitten* to race and painted "U.S. Navy Triplane" across the airplane's vertical tail, which might have been part of a Curtiss effort to charm the navy into financial support for the racer.[67]

According to Foxworth, Williams made one test flight in the Curtiss Triplane, and it proved "too high-spirited even for" Williams.[68] If Williams, a very able pilot, had that opinion, Clarence Coombs's accomplishment in finishing second in the 1921 Pulitzer is even more impressive. Finally, Williams was assigned to fly the second CR-2. No matter how good a pilot he was, Williams (and any other pilot) was surely better off in a CR-2 than in the triplane.

The fifth racer in Heat 2, the Wright Company's NW-1, for "Navy Wright-1," like the Curtiss racers, was built around an engine. Following on its success in the manufacture of liquid-cooled V-8s based on Hispano-Suizas, the Wright Airplane Company built more powerful V-12s. In May 1921, the navy had successfully bench-tested Wright's new V-12 Tornado-2 (T-2) engine. Navy officers and engineers, enthusiastic about the smooth-running 600+-hp engine, wanted to try it in the wide-open competition at the Pulitzer, but they were frustrated by the lack of an available airframe that could accommodate the new engine and its power.

The navy turned to Jerome Hunsaker, head of its Bureau of Aeronautical Design, to

Ensign "Al" Williams (USN) and the 1922 Curtiss Triplane, the former *Cactus Kitten* (courtesy National Archives and Records Administration).

design its racer. Hunsaker, who was to become a towering figures in American aviation and engineering, was already prominent, at age 36, in 1922. He had graduated number one in the Naval Academy class of 1908, attended MIT, where he earned the institute's first PhD in Aeronautical Engineering in 1916, participated in the design of the navy NC transatlantic flying boats, and championed aircraft carriers. His enthusiasm for dirigibles led him to the Goodyear-Zeppelin Company, which planned to build a fleet of transatlantic passenger-carrying airships in the 1920s. Subsequently convinced by the losses of the dirigibles *Macon* and *Akron* that passenger-carrying airships had no future,[69] he returned to MIT in 1933, where he headed the Department of Aeronautical Engineering, which he founded in 1939. He retired from MIT in 1951, but continued to walk to his office and work there through the 1960s (he died at age 98 in 1984). Among his honors, he received the Navy Cross and the Guggenheim Medal for contributions to aircraft design. A biographer summed up Hunsaker's career: "[it] might be considered the epitome of success in engineering."[70]

As great as his talent was and as great as his achievements were to be, he had no experience with racer design in 1922. Examining successful European designs led him to follow the example of Gustave Delange, who had designed the Nieuport-Delange sesquiplane that Joseph Sadi Lecointe, winner of the 1920 Gordon Bennett, had flown to a new world speed record of 330 k/hr (212 mph) on September 26, 1921. It was the first officially timed flight at over 200 mph, and no American racer, even with engines of twice the Nieuport-Delange's 300-or-so horsepower, had matched it. Sesquiplanes, intermediate between monoplanes and biplanes, usually had a normal-sized upper and a diminutive lower wing. In Delange's design, a shoulder-mounted wing was paired with a much shorter airfoil-shaped "wing" between the fixed landing gear wheels. Scoffers dismissed the design as "a dubious monoplane or halfhearted biplane."[71]

The lower wing of Hunsaker's design was proportionally larger than the Nieuport-

Navy Wright NW-1. The streamlined tip of the spinner has not been installed, as the engine is being run up on the ground. Note large size of the racer in comparison to the men, who appear reluctant to more closely approach the flailing propeller. Note also the Lamblin radiators, the inferior wing at wheel height, and the massive landing gear. An oil leak and engine seizure forced the racer down on its last lap, and its pilot, Lt. Lawson Sanderson (USMC), crash-landed near the shore of Lake St. Clair (courtesy National Archives and Records Administration).

Delange's and extended beyond the landing gear wheels. The wings of the big, powerful-looking airplane were braced with struts from the landing gear but there was neither wire nor strut bracing between the fuselage and the upper surface of the wing. The racer's two Lamblin radiators were tucked in between the upper and lower wings.

Hunsaker's racer was still under construction in September when Marine lieutenant Lawson "Sandy" Sanderson was designated as its pilot. (Sanderson had distinguished himself flying a Curtiss Jenny to dive-bomb guerrillas fighting Marines in Santo Domingo in February 1919. It can be assumed that the guerrillas had no effective antiaircraft weapons.) Earlier in the week of the Pulitzer, with victory in Curtiss Marine Trophy Race in sight, his Curtiss 18-T ran out of fuel forcing him down on his last lap. Although Sanderson was at the Wright factory on Long Island in late September, the racer was not ready to fly before it had to be crated for shipment to Selfridge. At Selfridge, Wright engineers, technicians, and mechanics assembled the airplane — minus the upper engine cowling and a few other small pieces that were still being fabricated.

Unofficially, Sanderson was clocked at 209 mph in the only flight he made before the Pulitzer. He was happy with the airplane, reported it was powerful, smooth, and steady, and easy to fly and land. He predicted still higher speeds in the race. The oil temperature had climbed to 160 degrees during the 30-minute test flight, but everyone was so busy and so pleased with the racer's performance that little attention was paid to it.

The Wright NW-1 and three other racers that Wright subsequently built for the navy were, at the time, and sometimes still are, referred to as "mystery planes" because the Wright Company released so little information about them. In the case of the NW-1, the company had little choice because only a single flight was possible before the NW-1 went on public display at the Pulitzer. The large NW-1 loomed over Curtiss racers. Its wingspan was one-

and-a-half times the wingspan of its principal competitors, the Curtiss R-6s (30 feet, 6 inches versus 19 feet). It was six feet longer (24 feet versus 18 feet), and about three feet higher. The Curtiss racers had been flown several times before being shipped to Selfridge, where they were flown again; the Wright NW-1 was flown exactly once at Selfridge. If all the racers were of equal performance, the ones that had been test flown most thoroughly and the ones that the pilots knew best — the Curtisses — would have to have been favored.

Heat 3 was a mixture of new designs — five monoplanes — and the oldest, the VCP-R biplane, now redesignated "R-1" (see Table 2). All were army entrants. The five monoplanes were from three manufacturers: One Verville-Sperry R-3, two Loening R-4s, and two all-metal Thomas Morse R-5s. The army had stored the Verville VCP-R, at McCook Field during 1921. In April 1922, the chief of the Air Service ordered McCook Field engineers and mechanics to spruce it up to fly the Pulitzer. The wings and fuselage were shortened; a new Packard engine, propeller, radiator, and smaller wheels were installed; and the racer was painted glossy white. Corliss Moseley, who had flown the VCP-R to victory in the 1920 Pulitzer, flew the newly designated R-1 racer in July, August, and September. On September 26, he reached 179 mph, according to hand-held stopwatches, and 176 mph on electric timers. The racer was crated and shipped to Selfridge.

The Air Service's Engineering Department at McCook Field, planned as the army's "experimental factory," had built the Verville-designed VCP-R and later modified it to its R-1 configuration. The Sperry Company built the three Verville-Sperry R-3s — two flying in Heat 1 and one in Heat 3 — and damning comparisons were made between the costs at McCook and Sperry.

McCook Field's role in 1920s aviation reached far beyond the army. Its engineers, designers, and mechanics played major roles in deciding which airplanes the army would buy and which modifications would improve their performances, dictating in some cases, which airplanes, parts, and accessories would be produced. McCook personnel made pioneering decisions about equipment, such as ordering Lawrance's first air-cooled engine. Some of McCook's suggested and directed changes and improvements made their way into the civilian as well as military market. McCook's central role was no surprise. In those days, "McCook Field was the nation's largest aviation facility. In the early 1920s, a nadir for aviation business, the workforce at McCook Field was roughly the same size as all manufacturing firms combined — approximately 2,000 each."[72]

As with all government activities, McCook came in for criticisms of inefficiency and profligate spending. Comparisons between the Sperry Company's apparently inexpensive building of the R-3s and the comparably exorbitant cost of McCook's construction of the VCP/R-1 appeared to confirm the criticisms. A writer for *Aviation* magazine reported that McCook engineers worked 33,000 hours on the VCP-R/R-1 and that the Sperry Company had designed and built the R-3s in one-tenth the time, 3,000 hours. The writer also estimated the cost of the R-3s at about one-tenth that of the R-1.[73] The design, manufacture, modification, and maintenance of the VCP-R/R-1 cost about $250,000. In contrast, the R-3s of 1922 cost about $30,000 each.

The comparison is unfair in some respects. The progression from VCP to VCP-R to R-1 involved redesign and modifications that produced performance improvements over a three-year period. In contrast, the costs for the R-3s were estimated after their initial development and manufacture. Adding in the costs of improvements and modifications to be made in the R-3s in 1923 and 1924, when an R-3 flew its last race, would have made a fairer comparison, but that was impossible in 1922 when the *Aviation* writer made his.

Grover Loening, the designer, and his company had designed and built three navy entrants in the first Pulitzer (see Chapter Two). None did well. One of two M-8 fighters collided with another racer before takeoff. The engines of the other M-8 and of the racer derived from the M-8, the M-81 S Special, had failed. Those experiences did not sour the navy on Loening. After the 1920 Pulitzer, it ordered 83 aircraft from the company.

All Air Service contracts for racers in 1922 specified a top speed of at least 190 mph and reduced the amount of money paid to the manufacturer by $2,500 for every mile per hour below that standard. Because of that stipulation, neither Loening nor any other manufacturer would receive a penny unless its racer exceeded 175 mph. Grover Loening's fortunes didn't fall that far, but they came close.

By the time Loening and the Air Service signed a contract in 1922, only three months remained to design, manufacture, and ready two racers for the Pulitzer. The Air Service provided 578-hp Packard engines and wood propellers, and Loening designed the smallest airframe that would accommodate the Packard engine. His design was called, maybe for the first time, "the flying engine," but many "flying engines," airframes barely large enough to hold an engine, were to follow.

The Loening R-4 emerged with thick, cantilever wings mounted at the bottom of a slab-sided fuselage. The continuous wing spars passed through the fuselage and bore the weight of the Packard engine's 1,118 pounds, which, combined with the weights of radiator, water, and plumbing came to 1,500 pounds. The clever use of the pass-through-the-fuselage wing spar to bear part of the engine's weight kept the airplane's weight down. The loaded racer, including pilot and fuel, weighed 3,050 pounds, with essentially half the weight in engine and accessories.

The wing spar that passed through the fuselage saved weight but passed engine vibration directly to the wings, contributing to the wing flutter that plagued the R-4. The racer sat on un-streamlined fixed landing gear, and the cockpit, over the trailing edge of the wing, was between the engine and the gas tank.

After being proudly displayed to the press on Long Island on September 18, the first R-4 and its sister ship were crated up and shipped to Selfridge. First Pursuit Group pilot Lt. Ennis Whitehead, who was to go on to command the 5th Air Force in World War II and retire as a major general, took the first R-4 up on its first flight. He reached 170 mph but his report of the 30-minute flight was damning. The flexing of the wings alarmed him, the racer yawed frighteningly back and forth, and use of the ailerons did not always produce the predicted motions. Overall, the airplane handled poorly and unpredictably, and wind blowing into and around the cockpit made it uncomfortable. In the afternoon, Lt. Lester Schulze flew the second R-4 for 18 minutes. His impressions were more negative than Whitehead's. His tail-heavy racer was impossible to control when he opened the throttle, and the wings fluttered and vibrated.

The wing flutter surely contributed to the cracks in the wing spars that Selfridge's chief engineer detected during his inspection of the R-4s. Loening received a multitude of suggested solutions for the problems, and he had to choose among them with no time for testing. To improve lateral stability, he had a larger fin installed. At the suggestion of Alfred Verville, who had observed similar wing flutter in the R-3, he reinforced the wing structures, stripped off the fabric wing covering and replaced it with wood veneer from the rear spar forward. Bracing wires were installed between the landing gear and outer wings and between the stabilizers and outer wings. A larger windshield reduced air flow into the cockpit, improving the pilot's comfort. All the modifications added about 700 pounds to the racers' weights without entirely solving the problems of directional instability and wing flutter.

An additional modification to control wing flutter disfigured Schulze's R-4. A kingpost was added in front of the cockpit and two wood braces were extended from it to each wing. The kingpost and braces looked like pieces of scaffolding hastily nailed to a failing structure, but did not eliminate wing flutter. Bill Wait described the R-4's wing problems in more human and dramatic words. He remembered hearing Lt. Whitehead "say that the wings on the plane banked in the wrong direction when he flew it. He was pretty upset, too, because the people who backed the design said he was too yellow to open it all the way up. Whitehead finally discovered the problem and he had been right all along. When the ailerons deflected the wing warped and twisted so badly the angle of incidence changed.... His plane was eventually wire-braced and flown in the race."[74] Indeed, both R-4s flew in the Pulitzer, but they flew poorly.

A contemporary magazine writer described the R-4s in more favorable terms. They "impressed the observer as having been designed solely for service purposes," in contrast to other racers that would require considerable modification for service use.[75] The reward of service adoption would be denied to Loening and all manufacturers but Curtiss. The Curtiss racers, after some considerable modifications to equip them with armament and service equipment, were the bases for a long line of biplane fighters powered by liquid-cooled engines.

Lt. Lester Schulze (USAS) and Loening R-4 with external bracing to reduce wing flutter. The engine was left uncowled because of overheating problems. This racer finished 9th in the 1922 Pulitzer (courtesy National Archives and Records Administration).

Capt. Frank Hunter (USAS) and the Thomas Morse R-5 he flew to 11th place in the 1922 Pulitzer (courtesy National Archives and Records Administration).

The Thomas Morse Company, which was about to disappear as an airplane manufacturer, had three racers at Selfridge. In addition to the navy's MB-7 monoplane, originally built for the 1921 Pulitzer, which flew in Heat 1, it had new army racers in Heat 3. The new racers, the "R-5s," were of all-metal construction and powered by 578-hp Packard engines. The "parasol" monoplane wing was mounted on struts extending from the fuselage sides to the underside of the wing at about one-third the distance between fuselage and wingtip, and the single residual of the Alula wing was a hump and broader chord in the wing's center section, giving the racer a peculiar hump-backed look. Smooth metal covered the fuselage forward of the cockpit, and dull-looking corrugated metal covered the wings and aft fuselage. The radiator was mounted in a bomb-shaped structure between the landing gear struts.

The wingspan of the first R-5 was 29 feet; the second 25 feet. Captain Frank "Monk" Hunter, who was credited with shooting down nine German planes during the war, flew the longer-span R-5 for the first time on September 29.[76] Cooling problems forced him to land. A new radiator and a new set of the shorter wings were ordered. One flight of 11 minutes was made with the new radiator, but the new, short, wings were not ready in time for the Pulitzer.

Lt. Clayton Bissell, credited with seven German airplanes, was selected to fly the short-wing R-5, which arrived at Selfridge during the tests of the long-wing version. Bissell had been assigned as an aide to General Mitchell and accompanied him on his inspection tour of post–World War I aviation in Europe. During World War II, he commanded U.S. Air Forces in the China, India, Burma (CBI) Theater, and served as assistant chief of intelligence to General Marshall later in the war.[77]

Little has been written about the R-5s, which played pioneering roles in all-metal construction but performed poorly. William Wait, the designer of the Curtiss R-6, recalled "a little Thomas Morse at the meet [Detroit].... It was an all-metal ship with a fuel tank in the upper wing above the pilot's head. Every time he taxied in there was a stream of gasoline running down from that center section. Why it didn't catch fire I'll never know."[78] It is tempting and, perhaps, too easy to modify a saying often attached to good-looking, good-performing airplanes: "If it looks good, it is good." The R-5 looked dull because of its corrugated metal covering; it looked awkward because of its wing. "If it looks dull and awkward, it is dull and awkward" comes to mind. The R-5s finished last and next-to-last in 1922.

All racers in the 1922 Pulitzer were army and navy aircraft, with ten army and five navy entrants. The army's entrants ranged from the updated two-year-old Verville VCP-R/R-1 biplane through the two well-tested one-year-old Curtiss R-6 biplanes to seven brand-new and barely test-flown R-3s, R-4s, and R-5s, all monoplanes.

The navy's five entrants included two updated, well-tested biplanes, the Curtiss CR-2s, which were improved versions of the 1921 CRs, and the Thomas Morse MB-7 monoplane, which had been built for, but did not race in, the 1921 Pulitzer. The other three were promising but little-tested new designs, the Bee-Line BR-1 and BR-2 monoplanes (the BR-2 was withdrawn before the race), and the large Wright NW-1 sesquiplane. All in all, navy pilots were confident about their chances in the Pulitzer. The day before the race, "when it was announced that the Navy's entries [in the Liberty Engine Builders Race] had been canceled, one of the leading Navy pilots had shouted at his brethren airmen in the Army, 'We'll get even with you tomorrow when the Pulitzer Trophy will come back to us.'"[79]

Anyone who read newspapers and magazines expected that the fastest racers in the Pulitzer would face off in Heat 2. The fact that the contest for fastest speed would likely come down to Wright and Curtiss, long-time rivals in aviation, served only to increase interest in Heat 2, but Heat 1 came first. The first pilot to take off in the Pulitzer, Francis Mulcahy in the MB-7, roared across the grass landing field at 1:00 P.M. (see Table 2). The other three racers in Heat 1 followed at 30-second intervals.

All who had read about the racers knew that the Bee-Line BR-1 and the two Verville-Sperry R-3s in Heat 1 had retractable landing gears, but "the cold mention of this fact does little justice to the emotion the onlookers felt when the two actually *did* [emphasis in original] draw their landing gears into the fuselage…. Selfridge Field was treated to a strange race between airplanes shorn of their landing gears."[80] (The number of racers visually "shorn of their landing gears" was three, but the author of the article from which this quote is taken evidently wrote about the Verville-Sperry R-3s as distinct from the Bee-Line BR-1.)

Spectators who wanted to see four airplanes jockey for position over five 50-km laps were disappointed. The navy's Bee-Line racer dropped out on the second lap when all the coolant drained from a broken line, which must have reinforced Admiral Moffett's opinion that the cooling system of liquid-cooled engines was an Achilles heel. The second plane to drop out, the MB-7, came to grief on the third lap because of lubrication problems.

Heat 1's two finishers, the two Verville-Sperry R-3s, sped around the course at speeds that would have won the 1921 Pulitzer. Eugene Barksdale's R-3 finished at 181 mph, and Fonda Johnson finished three miles per hour slower at 178 mph, both better than Bert Acosta's 177 mph at Omaha. For 20 minutes, until Heat 2 was completed, Barksdale held the world record for speed over a closed course. His speed over the course was sufficient to set a closed-course speed world record for monoplanes.[81] When all three heats were complete, Barksdale would be 5th and Johnson 7th.

Alfred Verville had been disappointed with having to settle on a Wright H-3 engine instead of a Curtiss, but he could not have known the havoc the vibrating Wright engine would cause. Foxworth put it bluntly: "Johnson [the pilot of one of the R-3s in Heat 1] had been lucky to land alive."[82] Engine vibration broke a water line, loosened motor mounts, sheared off engine hold-down bolts, and snapped internal fuselage bracing wires.

Spectators who had expected speeds of 190 or 200 mph or more in Heat 1 were disappointed but patient. Heat 2 offered better chances for such speeds, and it was flown an hour later. Army lieutenant Lester J. Maitland in his R-6 was first off in the second heat, followed by the other three Curtiss racers at half-minute intervals. The Wright was last off, 30 seconds after the last Curtiss. The Wright never gained on the Curtiss racers, making it possible to discuss the contests between the R-6s and the CR-2s as separate races.

The Curtiss airplanes climbed away, trailing lines of smoky exhaust, and turned toward the pylon at Gaukler Point. Some spectators along the shore could see the racers turn east at Gaukler Point to head across Lake St. Clair toward the marker boat anchored off the Canadian shore. Sooner than the spectators at Selfridge expected, they saw tiny specks appear over the lake to the southeast. Accompanied by roaring engines, the specks emerged as full-sized racers rounding the pylon on the field, turning south to start the next lap. The racers were spectacular: "Maughan and Maitland [in the R-6s] ... flew across Selfridge Field at a height of about 50 feet.... Once Maitland descended to within 20 feet of the earth and the crowd gasped in the belief he was about to land precipitately. But he merely 'nosed' his tiny craft upwards."[83] Bill Wait more colorfully recalled the army pilots' spectacularly low flying:

> [A]ll of a sudden he [Maughan] came skimming over Lake St. Clair about 15 or 20 feet above the water. He skipped over the shoreline bulkhead, went into a steep climb, rolled over on his side in a vertical back and flew in that position for a quarter mile to the last pylon. Then he yanked the plane around the pylon.... He flew the plane the last quarter mile on the lift of the fuselage alone. The wings weren't doing a thing. Then he flew back out across the lake, so low that some men sitting there in a boat raised their paddle to signal. He knocked it right out of their hands, and he won the race.[84]

More than one explanation can be offered for the paddle flying from the boater's hands. Maughan could have hit it; the onrush of air around the racer could have blown it from the boater's hands; the low altitude, speed, and noise of the racer could have so frightened the boater so that he tossed the paddle away or angered him to the point that he threw the paddle at the racer. And maybe Wait imagined it all. The story is too good for more inquiry.

Maughan, who led from the time of his takeoff, won the Pulitzer. The R-6 averaged 206 mph for the 50-km course, fully 38 mph faster (a 21 percent increase) over Acosta's speed in the Curtiss CR a year earlier. Maitland, in the other R-6, finished second, and navy lieutenants Harold Brow and "Al" Williams finished third and fourth in the CR-2s. Curtiss airplanes, all biplanes with conventional landing gear, had flown off with the first four places in the Pulitzer.

Maitland was initially credited with the breathtaking speed of 216 mph on his first lap: "[T]he grand stand and press box were in a turmoil. They could not believe it. The judges, however, insisted that was his time."[85] In excited postrace commentary, the *New York Times*[86] and the Associated Press[87] repeated the claimed speed, calling it the fastest man had ever traveled, beating by a fraction of one mile per hour the world-record speed, set during 1-km straightaway flights by Joseph Sadi Lecointe just weeks before. Astoundingly, Maitland had apparently flown a 50-km triangle course a bit faster than Sadi Lecointe had managed on a straightaway flight that benefitted from a diving start that drove his initial speed far above anything the racer was capable of in level flight.

It's easy to see why some people didn't believe the reported speed. Maughan took off first and led at the end of the first lap, and "right behind him, their props almost chewing his rudder, flashed Maitland and the two Navy CRs."[88] If Maitland were flying 11 mph faster

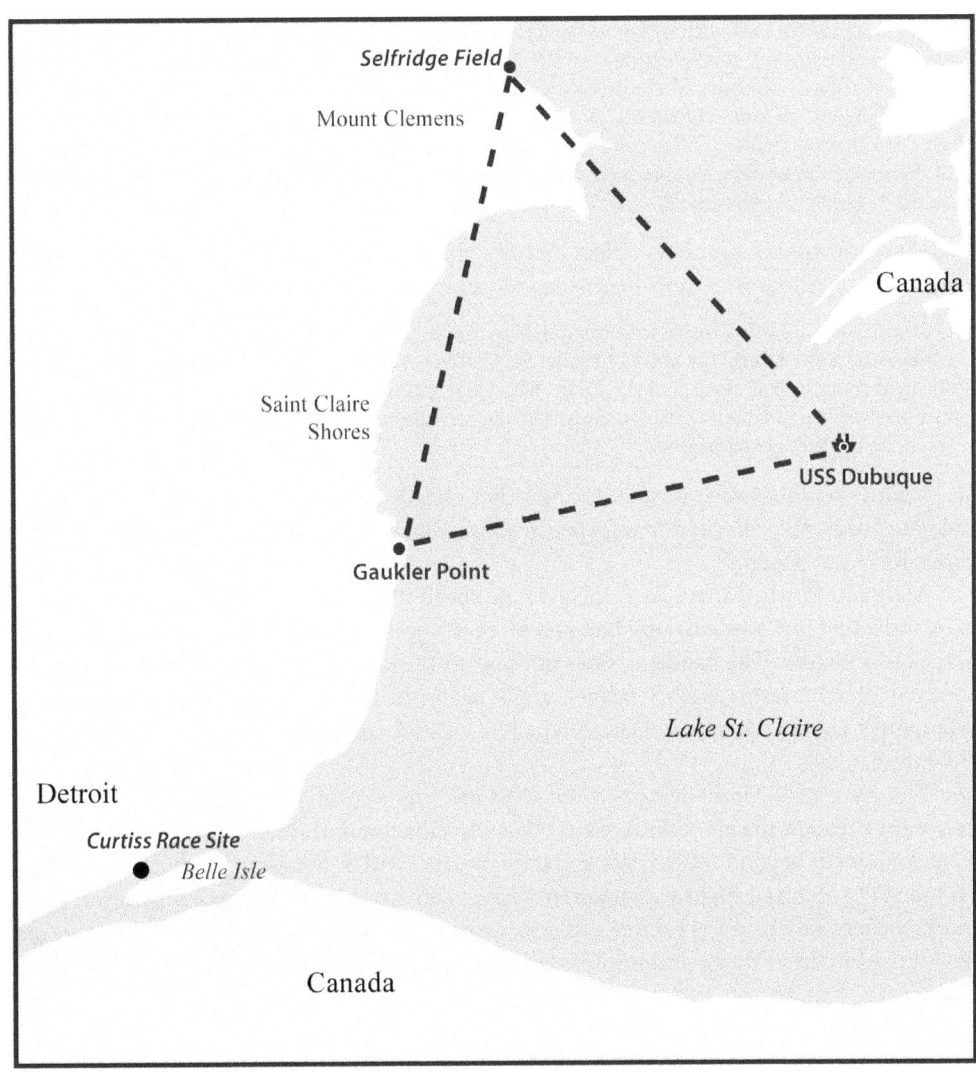

Figure 4. Map of the third Pulitzer Race, Selfridge Field, Michigan, 1922. The originally intended course was to have been over land to the west and south of Selfridge Field. The course was changed to fly largely over water to reduce dangers to people on the ground and to reduce dangers to pilots because a forced landing on water was thought less dangerous than being forced down on land. A tethered balloon was flown from a barge attached to the S.S. *Dubuque* to guide the pilots on the 11-mile or so crossing of Lake St. Clair. The Pulitzer Race was five laps of the 50-km course. The John L. Mitchell Race was four laps over the same 50-km course (map by Lynn Springer).

than the 206 mph recorded for Maughan, he should have been overtaking, or in front of, Maughan, who took off only a half minute before him. Clearly, he wasn't. Maitland followed Maughan all around the 250-mile course.

The claim of 216 mph for Maitland's first lap was quickly and quietly set aside. The official timers prepared a notarized document that verified Maughan's speeds for 100, 200, and 250 km around a closed course and stated that Maughan flew the fastest 50-km lap at 206.6 mph and averaged 205.9 mph for the five laps of the 250-km race. The official timer added,

It will be noted that this official report corrects the error in time and speed announced to the public and press during the course of the race for the first lap of #44 airplane piloted by Maitland. The incorrect time as announced credited Maitland with 8.37.16 [minutes, seconds, and hundredths of seconds] or 216 miles per hour.... The times printed by the timing apparatues [sic] clearly show the error of these figures.... This error in the times announced can only be attributed to the eagerness of the scorers in endeavoring to have the times announced to the public and posted on the score board as quickly as possible.[89]

Maughan's postrace comments reflect that the effects of high speed flight, turns, and other maneuvers on pilots were only beginning to appear, much less be understood, in 1922:

> On the 4th lap ... I was whirled into unconsciousness for a few seconds. When I regained my senses I was almost skimming the waves of Lake St. Claire.... I was lost four times in the haze.... I was stunned more or less at each of the turns. My worst moments, however, were at one turn when I lost confidence and then consciousness. On the straightaway I came to. Another problem I had was with my feet going to sleep.[90]

Many pilots reported such blackouts and other effects of turns at high speeds. Mysterious and dangerous, they merited newspaper headlines such as "Air Race Winner Made Unconscious By Great Speed."[91]

Maitland finished the race at the average speed of 199 mph, 7 mph behind Maughan. Maitland's fuel line was partially blocked after the first lap, and he had to hand pump fuel to the D-12 engine. The handicap does not appear to have affected his speed, which fell off hardly at all in his last laps when fatigue might have affected how vigorously he could pump. (Maitland's lap speeds were 193.4, 200.0, 196.1, 202.7, and 201.0 for an average speed of 198.9 mph.)

The *New York Times* described the R-6s as "tiny aircraft, wasplike in appearance as they went through the air" and reported that Maughan and Maitland, "in banking around the pylons, went beyond 180 degrees and for a brief second or two were partly upside down." The partially inverted flight was apparent rather than actual, but its reporting in the *Times* and elsewhere reflected the excitement generated by the races. The magazine *Aviation* discussed the low-level flying and impressively sharp, low, fast pylon turns in more analytical terms:

> The existence of a "ground effect," increasing the lift and decreasing the drag of wings at a low altitude as a result of the compression of the air between the wings and the ground has long been known to workers in aerodynamics, but there has seldom been evidence of such an attempt to take advantage of it in practise as was shown week before last. The turn at the field was made just above the roofs of the hangars in most cases, and immediately after completing the turn many of the pilots dove to cross the field itself at a height of 20 ft. or less climbing again to clear the trees on the far side.[92]

In the R-6s, Maughan and Maitland had a race between themselves, and, not far behind, Brow and Williams dueled in a similar one-on-one contest in their CR-2s. Brow had an equipment failure of little consequence, which could have been a disaster. The thin rubber strip glued on the underside of the gap between the horizontal stabilizer and the elevator to reduce drag came loose and blew away. The elevator drooped down, and the plane "shot for the sky." Brow brought it swiftly under control but later said, "Lucky it didn't go the other way,"[93] which would have pitched the nose down and the low-flying racer into the ground. Williams too had his problems. The fire extinguisher in his cockpit exploded on the second lap, shards of the extinguisher ripped through his helmet and goggles, and fumes made him violently sick. He escaped injury and landed 4th, behind Brow, who took 3rd.

The modifications made to the CRs had paid off. Brow's speed was almost 17 mph better than Acosta's 177 mph, and Williams, five mph slower than Brow, was still 12 mph faster than Acosta. Thirty seconds behind Williams' CR-2, the last of the Curtiss racers to take off, the Wright NW-1 started its takeoff. Bill Wait recalled Lawton Sanderson's takeoff run: "[He was] taxiing across the field, with several men running along of each side, dashing ahead, then flopping on the ground like a school of fish. We couldn't for the life of us figure out what they were trying to do. Later we learned they were checking the clearance of a little axle wing which was very low to the runway. They were afraid that if it hit the ground and tore the fabric, they'd have terrible problems in the air."[94] The NW-1 shot into the air. Because of its large size, it may have looked faster than the Curtiss racers as it headed south, and Sanderson's airplane had the power to close the gap. Was it up to the challenge? The answer emerged on the first lap. Sanderson's first-lap speed, 185.3 mph, was one mile behind Williams' 186.4 in the slower of the two CR-2s. Despite its power and (unofficial) 209 mph in a test flight, the NW-1 could not hold its own, much less make up any distance between it and the Curtiss biplanes.

Neither was the small oil radiator on the bottom of the NW-1's cowl capable of holding the oil temperature within normal operating limits. Disconcerted by the constantly rising oil temperature gauge, Sanderson crammed a handkerchief over its face. An excellent pilot, Sanderson might have managed not to look at the gauge, but he could not ignore the smell of burning oil and overheated metal and the thick trail of white smoke streaming behind the speeding silver racer as he turned at Gaukler Point to head across Lake St. Clair on his fourth lap. Realizing that the engine would not last through the remainder of the lap, and not wanting to have to ditch in Lake St. Clair's deep water, he made an abrupt turn back toward shore.

Spectators lay, sat, and stood on the shore before him. They might have scattered had they realized Sanderson's distress, but they weren't immediately aware that anything was wrong. Before he reached the shore and endangered anyone there, the overheated engine clunked to a stop, and Sanderson set the big sesquiplane down in about five feet of water. The racer flipped onto its back, as the small lower wing had plowed into the water and tossed the airplane over its nose. Thrown from the cockpit, Sanderson was plunged head-down into the water and the lake's muddy bottom. He emerged unhurt. The racer was a mud-clogged wreck. Foxworth's amusing story about Sanderson's water landing and emergence, like Bill Wait's story about the hit paddle, is likely too good to be true: "Though it had occurred to some witnesses that the 'durn fool aviator' had been under water 'a bit long,' many wide-eyed spectators were seeing aeroplanes for literally the first time in their lives, and unsure of just what to expect no one made a move to save Sanderson. He vividly remembered an old man drawling laconically to him as he waded ashore. 'I thought it was about time you came up, son.' The old man thought the whole display had been staged!"[95]

The much-anticipated race between Curtiss and Wright had turned out to be a race among Curtiss airplanes. Even if the Wright NW-1 had completed the course at 186 mph, the speed it achieved on its first three laps, it would have finished fifth, behind all the Curtiss biplanes and "out of the money." Sanderson had had a bad week. He had been leading in the Curtiss Marine Trophy Race and within sight of the finish when his yellow-painted Curtiss 18-T ran out of gas and he was forced down. Five days later, his engine failed in the Pulitzer. He was left with "Did Not Finish" (DNF) written against his name in the first and the last, the two most glamorous races of the National Air Races.

Maughan, who took off first, landed first in the Pulitzer. The other Curtiss pilots followed him in. Tallying up speeds was not really necessary to confirm that Maughan had flown the fastest race. Officially, Maughan set world records for 100 and 200 kilometers over a closed course. An *Aerial Age* writer wrote that the performance had been expected: "Hangar gossip had had Maughan a winner weeks ago."[96]

The pilots ended the race in crowd-satisfying fashion. When a racer crossed the finish line, the pilot pulled back on the stick and zoomed high into the air, leveled off, and descended to land from a gentle glide. The soaring zoom pleased race-watchers and served the important functions of getting out of the way of trailing, low-flying racers and spilling off the high speed of the race so that the landing would be made at the specified 75 mph or less.

As Maughan's racer rolled to a halt, General Patrick and navy secretary Denby rushed up to congratulate him. According to Foxworth,[97] the general was so excited that he patted Maughan's head rather than shaking his hand, and Secretary Denby broke into tears. It's impossible to tell from a photo of the three men and the R-6 if Secretary Denby has tears in his eyes, but General Patrick is definitely shaking Maughan's hand. It's possible too that the photo was taken after an earlier, more emotional meeting.

When getting ready for the race, Maughan had wondered about his wife, who was about to deliver a baby. After the race, a telegram informed him that wife and son were doing well.

The R-6s, as expected, had been fast. When all the racers had landed after Heat 2, the R-6s were 1st and 2nd. Surely galling to other designers and companies, all of their own racers finished behind the year-old CR-2s, which finished 3rd and 4th.

Lt. Russell Maughan leans against the fuselage of Curtiss R-6, racer number 43, in which he won the 1922 Pulitzer (courtesy National Archives and Records Administration).

Heat 3 remained to be flown. Did anyone imagine that any of its six airplanes would best Maughan's speed? Even more unlikely, who expected that the oldest airplane in the Pulitzer would be the fastest racer in Heat 3? Moseley and the R-1 were first off in Heat 3. No one caught them. The R-1's speed of 178.9 mph over the 250-km Pulitzer course with 15 turns was a surprising three mph faster than its best electrically measured speed over a straight one-km course. Either the racer performed better in the Pulitzer or the speed trial timing had been inaccurate. Moseley and the R-1 bettered his and the VCP-R's 1920 performance by 13 mph and took 6th place in the 1922 race.

The R-4s had rolled to the start line without engine cowlings because of cooling problems. Both finished. Whitehead finished 35 mph behind the winner, and the clumsy, drag-inducing bracing on Schulze's racer held his speed nine mph below Whitehead's. Whitehead finished 8th, Schulze 9th. The other Heat 3 racers, all monoplanes, disappointed. Foxworth characterizes the R-4s' performance as "deplorable"[98] but goes on to explain that Grover Loening had only three months to design his racers and two weeks after their first problem-plagued flights to modify them. With more time, he might have had winning airplanes. He didn't.

Deplorable they might have been, but the R-4s did better than the all-metal Thomas Morse R-5s. Clayton Bissell flew the short-wing R-5 to 10th place at 155 mph, some 50 mph slower than the winning speed. Hampered by a rough-running engine and disoriented in the haze on his first lap, Frank Hunter in the long-wing R-5 managed only 113 mph on that lap. He raised his lap speed to 165 mph and finished at just under 150 mph for the 250-km race.

Capt. Moseley flew Alfred Verville's VCP-R/R-1 to some sort of glory when he finished first in Heat 3. Lt. St. Clair Streett, flying Verville's far more modern R-3 in the same heat, had no such success. "Bill" Streett had flown 500 hours in combat over France in World War I and rose to command the 31st Aero Squadron. He had finished fourth in the 1920 Pulitzer, flying an Orenco D, and had commanded the epic New York to Nome, Alaska, and return flight in 1920. He remained in the Air Service, and early in World War II commanded several training units,. In 1944, he took command of the 13th Air Force in the Southwest Pacific and directed air raids against Japanese military installations during General MacArthur's "island-hopping" campaign. After the war, he commanded the Continental Air Force, which became the Strategic Air Command (SAC), and he commanded SAC before his retirement in 1952.

Streett took off last in heat 3 and averaged 164 mph, 15 mph slower than the R-1 and 40 mph slower than the R-6, on the first four laps. A broken oil line — likely another casualty of a vibrating Wright engine — ended his race, forcing him to make an emergency landing on his final lap. The engine died while he was over the lake. He managed to keep the powerless racer airborne until he reached land, but not long enough to find a safe landing place. He tried to slip the R-3 between a tree and a hedge, but the space was not sufficiently wide. The left wing was damaged but repairable.

Alfred Verville might have taken some comfort from the performance of the other two R-3s. Barksdale and Johnson, in Heat 1, had finished 5th and 7th, faster than the R-1 and faster than Acosta's winning speed the year before. Neither Verville nor anyone else could see ahead two years, when an R-3 would win the 1924 Pulitzer. Neither could have anyone seen that the R-3's winning speed would be a disappointment, the only winning speed slower than the previous year's winner.

Speed was uppermost at Selfridge in October 1922, and, on the 16th, two days after the Pulitzer, Maughan took off to set a world speed record. FAI rules required four passes over a one-kilometer straightaway course, two upwind and two downwind, with the average of the four speeds submitted for the record. The rules also required that the four passes be made during a single flight. Maughan's best single-lap speed was 248.5 mph, "a greater speed than any human being ever before attained."[99] "Aeronautical engineers and army and navy officers who witnessed the flight were astounded. The demonstration, they declared, proved that there was no limit to the speed that might be attained in the air."[100]

Maughan averaged 232.2 mph on four passes, good enough for the world record, but Maughan had not flown the four passes during a single flight. Surely few people knew of such mundane requirements for a record-breaking flight, and maybe no one who watched Maughan's last flight knew it. Apart from his speed, Maughan "added to the astonishment ... by flying approximately one mile with the machine on its side. The feat in reality was a flight without the aid of wings."[101] "Flight without the aid of wings" was part of the show at Selfridge. Bill Wait remembered Maughan's R-6 tipped over on its side for a quarter mile when he crossed Selfridge Field during the Pulitzer.[102]

An *Aviation* reporter wrote skeptically about Maughan's 248.5 mph fastest lap: "Some press dispatches even spoke of a maximum speed of 248.5 m.p.h., but this figure has not been substantiated."[103] The doubt was misplaced, but did not disappear until R-6s made official flights at higher speeds in 1923. Maughan made no attempt to better his speed in October 1922, the month of the Pulitzer. He left Selfridge to see his wife and new baby. "For family reasons, Lt. Maughan could not make the test and General Mitchell volunteered, even though he had never flown the racer before."[104] Any veteran of an armed service will raise an eyebrow at the idea of a general volunteering. Generals don't volunteer. They do whatever they want.

Bill Wait remembered General Mitchell's participation in preparing the R-6s for his flights:

> After the [Pulitzer] race General Mitchell told us he wanted to fly the airplane in the upcoming speed record tests and that we should get it ready soon. The plane had entered competition theoretically overweight, and in order to meet the technical requirements limiting landing speed to 75 miles per hour, we built a small wing on the landing gear axle. It added just enough lift to stay within the rules.
>
> With the kilometer speed test scheduled for the next day, I took crew chief Bill Irwin and some of my mechanics over to the hangar that evening to remove the little wing and replace it with a low-drag fairing. Well, the boys were really tearing and smashing that wing to pieces. Suddenly, Billy Mitchell and some of the brass walked into the hangar and saw what we were doing. He started screaming, "Call out the guard! Call out the guard!" They all came rushing up to us yelling, "What are you maniacs doing to that plane?" I told him, and they finally quieted down, but for a few minutes it was a pretty wild scene.[105]

Mitchell flew both Maughan's and Maitland's R-6s for the first time on cold, gray, and blustery October 18. He chose Maughan's slightly faster racer and flew the FAI-required two-upwind and two-downwind passes over the one-kilometer course in a single flight to satisfy the rules and set an officially sanctioned world record of 223 mph.[106] An Air Service publication stated that Mitchell had "further endeared himself to the flyers of the Air Service," with his record-setting flight. *Aviation* gushed with praise: "His [Mitchell's] brilliant piloting of the speed plane has brought to him the distinction of being one of the greatest fliers of the world and again gives Army aviators evidence that they are led by a man who will take any chance he demands of them."[107]

Not everyone joined in the praise. Rumors circulated that some pilots believed that

Mitchell had "pulled rank" to make the record-setting flight, and they drew attention to the general's falling well short of Maughan's best speed of 248 mph and his average of 232 mph for four passes. Mitchell would bristle at suggestions that he flew to burnish his own image, but it is impossible to believe his ego played no role. On the other hand, his ability as a pilot cannot be questioned. The R-6 was a "hot airplane" in 1922, and he flew both of them for the first time on the day he flew the faster one to a new speed record. He flew it well. Mitchell's world record became official in January 1923, with a letter from the secretary general of the FAI that announced his acceptance of the speed reported to him on October 19, 1922.[108] Mitchell did not fly another record attempt.

Either to end rumors about his possibly grandstanding or in possible but silent recognition that other pilots could fly hot racers faster, he asked General Patrick to request the National Aeronautic Association's Contest Committee to arrange speed trials to provide Maughan and Maitland opportunities to set a new record.[109] Those trials were undertaken in early 1923 (Chapter Five).

The Pulitzers were off to a great start. The 1920 event had been a good race, an unofficial world record was set in 1921, and world records were shattered in 1922. The first National Air Meet (1922) was a triumph. "Competition, speed, a week of intense activity ... attracted many thousands of people to Detroit and prompted newspapers in every state to ... [feature] as first page news each item of flying activity recorded during the air races."[110] *Aeronautical Digest* wrote: "It was the greatest exhibition of the kind ever held in the world and the results have justified every expectation."[111] The British magazine *Flight* praised the meet and its demonstrating "how Government departments can assist the cause of aviation ... by placing orders for machines which are not strictly service machines, although incorporating features which are as desirable in service craft as in sporting machines."[112] Combined with other achievements of 1922, "America now, for the first time since 1909, [held] supremacy alike in altitude, in duration, and in speed, with all three of the major records in the possession of officers of the Army Air Service."[113]

Fifteen racers had taken off in the Pulitzer; eleven finished; seven—Maughan, Maitland, three other army and two navy pilots—flew faster than the existing world record for 200 km (179 mph) over a closed course. Maughan set new records for both 100 and 200 km closed courses. Mitchell had set a world absolute speed record. Although not so splashy as the speeds, the safety record at the 1922 National Air Race was equally impressive. More than 292,000 miles were flown without serious injury or accident.

Kudos could be shared. The army could be proud of its pilots and racers. The Detroit aviation and automobile establishments had organized and hosted the first and greatly successful National Air Race. The town of Mt. Clemens had provided all kinds of assistance to everyone concerned with the large crowds and thousands of automobiles. The navy could see promise in the directions it was taking in airplanes and engines. Designers and engineers had produced impressive improvements in speeds and reliability. Successful race performance opened doors to commercial success. The Curtiss Company[114] and its suppliers[115] lost no time in advertising their triumphs in full-page advertisements in aviation magazines.

Unnoticed by almost everyone, the 1922 Detroit Air Races were the last major event to be sanctioned by the Aero Club of America. While the races were going on, "behind closed doors, a lawyer [William P. MacCracken, Jr.] with a passionate interest in aviation development made final revisions to the charter under which the NAA [National Aeronautics Association] would be incorporated."[116] Four general objectives called for the NAA to further the following:

1. Development of the Airmail Service to stimulate public interest and confidence in aviation.
2. Military purchases sufficient to keep airplane companies in business until markets of commercial airplanes developed.
3. Establishment of commercial air routes with proper navigation facilities.
4. All-encompassing federal regulation to govern licensing and inspection of pilots and aircraft.

The NAA did its part to recognize achievement on the way to those objectives. It awarded its Collier Trophy for "the Greatest Achievement in Aviation in America" to the Airmail Service in 1922 and 1923. It sanctioned the Pulitzer Race and other contests for army and navy aircraft at the National Air Races to bring attention to military aviation, with the expectation that the public would support additional funding. The NAA-sanctioned "On to Detroit Race" focused attention on civilian airplanes and pilots and their capabilities and laid the foundation for such events that continued to be popular for years. In other activities, the NAA publicized commercial aviation — passenger and freight — to encourage its use, and it encouraged NAA members to approach their communities to install navigation aids, beginning with such simple things as painting orange and white checkerboard patterns on high buildings and water towers and adding arrows pointing toward airports.

NAA and others had decided that federal regulations, with enforced standards and inspections, were necessary before insurance companies would insure aviation, and that availability of insurance was absolutely necessary before commercial aviation could really get off the ground. The NAA membership was split about support of federal regulations and especially about government licensing of pilots. Some members objected because of the restrictions it would place on "gypsy pilots" and "barnstormers," who had introduced much of the population to airplanes in the years after World War I. Whatever the objections, they were overcome. In 1926, Congress passed, and President Calvin Coolidge signed, legislation that established federal regulations of air commerce.[117]

The NAA's activities and plans were important but lacked the drama of high-speed flight in the three Pulitzer Races. Speeds had ascended from a good start with no-world-class performances on Long Island in 1920, to world-record competitive speeds at Omaha a year later, and to spectacular put-the-world-back-on-its-heels speeds in Detroit in 1922. What's more, the future looked bright for the next Pulitzer. Not all the new racers at Detroit — the Verville-Sperry R-3s, Loening R-4s, Thomas Morse R-5s, Bee-Line BRs, and Wright NW-1 — had proved their worth, and some never would, but they illuminated the energy, enthusiasm, and innovation of the American aviation industry. The Curtiss R-6s, in which the Curtiss Company combined its well-tried biplane layout with new engines and an innovative cooling system, resulted in world-class airplanes.

The R-6s bowled over a French air attaché who saw the 1922 Pulitzer: "I am writing my Government that in aviation they have two American achievements to surpass, one is the Curtiss motor and the other is the Curtiss wing-type radiator."[118] The first National Air Race ended triumphantly. Optimists could look ahead to a still better National Air Race in 1923. They were right to be optimistic.

CHAPTER FIVE

1923

The Greatest Show of All, Lambert Field, St. Louis, Missouri, October 6

> *"I have never seen such flying in my life. How do these youngsters accomplish such feats?"* asked the Secretary of Navy [Edwin Denby]. *"In no way can we determine as readily and as surely just what we need and what we must have in order to play our part in the national defense"* said the General [Major General Mason Patrick, Air Service chief], praising the Pulitzer Trophy winners.[1]

United States aviation passed through five remarkable years after 1918. At war's end in 1918, American designers and engineers had not designed state-of-the-art warplanes, and American manufacturers had struggled to build European designs to wartime standards. In early 1923, General Mason Patrick, chief of the U.S. Army Air Service wrote, "[O]ur Air Service stands supreme, as it leads the world in aviation records, possessing, as a matter of fact, all the official world's records possible of attainment in heavier-than-air machines."[2] From a United States perspective almost 90 years later, a French writer's discouraged comments about American gains and French losses in aviation elicit either sympathy or humor:

> One by one, The American Army pilots have broken the records which we, here in France, considered as our own for so long.
> Speed, endurance, long distance, as well as altitude records have passed out of our hands....
> We have been in the habit of considering France as the leader in the World's aeronautics, that it is not putting it any too strong when we say that there [*sic*] new American records struck us like a thunderbolt out of a clear sky....[3]

The French writer could have found scant comfort when he wrote about the records still held by his countrymen: "What is there left for us in matters of World Records...? We still carry the endurance and distance records for 600, 900, and 1200 c.m. balloons, as well as altitude record for dirigibles (3080 meters). We still have the altitude record for load-carrying of 250 and 500 kg. And last, but not least, we carry the endurance record for gliding flight." Another French writer identified General Mitchell as the reason for United States resurgence and dominance:

> WHO is at the head of the U.S. Air Service? A general, rather young, spirited, and full of "pep;" a superior officer, who himself has flown his ship more than 300 km. per hour, setting also the best example for his subordinates!

... we have no one in our Engineering Division who can compare for one instant with Brigadier-General Mitchell. Padded offices, adorned with reports and counter-reports, replace the cockpit. It is a lamentable stage of affairs.[4]

Unmentioned by the French authors, the record-breaking performance of United States racers was changing the face of military aviation. The Air Service's newest plane, the Curtiss PW-8 ("P" for "Pursuit" and "W" for "Water"-cooled engine) and its V-12 engine came from the same Curtiss designers and engineers who had produced the Pulitzer winners of 1921 and 22.[5] Similarities between the 1922 Pulitzer R-6 racers and the PW-8s were apparent to the most casual observer.

In 1923, the Pulitzer Race, which had provided the framework and foundation for the competitions that produced the Curtiss racers and their competitors, moved from its great triumph in Detroit, a city fast-growing in size and importance, to St. Louis, a Midwestern city of long-established prominence. The fourth largest city in the United States in 1900, St. Louis got its start when 30 French-speakers came up the Mississippi River from New Orleans in 1764, selected a site protected from flooding by high bluffs on the west side of the river and began clearing land and laying out a town. A year later, in 1765, France ceded land east of the Mississippi, which it had claimed, to Great Britain, and many French settlers on the east bank and further east moved across the river, increasing the population and shoring up the "Frenchness" of St. Louis on the west bank. In 1766, the population stood at about 300. Ownership changed in 1804, when the United States acquired St. Louis as part of the Louisiana Purchase. Two years later, in 1806, about 1,000 people lived in St. Louis, and a visitor lamented that fallen morals had accompanied settlers arriving from Tennessee and Kentucky.

Steamboats and the increased trade they brought spurred growth of the city's population and its importance as a transportation hub after 1821. In the 1800s, ships moved goods and people from the East Coast to New Orleans, and from there steamboats transported them to the bustling Missouri city. Moving west and north on the Missouri and Mississippi rivers, the steamboats served the fast-moving frontiers of the county.

Population jumped following the revolutions of 1848 in Europe that brought Irish and Germans emigrants in large numbers. The arrival of German settlers in St. Louis in mid-century paved the way for German veterans of the Civil War, including Joseph Pulitzer, who came to the city almost two decades later. The year 1849 was calamitous for St. Louis. A cholera epidemic killed hundreds, and a fire destroyed 15 blocks of the town and 23 steamboats. The city responded by draining and cleaning some polluted water bodies, beginning the first public sewers in the early 1850s, and rebuilding the burned areas. The importance of St. Louis to commerce spurred the city to recovery. In the middle 1850s, the Port of St. Louis, important to both the Mississippi and Missouri rivers, the mainlines for commerce in what is now the Midwest, was second only to the Port of New York in tons of goods entering and leaving. In 1860, St. Louis' population was 160,000, having more than doubled in ten years. At the same time, railroads were laying tracks to the south, west, and north, and St. Louis companies began manufacture of train engines, cars, and other railroad essentials. No railroad line ran eastward from the city; the mighty Mississippi River was too much a barrier. People and material arriving by train to the banks of the Mississippi had to change to boat or ferry to cross the river.

During the Civil War, beginning in 1861, Missouri was an exception to the generality that non-slavery states fought on the side of the Union and slave states fought on the side of the Confederacy. Slavery was legal in Missouri, but a special convention in January 1861 decided that the state would not secede with the Confederacy and would instead remain part of the Union. Some parts of the state strongly supported the Confederacy and constant tensions and incursions by Confederate troops and irregulars (guerrillas) produced violence and skirmishes and battles throughout the war. The war, which interrupted shipping between St. Louis and New Orleans, disastrously affected the commerce of St. Louis, and the stagnation caused St. Louis to lose status to Chicago as the most important Midwestern city.

Prosperity returned after the Civil War because St. Louis was essential to the transportation necessary for the country's westward expansion. Completion of the Eads Bridge over the Mississippi in 1874 finally linked St. Louis to the east by railroad and road, and the city became a hub for ground as well as river transport. The city increased its area by annexation and people continued to move in, and the population reached 311,000, not quite doubling between 1860 and 1870.

Forty-three years after President George Washington had watched the first balloon ascension in the United States at Philadelphia in 1793, Richard Clayton made the first ascension from St. Louis in 1836.[6] From that beginning, St. Louis, in keeping with its bustling image of a fast-growing, progressive, prosperous city, solidified its importance in ballooning. John Wise was probably the most important person in St. Louis' earning (or self-bestowing) the "title 'City of Flight.'"[7]

John Wise,[8] an avid balloonist, studied aeronautics and meteorology and was convinced that a steady wind blew west-to-east across the United States. Banking on that wind, he aspired to set up mail delivery by balloon from the United States to Europe. To test and publicize his idea, he made his most successful and epic ascension from St. Louis on July 1, 1859, expecting his balloon, the *Atlantic,* to drift from St. Louis to New York, and from there he would begin a transatlantic flight. A careful man, he supplied the *Atlantic* with ample food and drink and heavy clothing, and, for possible emergency alightings on water, a boat was suspended below the basket. On takeoff, Wise was in the basket with three other men; supplies were stowed in the boat.

The balloon drifted farther north than planned or hoped, and a strong wind swept it across Lake Erie. The gale continued "at a speed in excess of 100 miles per hour"[9] and drove the balloon, steadily losing altitude, across Lake Ontario. The crew tossed everything possible overboard and, after carrying the boat for hundreds of miles to be ready for a water landing, jettisoned it to remain airborne and avoid alighting on wind-blown and storm-tossed Lake Ontario. Finally across the lake, the balloon sank onto the top of a forest, the strong wind driving it along and through the treetops until the bag was punctured. The four-man crew, practically unharmed, was suspended in the basket 20 feet above the ground, from which they safely descended.

Wise had failed to reach New York, but he had established a world's record for distance in a balloon flight, and proved that a west-to-east air current blows across North America. In a straight line from St. Louis, the *Atlantic* drifted 826 miles, and its actual flight path was near double that, about 1,500 miles. The 826 miles stood as a world record for distance flown by a free balloon until 1900, and it was the longest flight in the United States until 1907. (Wise disappeared on a balloon flight in 1879, when he was 71 years old.[10])

Reading J.J. Horgan's *City of Flight: The History of Flight in St. Louis,*[11] makes it seem that balloons first and then balloons and airplanes filled the St. Louis sky for a decade beginning in 1904. Certainly, St. Louis, its Aero Club, the millionaire owner of the company that made Listerine, and many of its ordinary citizens were "air-minded."

Twenty million people attended the Louisiana Purchase Exposition, or St. Louis World's

Fair, to celebrate the 100th anniversary of the Louisiana Purchase; the fair opened on April 30 and closed on December 1, 1904. (Many people attended multiple times because the crowd of 20 million represented a fourth of the 1904 United States population of 82.2 million).[12] The fair's planners, while celebrating accomplishments of the past, also wanted to assure spectators that modern developments were not passing St. Louis by. They did not skimp on obtaining expert advice and contracted with Alberto Santos-Dumont, the inventor of the dirigible. In a bonus for all concerned, Santos-Dumont agreed to bring a balloon to the fair and ascend in it. What an addition to the fair! Men would go aloft, providing most spectators their first chance to see men fly.

Santos-Dumont and three assistants arrived in St. Louis in late June 1904 in anticipation of a July 4 flight. He pronounced his dirigible, *Number Seven,* in perfect shape when it was uncrated on June 27, and he insisted that the crate be left uncovered overnight at the fair grounds to let air circulate around the gas bag. A guard was posted, and he left his post one time but saw nothing amiss on his return. Next morning Santos-Dumont and his men discovered that the gas bag had been slashed with a knife. Oddly, Santos-Dumont, who "blamed the vandalism on an enemy," took no part in detectives' investigations. His damaged balloon was shipped back to France on July 2.

Disappointed fairgoers did not get to see the famous Brazilian airman and *Number Seven,*[13] but they could see a famous United States dirigible and a famous and popular dirigible pioneer in October. Thomas Baldwin[14] had made his living as a circus performer until he went up in a balloon in 1875 and turned to making exhibition balloon flights and parachute jumps. He built an "elongated balloon," a dirigible, called the *California Arrow,* powered by a Glenn Curtiss-built engine, and made the first circuitous flight in the United States in 1904.

Roy Knabenshue,[15] who, like Baldwin, earned money by charging spectators to watch his ascensions in balloons at fairs and other events, traveled to St. Louis in 1904 to sell rides on a captive balloon. He met Baldwin, and when the *California Arrow* proved incapable of lifting the hefty Baldwin into the air, Knabenshue agreed to pilot the airship. In the early afternoon on October 31, Knabenshue ascended about 30 feet in the *California Arrow* while the ground crew held onto ropes to keep him at that height. Even with that precaution, a gust of wind slammed the dirigible down and damaged a propeller. "Spectators scoffed" when the airship, apparently defeated by the very air that supported it, was pulled into a hangar, where Knabenshue repaired the damage.[16] Late in the afternoon, he took off again, climbed to 2,000 feet, and maneuvered over the fairgrounds for 45 minutes. He landed the *California Arrow* about 100 feet away from where he took off, having convincingly demonstrated his piloting ability and the *Arrow*'s controllability.

Aviation became the main attraction three years later when the second Coupe Aéronautique Gordon Bennett (Gordon Bennett balloon race) was flown from St. Louis. To this day, the FAI characterizes the Gordon Bennett balloon races as "the most prestigious event in aviation and the ultimate challenge for the balloon pilots and their equipment. The goal is simple: to fly the furthest distance from the launch site."[17] Unlike the Gordon Bennett airplane race, which was flown for the last time in 1920, the balloon race continues on a near-annual basis.

U.S. Army lieutenant Frank Lahm drifted 402 miles from Paris to Flying Dales, England, to win the first Gordon Bennett Balloon Race on September 30, 1906. In keeping with the rules of the Gordon Bennett Balloon Race, the aero club of the winning country was obligated to host the next year's contest. Midwestern cities in the United States, far from the Atlantic Ocean, which allowed long flights before the balloons had to land and had sufficiently large populations to organize and support international events, were preferred sites for starting balloon races. Chicago, Omaha, and Denver were interested in hosting the 1907 Gordon Bennett Balloon Race, but a newly established friendship between St. Louis businessman Daniel Nugent and Augustus Post, the secretary of the Aero Club of America, resulted in the race being awarded for St. Louis.[18] As it turned out, the Atlantic Ocean, 800 miles away, would end the race.

The most important person in St. Louis aviation over the next 40 years, Albert Bond Lambert, the millionaire president of the Lambert Pharmacal Company, and the Aero Club's honorary secretary in 1907, played a major role in organizing that year's International Aeronautical Tournament in St. Louis. Lambert's fortune, and, as a result, St. Louis aviation floated on a sea of mouthwash. In 1879, Dr. Joseph Lawrence and Jordan Wheat Lambert concocted an antiseptic mixture and dubbed it "Listerine" to suggest an association with Dr. Joseph Lister, who first used a chemical—carboxylic acid—as a disinfectant during surgeries. Six years later, Dr. Lawrence sold his interest in the product to the Lambert Pharmacal Company. The Lambert Company first sold Listerine as a surgical antiseptic and, looking for other marketing opportunities, later offered it as a floor cleaner and as a cure for gonorrhea (!). In the 1910s, a diluted mixture was marketed to dentists for use as a mouthwash and its regulatory status was changed from a prescription drug to an over-the-counter product. In 1921, the company began a marketing campaign that made the company rich and Listerine a household word.

The company popularized "halitosis"—a far more "scientific" and scarier word than "bad breath"—as the cause of lost jobs, friendships, and love because bosses, friends, and potential lovers shunned its victims. Featured in Listerine's print ads, perplexed people like Edna, pretty, stylish, almost 30, and unmarried, had no idea why she was "often a bridesmaid but never a bride." How was she to know? Halitosis was ruining her life.[19] A Lambert Company "survey" of hairdressers found that one-third of their clients, some from wealthy families, were afflicted. One can only wonder about the incidence of halitosis among hairdressers, few of whom were from wealthy families, and admire the tolerance of Edna's friends who kept inviting her to be a bridesmaid despite her affliction. Company revenues jumped from $115,000 to $8 million in seven years.[20] (This author remembers Edna appearing in Listerine print ads in the 1940s.)

"Listerine" became a generic name for mouthwash and appears in unexpected places. In his 2009 novel *Rhino Ranch*, Larry McMurtry writes that the inventor of Listerine gave a property containing a lake to Princeton University and that Albert Einstein rowed his wife about on "Lake Listerine."[21] Alas! An e-mail from Erin A. Metro at Princeton explains that McMurtry was taking artistic license: "Our University archivist is not familiar with a 'Lake Listerine' in the Princeton area but did share the below information. 'We have photos of Einstein sailing (not rowing) a small boat on Lake Carnegie and perhaps the author [McMurtry] swapped a steel magnate's name for that of a mouthwash?'"[22] It is too bad that Lake Listerine doesn't exist. It makes a good story for a book that brings attention to Mr. Lambert's importance to St. Louis aviation and to aviation in general.

In the early 1900s, before Listerine's discovery and exploitation of halitosis, Albert Bond Lambert made frequent marketing trips to Europe. In 1906, he attended the Gordon Bennett Balloon Race in Paris, and, along with his wife, made his first balloon ascension. Two years later, he received balloonist's license #18 from the FAI. For the next 40 years, he was St. Louis' "Most Air-Minded Citizen."[23]

On October 21, a total of 150,000 spectators watched nine balloons drift into the air to start the 1907 Gordon Bennett, a featured event at the St. Louis Aeronautical Tournament. The German balloon *Pommern* and the French *Isle de France* both came down in New Jersey within sight of the Atlantic Ocean on the next day.[24] The U.S. Geological Survey, called upon to determine distances flown because of the closeness of the landing places, declared that the *Pommern* had flown 873 miles, the *Isle de France* 867 miles, six fewer than the *Pommern*. Both the German and the French balloons had flown farther than John Wise's American record flight of 826 miles, set 48 years earlier on July 1–2, 1859.

In addition to the Gordon Bennett, the St. Louis Aeronautical Tournament featured a dirigible race and a planned airplane race. A crowd of 100,000 watched five dirigibles make up to three attempts each to fly the fastest round-trip around a captive balloon moored about 3,400 feet from the start line on October 23.[25] Lincoln Beachey, who was to go on to a brilliant career as an airplane pilot before crashing to his death in 1915, won the dirigible race.

Few people in 1907 had seen an airplane, and the entry of seven airplanes[26] in the airplane race promised a lively and brand-new kind of contest. A writer for a contemporary magazine underlined the event's importance: "For the first time in history [given that this was written less than four years after the Wright brothers' first flight, "for the first time in history" may seem a peculiar choice of words], as far as we know, there is expected to be actual 'races' between gasless flying machines at St. Louis."[27] Two airplanes arrived at the racecourse. Neither managed to take off. So much for the "gasless flying machine" race.

The featured attraction, the Gordon Bennett Balloon Race, had been a great success,

Ascension at the 1907 Gordon Bennett Balloon Race, St. Louis (courtesy National Archives and Records Administration).

and St. Louis remained a center for ballooning for years to come. Several St. Louis citizens manufactured balloons, and more made balloon flights, some of the flights directed at winning the Lahm Cup, which the Aero Club of America would award to its first member to fly farther than Frank Lahm's 402 miles that won the first Gordon Bennett in 1906. The winner of the cup would retain it until another Aero Club of America member exceeded the distance flown, and a balloonist who held the record for the longest flight for three years would take permanent possession of the trophy. In addition to being Aero Club members, Lahm Cup competitors had to inform the Aero Club at least 24 hours in advance of an attempt and pay a one-dollar fee.

Albert Bond Lambert and H.P. Honeywell, the manufacturer of the balloon *Yankee*, made their attempt for the Lahm Cup on November 28, 1908. It was Lambert's 13th ascension, Honeywell's first. They flew about 450 miles — 23 miles short of the then-current Lahm record — and landed in Alabama.

A year later, in 1909, the St. Louis Aero Club organized a mass assault on the Lahm Cup record. Seven "large" (80,000 cubic-foot capacity) balloons entered the contest at that year's centennial celebration of the incorporation of St. Louis as a town. The first- and the second-place balloons took off within five minutes of each other. Underlining the capriciousness of winds and ballooning, they drifted off in wildly different directions. The winner drifted 540 miles in a generally northward direction to land in Minnesota; the second-place finisher went 448 miles southeast to land in Alabama. The winner earned the Lahm Cup and $600 first prize; the second-place finisher won $400. A third balloon, which traveled only 146 miles to Edina, Missouri, spent its flight in crosswinds and remained aloft for 48 hours to win the trophy for longest endurance. Dirigible pilots flew several demonstration flights during the centennial celebration, but high winds canceled the scheduled dirigible race.

Centennial celebration organizers realized that the glamour and prestige in aviation was shifting from lighter-than-air to heavier-than-air flight and included airplanes in the event. Who better to demonstrate an airplane than its inventors, Orville and Wilbur Wright? Albert Bond Lambert invited them. They declined, saying they were not interested in exhibitions, and stated that they would consider legal actions against anyone who made money from flying an airplane that they had not manufactured or licensed. The Wrights repeatedly made good that threat through 1917, until the federal government negotiated an agreement between them and other manufacturers so that American companies could build airplanes for the huge market that emerged during World War I.

Rebuffed by the Wrights, Lambert turned to the other "big name" in American aviation, Glenn Curtiss. Only weeks earlier, Curtiss had been very much in the news because of winning the first Gordon Bennett airplane race at Reims, France. When Lambert's cables reached him, he was winning additional trophies and money at an air meet in Brescia, Italy.[28] Lambert and Curtiss negotiated a contract under which Curtiss would make four flights at St. Louis during "Centennial Week" for $6,000. Curtiss brought two airplanes, identical to his winning *Reims Racer*, to St. Louis.

"The first aeroplane flight west of Dayton, O. took place ... [on October 7, 1909] when Glenn Curtiss, watched by not more than fifteen persons, made two short trips."[29] After a stiff wind lay down later that day, Curtiss tried again at dusk. The crowd of spectators had grown: "Four hundred thousand people [!] had waited for hours in Forest Park to see the performance."[30] Curtiss flew only about 220 yards before a stopped engine brought him back to earth. Two days later, he was more successful. "Under the arch of a brilliant rainbow,

Curtiss Reims Racer, winner of the first Gordon Bennett Airplane Race, 1909. Caption for notations on original photograph: "Side view of Latest Curtiss Biplane. This machine, which won the world's speed record at Rheims, France, in August, 1909, is one of the lightest machines ever built, weight less than 500 pounds. Its lateral balance is maintained by the ailerons at *a* [the letter is barely visible at horizontal surface above engine cylinders and between the wings], and its longitudinal balance by the elevator surfaces *h h h*. The wheeled running gear *g g g* is without springs.... Propulsion is by a single propeller at the rear, revolved by an eight-cylinder, V-shaped, 50-horsepower motor." Letters *o* and *j* are unidentified (courtesy National Archives and Records Administration).

Glenn H. Curtiss thrilled thousands of rain-soaked spectators here to-day by an aeroplane flight of more than a mile over the tree tops of Forest Park."[31] He flew at about 60 feet and was in the air for one minute and 50 seconds.

In all, Curtiss made seven flights, totaling about that number of minutes in the air. It's tempting to ask, "A minute or so in the air? That's a flight? What was the big deal?" Even so, Horgan writes that Curtiss was "the most outstanding performer during Centennial Week, 1909."[32] Based on monetary rewards, Curtiss was the big winner, earning $6,000, or almost a thousand dollars for every minute of flight. In comparison, the huge balloons that drifted for hours covering hundreds of miles were competing for prizes of less than $1,000.

In 1910, the Aero Club of St. Louis, now with 700 members, offered total prizes of $25,000 in an "international tournament" at its own airfield just to the west of the city limits. Two foreign airplane pilots, Alexander Ogilvie, manager of the Wright Company in England, and Alfred Le Blanc (sometimes written as Leblanc), Louis Blériot's "right-hand man,"[33] journeyed from the site of the second Gordon Bennett Airplane Race at Belmont Park, New York, to St. Louis. The 1910 Gordon Bennett Balloon Race, scheduled for the next-to-last day of the St. Louis air meet on October 17, had more international competitors. The entrants consisted of two French, three German, two Swiss, and three United States balloons.

Well before the 1910 international tournament, Pulitzer's *St. Louis Post-Dispatch* had offered the Wright brothers $2,500 to make a flight along a stretch of the Mississippi River and across it. Saying "no" again, the brothers replied that they were too busy. The paper had success when it invited Thomas Baldwin to make the flight.

In September, Baldwin, the noted balloonist, dirigible designer and pilot—and at 205 pounds, "the heaviest successful aviator"—made the flight. He flew his homebuilt *Red Devil*, a Curtiss design that could probably manage about 50 miles per hour, from an airfield a mile away, to the river, turned south and flew down the river about six miles, crossing over to land on the Illinois side. There he rested, filled the airplane's radiator, and took off to make the northbound flight over the river. This time he flew under the two bridges—the Eads and the Merchants—that spanned the river. About 200,000 people saw what Baldwin called "his first flight of consequence."[34]

The Wright brothers had repeatedly said they were not interested in exhibition flying, but, as businessmen, they saw there was money to be made. To cash in, they organized the Wright Exposition Company, paying their pilots $20 per week plus $50 for each flight. The Wrights took home any prize money, which, in 1910, provided a profit of about $100,000.

Walter Brookins, Arch Hoxsey, and Ralph Johnstone—the most famous of the Wright pilots and none a licensed pilot—were in the air throughout the 1910 international tournament week competing for various prizes and awards. Hoxsey made a historically important passenger-carrying flight when he took former President Theodore Roosevelt on a three-minute flight on October 11. Horgan quotes Roosevelt as saying, "Bully.... I wish I could have stayed up longer."[35] Albert Bond Lambert remembered the former President as not so "bully." "'He had to [take the airplane ride],' Lambert says with a sly grin. 'There were 15,000 people out there and two bands of music. He couldn't back out.'"[36]

Superstitious Alfred Le Blanc, whose Blériot XI was clearly the fastest airplane at St. Louis, refused to fly a time trial over a measured-mile course on October 13. The next day, he posted a speed of 68 mph, good enough for a new American record.[37]

The opening-day crowd of only 8,000 was hardly visible in grandstands built for 100,000. The low attendance imperiled the finances of the Aero Club, whose officials were counting on admission fees of $0.50 and $1.00 to offset expenses. With only two days of the tournament remaining, Alfred Bond Lambert, newly installed as Aero Club of St. Louis president, appealed to the citizens of St. Louis: "Pay your way ... through the gates and discourage the parasites who seek adjoining fields to deadbeat their way."[38] Crowds of 18,000 on October 15 and 35,000 on October 16 paid their way in, bringing the total paid attendance to 63,000. An uncounted number of "parasites" took in what they could for free, watching from outside the festival grounds, but the paying customers moved the Aero Club's finances into the black. The much smaller crowds than the mobs that had flocked to earlier St. Louis air meets probably reflected that aviation, if still not quite an ordinary part of life, was no longer especially extraordinary. Seven years after the Wrights' first flight, the novelty of seeing an airplane flying was wearing off.

Anyone who has spent a summer in the Midwest probably remembers heat, humidity, and grasshoppers. The insects sabotaged two United States entries in the biggest event of the 1910 aviation calendar, the Gordon Bennett balloon races, which was again in St. Louis. On the night before the race, grasshoppers discovered the uninflated, uncovered gas bags of the *Million Population Club*[39] and the *St. Louis IV* on the ground and chewed holes in their linen fabric. The holes were patched, and the two repaired balloons joined the other eight to start the race.

Well before the grasshopper attack, grievous injury had been done to the perception of courteous gentlemen balloonists floating in their balloons for the joy of the sport. The pilot of one of the German balloons announced that he would not race because no cash prizes were to be awarded, and several other pilots "intimated they were seriously considering withdrawing."[40] The same German airman also complained he had not been provided navigation instruments and that the order of starting (which was done by drawing lots) had been posted only four days before the race and not 14 days as required by the rules. Prize money was found for the first five finishers.

All ten balloons got away cleanly except the United States' *Million Population Club*, which avoided the grandstand at the last minute only by frantically dumping sand ballast. Stocked with ample supplies of food and drink, the balloons drifted off to the north and northwest, where, according to a meteorologist, they would rise into the west-to-east-blowing winds at higher altitudes,[41] as John Wise had predicted.

The *Million Population Club* and the *St. Louis IV*, probably suffering from inadequately repaired grasshopper damage, and the French *Condor*, which like the two United States balloons had a heavily lacquered fabric gas bag, came down south of the Great Lakes. The other seven balloons, all of which had rubber gas bags — clearly a technological breakthrough — drifted on over Canada. Flying over and landing in Canadian forests brought balloon crews unexpected dangers and thrills. A "huge bison" greeted Alfred Le Blanc, pilot of the French *Isle de France*, and his aide when they landed in a clearing. Within a day, "afraid of stepping on snakes of which I had read," and seeing moose and deer, they reached a telegraph station and cabled the Aero Club information about their landing and trek.[42] (Seven days after emerging from the Canadian wilderness, Le Blanc, flying a Blériot XI, was leading in the Gordon Bennett Airplane Race at Belmont Park. He did not win or finish in the money. For one reason or another, he collided with a telegraph pole on the last of 20 laps.)

The American crew of Alan Hawley and Augustus Post was the last to emerge from the Canadian forests. A snowstorm forced the *America II* to land 1,500 feet up on the side of an unnamed mountain, and Hawley and Post informed St. Louis about their flight and location[43] a week later after four days' walk and a canoe ride with French-Canadian trappers brought them to a settlement with a telegraph. Hawley and Post's flight of 1,173 miles was a new American record and enough to win the Lahm Cup[44] and the Gordon Bennett Trophy. It was the second consecutive United States victory in the Gordon Bennett, and if a United States balloon could win the next Gordon Bennett, the Aero Club of America would take permanent possession of the trophy.

Two months after the race, a German aeronautical association protested the award of the Gordon Bennett Trophy to the United States because the United States had not posted its entry two months before the race and Hawley had failed to produce documents that attested to *America II*'s reaching the claimed landing spot. Hawley dismissed the complaints as groundless. After all, the FAI had already accepted his statements. He added, "We were provisioned and equipped to stay in the North a month longer…. Most of the foreign contestants left helter-skelter and were forced to drop out by reason of their poor equipment."[45] Hawley knew from his own experience that unless a downed crew elected to stay with its balloon, hoping for rescue, the month's provisions would have done little good. If the men started walking, most provisions would have to be left behind. Hawley's comment about "poor equipment" could be interpreted to bolster the German claim that German crews were not provided with appropriate navigational instruments. Both the protest and Hawley's gratuitous statement were additional assaults on the picture of gentlemanly balloonists, good sports all-around, happily floating at the mercy of the wind.

Balloon races were continued in St. Louis until suspended during World War I. They resumed after the war, in 1919, but huge paying crowds of tens of thousands were things of the past. Worse, the crew of a balloon in the 1919 National Balloon Race from St. Louis perished in a crash in Lake Huron.

Business slowed in St. Louis during World War I. It picked up after the war — the city was prosperous and progressive — but growth never again reached the pace seen between the 1904 World's Fair and the war. In 1920, St. Louis slipped from fourth to sixth largest city in the country (773,000 residents) as other cities annexed additional suburbs[46] and attracted more new residents. The population slid farther in the censuses of 1930 and 1940, recovered during World War II to peak at 857,000 in 1950, and has continued generally downward since. (St. Louis is now the second largest city in Missouri, behind Kansas City, and 53rd in the nation.)[47]

In early 1922, when Detroit was preparing for that year's National Air Races, the Flying Club of St. Louis, largely composed of World War I veterans, adopted a resolution urging that St. Louis host the 1923 races.[48] On October 19, 1922, Colonel C.G. Hill, the commander of Scott Field in Belleville, Illinois, 35 miles east of St. Louis, thrilled the St. Louis Lions Club with a speech about the recently completed Detroit National Air Races. Two days later, Randall Foster, president of the Flying Club of St. Louis, visited Joseph Pulitzer, Jr., publisher of the *St. Louis Post-Dispatch*, and one of the donors of the Pulitzer Trophy, to talk about bringing the 1923 National Air Race to St. Louis.

Good reasons abounded for selecting St. Louis: The city was home to the Pulitzer newspapers that had endowed the Pulitzer Trophy; it had a rich aviation history, if concentrated in lighter-than-air activities; it had an industrial base and good transportation; city leaders saw their home as a city on the move, and some city leaders saw a national air race as a way to hasten construction and use of a major commercial airport in the city.

Pulitzer suggested the formation of a citizens' committee which, when organized, agreed to underwrite the costs of the air meeting. In February 1923 the NAA informally informed St. Louis that it had been selected to host the 1923 contests and made the formal announcement on June 25. The citizens' committee dissolved itself and was reconstituted as two other organizations with overlapping membership and officers— the St. Louis Aeronautical Corporation and the St. Louis Air Board. The corporation raised $200,000 through the sale of 4,000 shares at $50 each, and raised a total of more than $300,000.[49] The board used the money to pay for preparing, maintaining, and operating a flying field.

The organizers first considered holding the meet at Scott Field, but it was far from St. Louis, and, as an Air Service field, it had no future as a municipal airport. They settled on the use of the St. Louis Flying Field in Bridgeton, six miles west of St. Louis. The Missouri Aeronautical Society had leased the land in 1921, and Albert Bond Lambert,[50] who was maintaining and improving it at his own expense, offered its services to everyone interested in aviation.

At the time, and throughout the 1920s, Lambert was frequently identified as "Major Lambert" in honor of his World War I service when he had donated a large balloon to the Missouri Aeronautical Reserve Corp and was commissioned a captain in the Air Service. Actively participating in the training of 350 balloon pilots during the war, he was subsequently promoted to major.

In preparation for the 1923 National Air Races, the Missouri Aeronautical Corporation purchased the 183 acres used as the flying field and leased an adjoining 316 acres.[51] More than $130,000 was spent on grading and improving the grounds to enable "ships landing at 150 miles an hour to glide safely over the field as if they were on a billiard table."[52] The "150 miles an hour" seems a bit overstated and unnecessary; the Pulitzer racers were required to have a landing speed of 75 mph or less. Two steel "twin hangars," each being two hangars joined together lengthwise and each capable of housing 25 airplanes, were built. A 10,000-gallon gasoline tank[53] was buried and a 10,000-gallon water tank installed.

Bleachers and box seats were provided for 45,000 spectators, and space was made to park 20,000 cars.[54] The seats looked directly north across the airfield. Racers rounded a pylon on the west side 200 yards in front of bleachers, providing spectators a close-up view for a distance of at least one mile. "But that will not be all," J.J. Horgan points out. "The terrain for miles in each direction out there is comparatively level, and the planes can be seen with naked eye a distance of about six miles coming and about the same distance going over the triangular course, thus affording all spectators unobstructed view of the races for a third of the time of flight around the course."[55] In addition to being able to see, spectators would be safe and comfortable:

> All the roads to and from the field will be adequately policed during racing hours, and the traffic regulations will be such as to avert confusion and delay....
>
> [O]ne may go to the races, either by motor or street car; take one's lunch, or purchase it at one of the innumerable refreshment concessions ... near the field, and spend the day in comparative comfort....
>
> [C]omfort stations for men and women have been erected, so that anyone may spend an unlimited time at the field without discomfort. Maids will be in charge of the stations for women.[56]

One hundred eighty policemen and 70 detectives from St. Louis and other cities would provide information and directions and protection of the spectators from pickpockets and confidence men.

Lambert Field, or "St. Louis Flying Field" as it had been called and as it is identified on the maps included in the race program, is located northwest of the city. As at Omaha two years earlier, the Missouri River meandered through the racecourse. From Lambert the racers flew north on the first leg of the nearly-equilateral triangle course. Almost as soon as they took off, they could see the stretch of the Missouri River that lay under the last half of the first leg. A pylon marked the first turning point, and the racers swooped left to fly west-northwest and parallel to a train track to another pylon, where a turn to the southeast headed them back to Lambert. They crossed the Missouri River, almost a mile and a half wide, at about midway along the last leg of the race.

Pilots liked the airfield. For instance, navy lieutenant Steven Calloway, whose picture and a picture of him test-flying a navy racer were featured on the front page of the September 30 *Post-Dispatch*, praised the new airport: "It is three times as good as Mitchel Field [the newspaper explained that Mitchel, on Long Island, had been the site of the first test flights of the Curtiss racers].... I'm convinced that St. Louis has one of the finest fields in the county."[57]

Major Lambert had laid the airport's cornerstone in 1922. With his friend and aviation enthusiast William B. Robertson piloting, he tossed the cornerstone, bearing the inscription, "St. Louis—The Aerial Crossroads of America," from the rear cockpit of a biplane. The Flying Club of St. Louis, to honor the man most important to St. Louis aviation, voted to name the field Albert Bond Lambert Field.[58] On the first day of the 1923 air meet, a young woman from St. Louis "came down out of the sky" and made a speech of "military terseness" to General Mason Patrick: "The field in finished. I dedicate it as the Lambert St. Louis Flying Field. I turn it over to your command for the races, sir. All contestants are present

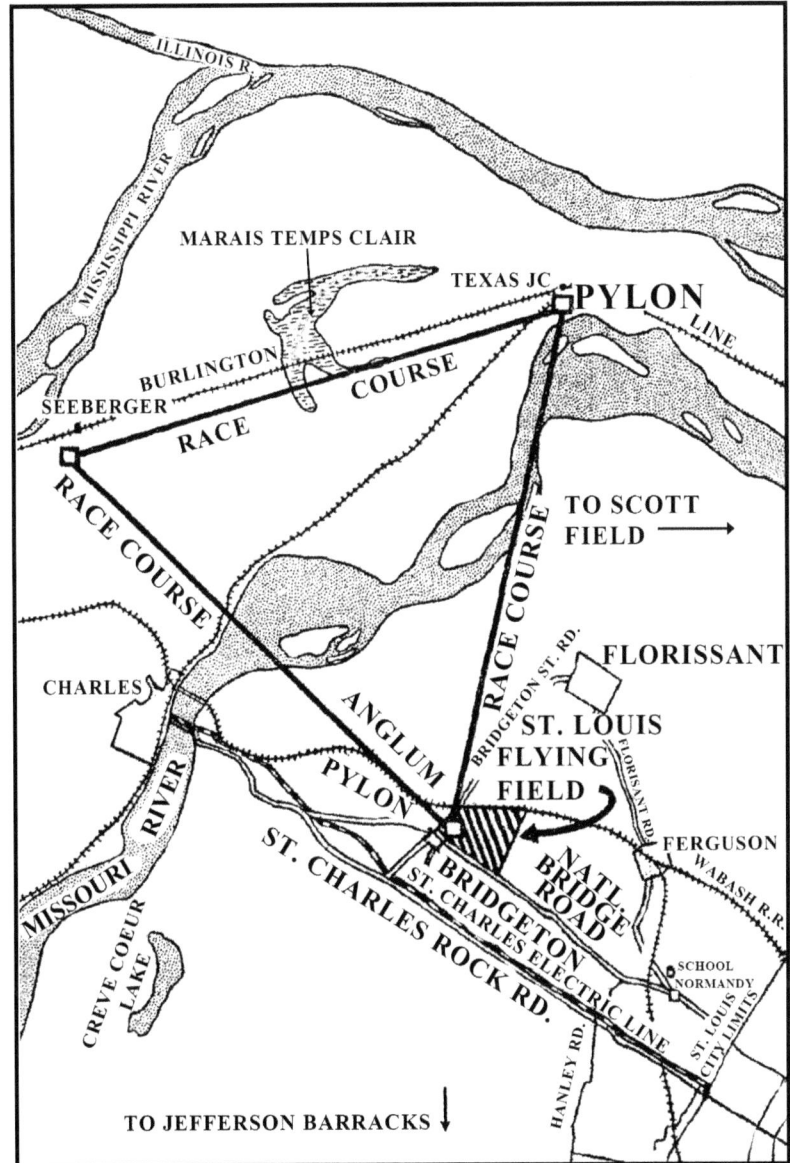

Figure 5. Map of the fourth Pulitzer Race, St. Louis Flying Field, then and now "Lambert Field," St. Louis, Missouri. The Pulitzer required four laps of the 50-km triangular course. The John L. Mitchell Race of the same length was flown over the same course (courtesy Missouri History Museum, St. Louis).

and accounted for."[59] The man who had done and continued to do so much for St. Louis aviation was honored, and the St. Louis International Airport remains "Lambert Field" to this day.

For the races, 60-foot pylons were erected at the corners of a 50-km equilateral triangle. Initially, the pylons were doubled because Jerome Hunsaker, the designer of the Wright NW-1 sesquiplane for the 1922 Pulitzer, told the NAA's contest committee that recent tests in England had shown pilots could not withstand turns of more than four G's. The double pylons were intended to force pilots to make more gentle turns. Before the races, the "extra" pylons were removed when the worry about too-tight turns proved overblown. All the speed competitions at the National Air Race would be flown around the 50-km course. Improved performances were seen in all but two contests for which comparisons can be made between 1922 and 1923. The exceptions were the John L. Mitchell Race and the Aviation Country Club of Detroit Race.

TABLE 3
1923: Closed-Course Competitive Events at the Second National Air Race, Fourth Pulitzer Race, St. Louis, Missouri, October 4, 5, and 6[a]

Event	Entrants	Course	Pilot, Airplane (hp) Speed
Flying Club of St. Louis Trophy Race	Civilian pilots/two-seat airplanes, 90-hp or less, 340-lb load	150 km, 3 laps of 50-km course (ca. 93 miles)	1. Walter Lees,[b] Hartzell biplane (90) 89 mph 2. Perry Hutton, Laird Swallow (90) 87 mph 3. "Casey" Jones, Curtiss Oriole (90) 85 mph
Liberty Engine Builders' Trophy Race	Military pilots/two-seat airplanes, 340-lb crew + additional load based on engine hp	300 km, 6 laps of 50-km course (ca. 186 miles)	1. Lt. C. McMullen, USAS, Fokker CO-4 (400) 139 mph 2. Lt. H.K. Ramey, USAS, De Havilland DH-4 (400) 138 mph 3. Lt. L.H. Smith, USAS, Fokker CO-4 (a USAS-designed DH-4 derivative) (400) 135 mph 4. Lt. G.B. Hall, USMC, Vought UO-1 (200) 132 mph. *All the first four finishers bettered the 1922 winning speed of 129 mph over 257.7 miles.*
Aviation Country Club of Detroit Trophy Race	Civilian pilots/engine of less than 220 hp, load of 340 lb	250 km, 5 laps of 50-km course (ca. 155 miles). Contests for speed and efficiency calculated: *payload (lbs) x mph / hp*	Speed: 1. Jack Atkinson, Bellanca Sedan (95) 94 mph 2. Perry Hutton, Laird Swallow (98) 87 mph 3. "Casey" Jones, Curtiss Oriole (98) 83 mph Efficiency: 1. Jack Atkinson, Bellanca Sedan, 675 points. 2. Perry Hutton, Laird Swallow, 300 points. 3. "Casey" Jones, Curtiss Oriole, 286 points. *Speed winner in 1922 was a modified DH-4 powered by 440-hp engine; speed 135 mph. Efficiency points were not found for 1922 contest.*
Merchants Exchange of St. Louis Trophy Race	Civilian and military airplanes and pilots/ to develop airplanes capable of 2,000 lb payload. Only military types flew.	300 km, 6 laps of 50-km course (ca. 186 miles)	1. Lt. H.L. George, USAS, Martin Bomber (2 x 440 hp) 114 mph. 2. Lt. M.A. Shur, USN, Douglas DT-4 torpedo plane (650) 108 mph 3. Lt. W.S. Hallenberg, USMC, Martin Bomber (2 x 440) 105 mph *Winner of similar 1922 Detroit News Aerial Mail Trophy Race, 105 mph over 257.7 mile course.*

Event	Entrants	Course	Pilot, Airplane (hp) Speed
Detroit News Air Mail Trophy Race	U.S. Air Mail pilots/ U.S. Air Mail De Haviland DH-4 airplanes/all powered by 400 hp Liberty engines.	300 km, 6 laps of 50-km course (ca. 186 miles)	1. J.F. "Dinty" Moore, 125 mph 2. D.C. Smith, 121 mph 3. P.F. Collins, 119 mph
John L. Mitchell Trophy Race	Pilots of 1st Pursuit Group flying standard MB-3A pursuits	4 laps of 50-km course (ca. 124 miles)	1. B.E. Skeel, MB-3A (340) 146 mph 2. G.P. Tourtellot. MB-3A (340) 143 mph 3. T.W. Blackburn, MB-3A (340) 141 mph. *1922 Winner, 148 mph.*
Pulitzer Trophy Race	Civilian and military pilots/airplanes w/ landing speed 75 mph and top speed greater than 140. Only military planes flew.	4 laps of 50-km course (ca. 124 miles)	1. A.J. Williams, USN, Curtiss R2C (507) 244 mph. 2. H.J. Brow, USN, Curtiss R2C (507) 242 mph 3. L.H. Sanderson, USMC, Wright F2W-1 (680) 230 mph. *1922 Winner, 206 mph over 250-km course.*

[a]From *Aircraft Year Book* (New York: Aeronautical Chamber of Commerce of America, 1924) 152–158 and 326–328, except for information about Mitchell and Pulitzer Races from T.G. Foxworth, *The Speed Seekers* (New York: Doubleday, 1974) 456–457.
[b]Lees is credited with training General Mitchell to fly.

The Mitchell had not been included in the race organizers' programs issued two months before the air races. General Mitchell intervened, and the Mitchell Race, designated as event 8a, appeared on updated programs. The Mitchell was flown on October 4, the second day of the meet. The commander of the 1st Pursuit Group, Captain Burt Skeel, won the race, but there was no improvement in the winning speed. Skeel finished at 146 mph, two mph slower than Lt. Donald Stace's 148 mph at Selfridge the year before. No new record, but the race had an orderly and exciting start and an unlikely and crowd-satisfying finish.

At 12:45, Air Service enlisted men swung the propellers of six MB-3As of the 1st Pursuit Group. When the engines were running smoothly — or as smoothly as the MB-3As' Wright H-3 engines would run — mechanics and other enlisted men lifted the pursuits' tails high enough to get the tail skids off the ground, and others hung on wingtips to pivot the airplanes and line them up for the start. The first pilot off, Lt. Thomas W. Blackburn, roared down the runway at 1:00, and the other five pilots followed at half-minute intervals. Skeel was last off. Blackburn led them in a climb to 1,000 feet then peeled off and dived across the start line. The others followed, in the same order in which they took off, and the timing for each plane was begun when it crossed the start line.

Skeel, starting last, had moved up to third and was pressing close to Lt. George Tourtellot in the last few hundred yards of the last lap. Tourtellot's engine sputtered, but he kept his racer airborne as the engine stopped completely, out of fuel. Realizing he would be unable to fly across the finish line, he "pancaked hard on the graded earth, and the resulting bounce carried him across the finish line in full flight."[60] Tourtellot's bouncing finish must qualify as unlikely in racing history.

Blackburn, the first off, Tourtellot, and Skeel crossed the finish line in a blur; two other pilots, lieutenants Vincent Dixon and J. Thad Johnson were seconds behind. The sixth starter, Lt. Thomas Matthews, had landed after three laps with fuel problems. The timers determined that Skeel won, finishing in the shortest length of time.

The 1923 Mitchell Race would be the last for Thomas Morse pursuits. In 1924, the 1st

Pursuit Group reequipped with Curtiss PW-8s, which had "all of the characteristics of the Curtiss Racer [the 1922 R-6s],"[61] and the MB-3As were gone.

The air-cooled engine, championed by the navy, continued to impress. A 200-hp air-cooled engine powered a navy Vought UO-1 observation plane to 4th place in the Liberty Engine Builders' Trophy Race, only 7 mph slower than the winner, a Fokker CO-4 powered by a 400-hp water-cooled Liberty engine. *The Aircraft Year Book* called its performance "especially interesting."[62]

In 1922, three of the four entrants in the Aviation Country Club of Detroit race had been military types. A year later, civil aviation had progressed to the point that six of the seven entries were purely civilian designs, and the winner, the Bellanca Sedan (Bellanca CF), was a well-streamlined, high-wing monoplane with airfoil-shaped wing braces—a standard feature on Bellancas to come. The Bellanca was faster than the other 1923 entrants and more than twice as efficient as judged by the speed it achieved while carrying a load with an engine of specified horsepower. It flew slower than the 1922 winner, which had about twice the horsepower. The National Air and Space museum quotes *Jane's All the World's Aircraft*, which commented that the Bellanca CF was "the first up-to-date transport aeroplane that was designed, built, and flown with success in the United States."[63] United States aviation was progressing in civil aviation as well as in high-speed flight.

Two disappointments marred the Pulitzer Race before it was flown. Three well-known foreign pilots had mailed entries, but all withdrew. The Pulitzer, despite great efforts on the part of organizers to attract foreign racers, remained a United States affair. In addition, the total of seven racers—products of three design teams—was small in comparison to the 15 from seven design groups of the year before. Those were surely the only disappointments associated with the 1923 Pulitzer.

The 1923 racers set new, seemingly impossible, speed records. Three of the 1923 starters—a Sperry-Verville and two Curtisses—were retreads from the year before. Four—two Wrights and two Curtisses—were new.

The retreads, the Sperry-Verville R-3 and the two Curtiss R-6s, had flown speed trials in the days after completion of the 1922 Pulitzer. The R-3's speed run had an air of necessity, if not desperation, about it. In order to avoid the penalty for below-specified speeds built into the army's contracts for 1922 Pulitzer racers, an R-3 had to reach 190 mph. Lt. Eugene Barksdale flew the R-3 he'd flown in the Pulitzer over a measured course two days after the Pulitzer. The R-3 managed 191 mph, by one mph saving the Sperry Company from paying a penalty.

In contrast to the R-3's barely meeting its contract specification, General Mitchell and an R-6 set a new world record of 233 mph. The time trials, making the superior speed of the 1922 thin-wing biplanes glaringly apparent, were enough to convince some aviation commentators that thin-wing biplanes would always fly faster than monoplanes with the thick wings necessary for their cantilever structure.

Two of the three R-3s had been in bad shape when they left Selfridge after the Pulitzer. One had been damaged when an oil line break had forced Capt. St. Clair Streett to land it among trees and shrubs on his last lap in the Pulitzer. After repair of damage to water lines and motor mounts caused by the vibrating Wright engine on Lt. Fonda Johnson's R-3 during the Pulitzer, Johnson had made another flight from Selfridge. The engine still vibrated and broke an oil line. With oil pressure falling, Johnson switched off the power, but the engine "dieseled" and pulled him back over the airfield. He lined up to land, forgot to lower the landing gear, and skidded 75 yards across the field. The two damaged R-3s and one whole R-3 were loaded onto trucks and taken to McCook Field. Barksdale's undamaged R-3 was given the McCook Field number P-269 and readied for engine and airfoil research. The other two were put in storage.

Alfred Verville was unhappy with the army's decisions about the R-3s. Verville said that experts at the National Advisory Committee for Aeronautics (NACA) technical facility at Langley, Virginia, who had requested an R-3 for evaluation "probably could have straightened out the R-3 within a year and the military could have had monoplane fighters ten years before the P-26."[64] The army didn't comply with NACA's request, asserting that a year's flight testing would reveal more about the airplane, its capabilities, and its shortcomings than a year's examination by engineers.

Verville thought the army incompetent (after all, McCook Field had, according to some accounting, spent 10 times as much time and money on the VCP-R/R-1 as the Sperry Company had spent in preparing three R-3s to race in 1922). He also thought that a cabal of established manufacturers, Curtiss and Wright, with vested interests in making and selling biplanes didn't want to give the R-3 a fair shake.[65]

Lt. Alex Pearson was assigned to test the flyable R-3, at McCook, and he flew it often. Gear-up, the R-3 could fly 185 mph, gear down 162, 23 mph slower. Pearson put the R-3 into the record book by setting a world record speed of 167 mph over a measured distance of 500 km. Two days after Pearson's flight, and before any documentation could be completed at McCook, a cable from France notified the NAA that a Frenchman had set a new speed record of 114 mph for 500 km. Pearson and the R-3 had flown 53 mph faster.

In spring 1923, the army planned to modify and enter two of its R-3s in that year's Pulitzer, but it scaled back to one when Sperry estimated the cost at about $15,000 each to prepare the racers. The army choose Pearson's plane and shipped it to the Sperry factory on Long Island where it was reengined with a Curtiss D-12 engine and fitted with wing radiators—the combination that Verville had wanted the year before. Other changes were made: the wing was completely plywood-covered to increase its rigidity, and the length and wingspan were both increased about a foot. In keeping with Curtiss practice, a new oil cooler was fitted in the radiator outflow.

The modifications were complete on Labor Day, and a metal Curtiss-Reed propeller was ordered for the racer. Before it arrived, Pearson test-flew the R-3 and saw the wings twist alarmingly when the ailerons were used. Although Verville had stiffened the wings, it was not enough. Pearson saw the wingtips flex in the direction opposite to the direction of aileron movement. For example, as an aileron was raised, the wingtip flexed downward, counteracting the expected aileron effect and making control of the airplane uncertain. It was not the R-3's only problem. It also had a tendency to roll left when the controls were in the neutral position for straight-ahead flight. The effort needed to hold the airplane on a straight course was so great that shock-absorber cord was strung from the cockpit wall to the control stick to assist the pilot in holding the stick to the right.

Before shipping the R-3 to St. Louis, Verville studied the roll problem and decided to glue plywood sheeting on the outside four feet of the top of the left wing and on the outside four feet of the bottom of the right wing to increase lift on the left wing and decrease it on the right. The hasty modification worked perfectly; the rolling tendency disappeared, and the shock-absorber cord was removed. When a Curtiss-Reed metal propeller arrived, it was hastily installed, and Pearson and the R-3 clocked 233 mph, a 15-mph improvement.

The mechanics, engineers, and pilots who had nursed the R-3 though its teething problems may have considered the improvements too good to be true. They were too good to last. On a test flight, the spinner of the Curtiss-Reed propeller broke free after five minutes at full throttle. The engine vibrated greatly. Pearson was forced to land. Worse was to come. From somewhere, a decree came down that the R-3 could not race with a metal propeller. Verville blamed the Curtiss Company for instigating the rule. Whatever the facts, past events provided Verville reasons to suspect Curtiss. Curtiss had prevented Verville and everyone else from obtaining Curtiss engines from the army to power their 1922 racers, and it was to the company's advantage not to allow other companies to equip their airplanes with Curtiss-Reed metal propellers. Verville also claimed the rule was announced very late, only a short time before the Pulitzer, making it too late for General Mitchell, recuperating from injuries suffered in a fall from a horse, to overturn it. A wooden propeller was found, and a new spinner was fashioned for the R-3 an hour before the start of the Pulitzer.

The other two retreads in the 1923 Pulitzer, the Curtiss R-6s, had more glamorous post–1922 Pulitzer histories than the R-3s, but a tiny bad smell had become associated with their record-breaking post–Pulitzer flights in 1922. General Mitchell had flown Lt. Maughan's racer to 223 mph, a world record, on October 22, but Lt. Maughan had flown faster only days before in carefully timed but unofficial flights. No matter how muted, criticism was directed at the general for "grandstanding," of taking advantage of his rank to make the world-record flights himself rather than giving the honor to one of the race pilots.

Whether or not he was responding to muttered and rumored criticisms, in December 1922 Mitchell arranged for lieutenants Maughan and Maitland to make attempts at new speed records in the spring. Wilbur Wright Field engineers made changes to the windshields, engine cowlings, and air intake scoops to reduce drag. For the same reason, the exhaust stacks were cut off, which must have made already-loud engines louder. Engineers streamlined everything possible and installed new carburetors and Curtiss-Reed metal propellers.

On March 29, lieutenants Maughan and Maitland attempted new records. Thirty times, Maitland climbed to 7,000 feet, screamed down in a full-speed dive, leveled out, and sped across the one-km course. His fastest four tries averaged 240 mph, and his fastest single pass was clocked at 281 mph. He credited the starting dive for his speed: "I guess there is nothing to it except which one is the biggest fool and takes the highest dive. That's where I got most of my speed."[66] Because Maitland did not fly a completely level flight over the measured distance, his effort at the world's record was disqualified.

Maughan followed Maitland into the air. He too dived down from several thousand feet and leveled out at 60 feet. He flew level courses, made a total of 24 tries, and his four best flights averaged 236.6 mph.[67] His world record for fastest official speed over a one-kilometer course, set on March 29, 1923, will stand forever. On April 1, the required distance to be flown to set absolute world records was changed from one to three kilometers.

The R-6s' assaults on a record over the new three-kilometer course did not turn out well. On April 9, Maitland taxied out to make an attempt, but the R-6 lurched into a rut; the left wheel was wrecked, and the racer ground looped.[68] Repairs were made to the propeller, left wings, one wheel, and one bulkhead, and Maitland made another attempt on April 25 (or 26). The racer ground looped to the right on takeoff and ended up, tail in the air, canted up on its nose with damage to its left wings. The aborted attempt at the speed record over three kilometers was Maitland's last flight in an R-6. In all, he made 13 flights and logged about six hours. As Foxworth notes in several places, most racer pilots in the early 1920s had few hours in their airplanes.

Both R-6s were repaired and moved to McCook Field, and all work on them ended in May. Two months later, in July, work re-started to ready them for the Pulitzer.

The army had muted expectations for the 1923 Pulitzer. In the year following the 1922

Pulitzer, the army had focused its expenditures on observation planes and development of the "huge" six-engine Barling bomber and smaller two-engine Martin bombers, and it purchased no new racers. What were the army's chances? As in every race, there was the possibility that faster airplanes would have mechanical problems and drop out. The R-6s had been remarkably reliable, and they were bound to finish the race. There was a chance.

The army decided that other pilots needed a chance to shine in races, and Maughan and Maitland would not fly the 1923 race. Lt. John Corkille was assigned as the first pilot on Maughan's racer, and Lt. Harry Mills was named alternate pilot. The R-6's on-the-ground problems didn't go away. On September 4, Corkille ground looped and damaged the left wingtips. Repairs were quickly made — mechanics must have been very familiar with R-6s' left wingtips, with their recurring damages — and Corkille made two successful flights before Mills took off on his first flight on September 10. Running over a rock on landing marred Mills' otherwise uneventful flight. The R-6 tipped over and again damaged the left wingtips. Corkille and Mills' flights went smoothly after that.

Corkille's R-6, Maitland's old mount, was re-engined with a new Curtiss D-12A engine. With a larger engine volume than the D-12, the D-12A nominally generated 507 hp and, at maximum, 520 hp. The new engine was mated to a Curtiss-Reed metal propeller. Lt. Walter Miller and Lt. Lucas Beau, the pilot and alternate for the other R-6, still powered by the 1922 race engine of nominal 450-hp, made daily flights. Their P-6 had a new propeller, but it was of wood, not a metal Curtiss-Reed.

In contrast to the army's depending on last-year's racers — spectacular as they had been — the navy bought new racers from both Curtiss and Wright for 1923. *Aviation* summed up expectations: "The Navy will have the new machines. Barring accidents, the laurels will probably pass from the Army to the Navy.... [T]he race will be in reality a contest between the Wright and Curtiss designers."[69] The magazine and everyone else who picked navy got it right.

Curtiss and Wright had squared off in the 1922 race, but the Wright Company, at that time, was primarily an engine manufacturer. That focus had changed by early in 1923. F.B. Rentschler, president of Wright Aeronautical Corporation, announced, "Sometime in March we expect to have ready a new plant ... which will house our plane activities.... Wright should eventually resume the development and manufacture of complete airplanes."[70]

Wright had developed a new liquid-cooled, V-12 engine, the T-3, and made it in two versions. The low compression, heavy-duty engine was intended for patrol planes. The high compression engine, intended and tuned for racing, churned out a nominal 680 hp and maximum 780 hp. The Wright engine and the Wright announcement that it would build "complete airplanes" threw down a gauntlet before Curtiss. Wright had new powerful engines and intended to build airplanes to exploit that power.

Even so, Wright's 1923 situation paralleled its situation of the year before. It had a new engine. It needed a racer for it. To design the new racer, Wright hired Harry Booth and Arthur "Mike" Thurston, who, the year before, had designed and built the Bee-Line racers. As long ago as 1920, both men had been central to Curtiss' racing program. In a hectic four days in 1920 in France, Thurston had designed and supervised the building of new biplane wings that were necessary for S.E.J. Cox's *Texas Wildcat* to fly from short French airfields. Both he and Booth had made major contributions to the design of Curtiss' 1921 CR racers. Working for their new employer, the Wright Company, they pushed ahead on a new racer in the summer of 1923. Somewhat surprisingly, given their having designed the Bee-Line BR monoplanes the year before, their 1923 racer resembled the Curtiss biplanes. It is possible that their switch from monoplane to biplane was influenced by the then-common opinion that thin-wing biplanes were inherently faster.

Because of complaints that the army and navy were "wasting money" on racers, the navy did not give its racers "R" for "racer" designations. The navy called its 1923 Wright racer "F2W-1," the "F" standing for "fighter." The F2W-1 was a single-bay biplane with I-shaped interplane struts and a thin carbane pylon from the center of the fuselage supporting the upper wing. An expansion tank for the cooling water, which flowed through wing radiators, was mounted on top of the upper wing at the centerline. The Wright racer had a metal strut behind each interplane strut that interconnected the ailerons on the upper and lower wings and provided a "good distinguishing mark" to tell the Wright racers from the Curtiss racers that did not have them.[71]

The F2W-1 was all wood except for metal in the upper engine cowing, one engine bearer, and metal frames for the cloth-covered elevators and ailerons. The bulky, heavy landing gear of the 1922 NW-1 sesquiplane was gone, replaced with a trim "V" arrangement. The tail of the racer was a mixture of old and new. It was externally braced with wires running from the fin to tops of the stabilizers and from the bottoms of the stabilizers to the fuselage, but stabilizer incidence was adjustable on the ground, an innovation at the time.

Unlike the 1922 NW-1, which had been a huge airplane compared to the Curtiss racers at Detroit, the wingspan and the length of the F2W-1 were similar to those of the new Curtiss racers, the Curtiss R2Cs that would race against the F2W-1s (see appendix). Nevertheless, in photos of both racers on the ground, the somewhat boxy-looking Wrights loom over the greyhound-looking R2Cs. All in all, the F2W-1 was smoothly finished and looked fast. Nor did it lack for famous admirers.

The Wright designers, calculating that the 200-km Pulitzer Race would be flown in 30

Curtiss R2C-1s (Race Numbers 10 and 9) and Wright F2W-1s (Race Numbers 8 and 7), at the 1923 Pulitzer (courtesy National Archives and Records Administration).

Navy Wright "Fighter," F2W-1, actually a racer, with Orville Wright and (Marine) Lt. Lawson "Sandy" Sanderson at the 1923 National Air Race, St. Louis (courtesy National Archives and Records Administration).

minutes, installed a small gas tank — 31.7 gallons for 31.7 minutes at full throttle — and only enough cooling liquid and oil to keep the engine within safe operating temperatures for that length of time in the racer that Marine lt. Lawson "Sandy" Sanderson would fly. The second Wright racer, to be flown by navy lt. Steven Calloway, had a gas tank twice as big — 60 gallons. The weight differences had an almost immeasurably small difference on the racers' top speed. In the Pulitzer, the Wright with the small tank was 0.065 mph faster.

"Sandy" Stevenson, who had flown the NW-1 at Detroit the year before, flew the first flight in the light blue-painted F2W-1 on August 27. Equipped with a two-bladed wood propeller, it was not race ready, but Sanderson reported it was responsive and fast. His speed in the first time trials on September 10 at Curtiss Field was good, 238 mph. But it was hardly good enough. Three days later, a new Curtiss racer flew 244 over the same course. And three days after that, on September 16, Sanderson cranked up the competition by flying 248. The already fierce competition between the Curtiss Company, which had dominated the 1921 and '22 races, and the Wright Company was heating up.

The second F2W-1 flown by Calloway was delivered with fire engine-red fuselage and white wings. Sanderson's F2W-1 was quickly repainted red and white, and "Navy Wright Fighter" was lettered across the red, white, and blue vertical stripes on the airplanes' rudders. The labeling was intended, surely, to convince anyone who saw it, that the F2W-1 was a fighter — a warplane — and not a racer that could be criticized as a frivolous and unnecessary purchase. Who was convinced? Most likely no one.

Stung by the army's runaway victory in 1922, the navy was determined to turn the tables in 1923 and did not stop with the contract with Wright. What better chance existed to fly off with the Pulitzer Trophy than to turn to the company that had produced winners for two years in a row? The navy contracted with Curtiss as well as Wright. The navy's reluctance to label its racers as racers did not extend to the two Curtiss machines, designated "R2Cs."

The Curtiss design had at least two features in advance of the Wright F2Ws. The new Curtiss racer had its landing gear suspended on single struts, reducing the drag from "V" struts, and its upper wing was mounted directly to the fuselage, eliminating the drag of carbane struts or pylons used to support the wing above the fuselage. Still, whatever the F2W-1's design deficiencies, its Wright engine, swinging a three-bladed metal propeller, was rated 200-hp higher than the Curtiss. The Wright was a contender.

Aviation summarized the visual difference between the R-6s readied for the 1923 Pulitzer and the R2Cs. They "differ very little ... [the R2Cs having] a somewhat smaller and better streamlined body, the placing of the top wing flush with the top of the fuselage and a simplified landing gear."[72] There were other changes; the new D-12A engine powered the R2Cs, the wings had a new airfoil, unequal spans (a change from the CRs' and R-6s' equal upper and lower wingspans) and steel tubing was used in the engine mounts. Wing radiators were installed on both upper and lower wings, and cooling water was circulated through them about five times a minute.[73] A wood propeller was initially fitted to the first R2C, but

Curtiss R2C and Lt. Alford Williams, USN. The relative sizes of the Wright and Curtiss racers are apparent from comparing the F2W that looms over Orville Wright and Sandy Sanderson in the previous photograph and the R2C at Al Williams' side in this photograph (courtesy National Archives and Records Administration).

it was replaced with a metal Curtiss-Reed propeller and a 10-mph increase in speed. The second R2C had a metal propeller from the time it was manufactured. For all the changes, the R2Cs were clearly derived from the R-6s, which had been developed from the CRs.

St. Louis' major dailies, the Pulitzer's *Post-Dispatch* and the *Globe-Democrat*, splashed across their front pages news of arrivals of civilian and military airplanes in the week before the scheduled start of the three-day air meet on Monday, October 1. The air meet organizers had dovetailed an old St. Louis festival, the arrival of the Veiled Prophet, with the kickoff for the races.

The Veiled Prophet ("VP" among the citizens), a "wise and kindly ruler" from "the faraway land of Khorassan," "a land of plenty," has visited St. Louis once a year since 1878 to share his land's happiness with the city. The identity of the VP, drawn from the ranks of the organizers of the VP pageant, is kept secret. His visit is the occasion of a parade, a debutant ball, and the selection, again by secret means, of a Queen of Love and Beauty.[74] The whole idea, modeled after New Orleans' Mardi Gras, had been dreamed up by St. Louis merchants who wanted public events to draw attention to harvest festivals in the city.

In 1923, the VP was scheduled to arrive by air at 10:30 A.M. on Monday, October 1. Socially important people of St. Louis would greet him when he landed, and his escort would, somehow, include the 300 to 400 airplanes in St. Louis for the air meet. The headline on page one of the September 30 *Post-Dispatch* read, "100 Maids in Purple Knickers to Greet Prophet Tomorrow."[75] "Knickers" may have caught male readers' attention, but the word referred to short trousers that reached about mid-calf and were worn with socks, as worn by many schoolboys of the time. They were not undergarments. To complement the purple knickers, the maids wore heavy satin yellow jackets, yellow stockings, and flying caps of "the same bright hues." Katherine Lemoine Perkins, one of the maids, was to ride in the airplane that would meet the one "occupied by the Prophet." The *Post-Dispatch* said all the maids would "look perfectly stunning" in the Prophet's motorcade from the airport to his den, where he resides during visits to St. Louis. The competitions were set to start after the Veiled Prophet's departure from the airport at about noon on Monday, October 1.

On Sunday, the day before the scheduled start of the air meet, the new airport opened at 11:00 A.M. (admission $0.50 for adult, $0.25 for child), and 9,000 people came to see displays of airplanes, aviation accouterments, equipment, tools, models and artifacts worth an estimated $5 million:

> Although the gates were not to open until 11:00 A.M. yesterday [Sunday, September 30] spectators began arriving by street car and in automobiles before 9 A.M., and before the gates were opened several thousand people had assembled.
> The field was in deplorable condition from a downpour at midnight. Early yesterday morning it appeared that the sun which then was shining, would remain out and put the field in condition for racing today [Monday, October 1], but a dense cloud came from the southwest about 10:30 o'clock, and at 11, when rain fell, the field was as dark as night....

The rain, which lasted for an hour, drove everyone to shelter.[76]

Standing pools of water on soggy ground made for a terrible and dangerous surface for high-speed airplanes taking off and landing.[77] Air meet officials with long memories may have recalled a 1914 air meet in St. Louis when the weather forced day-to-day decisions about whether to fly or not. What had been already-poor attendance suffered even more from the weather-related daily uncertainty.

The 1923 organizers acted swiftly and decisively. They pushed back the opening of the air meet from Monday to Thursday, allowing time to restore the airfield to tip-top shape, which cost an additional $40,000. Events originally scheduled for Monday, Tuesday, and Wednesday were rescheduled for Thursday, Friday, and Saturday, respectively. The Pulitzer would be flown on Saturday, a nonworkday, which increased the number of people who could attend.

The air meet could be delayed. The Veiled Prophet couldn't. He made his grand entrance by automobile — not airplane — on Monday, and "without any gay bevy of maids to greet him."[78] The rest of it — the parade, the ball, enlivened by the splendor of many army and navy aviators in full-dress uniforms displaying all the metals, ribbons, and badges to which they were entitled, and the crowning of the "Queen of Love and Beauty"— went off as planned.

Two "giant" aerial visitors flew to the air meet. The "largest aircraft in the world," the navy's Dirigible ZR-1 (to be renamed *Shenandoah* on October 10, 1923, and crash under that name) flew to St. Louis from Lakehurst, New Jersey.[79] It arrived over St. Louis at 3:30 A.M. on Thursday, cruised over the city until after six, and then flew over the airport grounds. Lines were dropped from the ZR-1 to about 120 soldiers and sailors who were serving as ground crew, and the men held the lines to keep the dirigible in place during its visit.

The navy had planned that the airship would remain for most of the day, but gusty winds and a far from ideal mooring arrangement — all those soldiers and sailors holding

The navy airship ZR-1 (later renamed *Shenandoah*) is pulled to earth and moored by 120 soldiers and sailors at St. Louis in 1923. The comparative size of the soldiers and sailors and the dirigible provides a scale to judge the ZR-1's size. The two taut lines that are barely discernible against the background of trees were passed through winches to draw the airship to the ground (photograph from the *St. Louis Globe-Democrat*, October 2, 1923, now in the collection of Mercantile Library at the University of Missouri–St. Louis, and available at Wikipedia, http://en.wikipedia.org/wiki/USS_Shenandoah_%28ZR-1%29).

the dirigible to the ground—convinced the navy to shorten the visit. The ZR-1 cast off after about three hours, but it took time to fly over and around the city, an event remembered by many of the city's inhabitants years later.

The navy had the "largest aircraft" in the world, but the army had the "largest airplane," and it came to St. Louis from Dayton, Ohio, where it had been built. The Barling Bomber, a triplane, was powered by six 400-hp Liberty engines, four mounted in tandem "tractor-pusher" nacelles and two in single tractor nacelles, all between the lower and middle wing. The upper and lower wings spanned 120 feet; the middle wing 111 feet.[80] It was so big, as judged against other aircraft of the period, that *Aviation* magazine called it "monumental."[81] Monumental yes. Speedy, no. It travelled at little more than its cruising speed of 61 mph on its Dayton to St. Louis flight.

In 1923, *Flight* magazine, surely depending on information supplied by the army, reported that the Barling could lift a normal payload of 5,000 pounds and fly at its top speed of 90 mph for 12 hours.[82] The National Museum of the United States Air Force Fact Sheet on the Barling paints a different picture. It agrees with *Flight* about the maximum bomb load and top speed, and even raises the speed to 96 mph, but the range of the loaded airplane was only 170 miles and its cruising speed was about 60 mph.[83] The Barling flew the round-trip from Dayton to St. Louis with only a little trouble from some of its six engines, but a planned flight from Dayton to Washington, where it was to be admired by awestruck citizens, revealed its limitations. The flight was canceled when the airplane proved unable to climb over the Appalachian Mountains. But that failure was in the future, and the huge bomber was a hit at St. Louis.[84]

The chairman of the air meet's ticket committee estimated that 85,000 people paid for

The Barling Bomber. Six engines, three wings, biplane horizontal tail, four fins and rudders, eight main wheels and two to prevent nose-overs, a tail skid and a crew of six. Range with full load: 170 miles at cruising speed of 61 mph. It was unable to climb over the Allegheny Mountains between Dayton, Ohio, and Washington, D.C. (courtesy National Archives and Records Administration).

admission on Saturday, October 6, to see the Pulitzer and that those who "stood in spaces to which no admission was charged" brought the total number of spectators to about 100,000. Over the three days of the meet, paid attendance reached a total of 140,000.[85]

The length of the Pulitzer Race had been reduced from 250 to 200 km because official speed records were recognized for 100- and 200-km but not for 250. In 1922, the 15 competitors in the Pulitzer had been divided into three heats; in 1923 the race was again flown in three heats. Takeoffs in the Pulitzer started at 2:30 P.M. By then army and navy enlisted men had pulled the propellers of all the racers and the racers were ready to take off. Marine Lt. Sanderson took off first in his red and white Wright, climbed to 4,000 feet in the cloudless bright blue sky, tipped downward, and dived across the start line and timers at 2:31 P.M. He was quickly a speck in the distance, gray, smoky exhaust trailing behind.

A minute after Sanderson left the ground, Air Service lt. Johnny Corkille lifted off in an R-6 with black fuselage and gold wings and horizontal tail. He climbed to 3,000 feet, came screaming down to cross the start line, and banked so closely to the pylon that some observers thought he had cut inside it. The violence of either the dive or the bank loosened the glass in Corkill's goggles, interfering with his vision. He was slightly off course when he completed his first lap, and, on his second lap, he was unable to see the railroad line that bordered the second leg of the course. On the third lap, he managed to wiggle the glass into a secure position in the goggles.

Navy lieutenant Al Williams, in his blue R2C, was first off in Heat 2. He climbed a few

Three sailors pull a Wright F2W-1's propeller to start the engine. Note the splendidly uniformed, relaxed spectator, and the distracted man in a hat. Who were they? What were their jobs? (courtesy National Archives and Records Administration).

thousand feet and "with an engine bark that was loud beyond belief ... dived past the starting line with a whine like that of a high explosive shell. The terrific roar of a D-12 engine, either in a climb, level flight, or dive, has never been equaled to this day."[86] The speed of Williams' start may have equaled its noise: "His speed at the starting line was so apparently greater than that of the contestants in the first section that a loud gasp went up from the crowd. He said later he was doing 280 miles per hour when he dived at the line."[87] The cheers of the "gobs," as navy enlisted men were called, in the navy's hangar drowned out loudspeaker announcements of record-breaking speed.

The other racer in Heat 2, the slate-colored Verville-Sperry R-3, was strikingly drab in the race with the red and white Wrights, the blue R2Cs, and the black and gold R-6s. Alex Pearson, the R-3's pilot, climbed to 4,000 feet, dived down, and headed for the start line. The R-3 weaved drastically from side to side. Pearson "explained his failure to pass the starting line after hopping off as being due to the extraordinary vibration of the metal propeller[88] installed on his plane last Friday: "The new 'prop' vibrated so that I knew I would never get around."[89] Pearson's landing ended the only R-3's participation in the 1923 race. In 1922, two R-3s had finished fifth and seventh, and the third had been forced down with a broken oil line. The R-3 flown by Pearson would be back for one more try in 1924.

TABLE 4
1923: Pulitzer Racers, Pilots, and Results, St. Louis, Missouri, October 6[a]

Heat 1

Aircraft	T/O[b] Order	Pilot	Place	Speed mph[c]	Laps Completed
Wright F2W-1	1	Lt. Lawson Sanderson, USMC	3rd	230.1	4
Army Curtiss (R-6)	2	Lt. Johnny Corkille, USAS	6th	216.4	4

Heat 2

Aircraft	T/O Order	Pilot	Place	Speed mph[c]	Laps Completed
Curtiss R2C-1	3	Lt. Alford Williams, USN	1st	243.7	4
Verville-Sperry R-3	4	Lt. Alex Pearson, USAS	-	-	0[c]

Heat 3

Aircraft	T/O Order	Pilot	Place	Speed mph[c]	Laps Completed
Wright F2W-1	5	Lt. Steven Calloway, USN	4th	230.0	4
Curtiss R2C-1	6	Lt. Harold Brow, USN	2nd	241.8	4
Army Curtiss (R-6)	7	Lt. Walter Miller, USAS	5th	218.9	4

[a]The nearly identical speeds of the pairs of airplane—the two R2C-1s, the two F2W-1s, and the two R-6s—are noteworthy. The close speeds of the R-6s were most remarkable. A new Curtiss D-12A engine (507 nominal horsepower) powered Miller's R-6, and Corkille's was powered by a D-12 of about 480 horsepower. Miller reported that his engine did not operate smoothly, which would have held his speed down. On the other hand, Corkille had trouble with his goggles, which interfered with his vision and must have slowed him.
[b]Takeoff.
[c]Unbalanced propeller forced Pearson to land before crossing the start line.

Al Williams did not see any other airplane: "It was rather lonesome out there on the course.... I looked around sometimes but I can't say that I saw much."[90] *Aviation* magazine reported that he had looked for Sanderson in the Wright but did not see him.[91] Williams was still in the air, having logged 245 mph on his first lap, when the judges—perhaps wanting to add drama by having more racers in the air or wishing to end the race as soon as possible—sent off the three airplanes in Heat 3. Navy lt. Steven Calloway took off in the second red and white Wright, followed by navy lt. Harold Brow in a blue Curtiss R2C, and army lieutenant Johnny Corkille in a black-and-gold Curtiss R-6.

When Sanderson crossed the finish line, the judges announced his speed as 230 mph, a new world record. Across the finish line, Sanderson pulled the nose of his Wright racer up and zoomed to 2,000 feet, and "as he then circled lazily away to avoid the next heat of racers his crimson plane began to plummet straight for the ground."[92] The plane disappeared, smashing into the ground out of spectators' sight behind a railroad embankment.

The crash knocked Sanderson out. When he regained consciousness, he hastily tried to unbuckle his seat belt, thinking to get out of the airplane in case it caught fire. His haste was unnecessary. When rescuers ran up, he shouted, "Help me get out!" They yelled back, "But you are out. There's no plane here."[93] The seat, with Sanderson firmly belted to it, had been tossed clear of the airplane. The leather cockpit ring, which caught on his body as he was tossed out, had accompanied him and hung "around his neck like a garland of flowers,"[94] or, perhaps, a horse collar. He escaped with cuts and bruises and a sprained ankle and was able to greet and congratulate Williams when he finished his flight. Sanderson later said, "I had sufficient altitude to get back to the field but with a dead stick [no power]. I was afraid I might somehow come down in the crowd. I saw two haystacks and a fence. I headed for one stack but saw I was due to strike the fence, so I turned her into stack number two, and when I struck it I was thrown out through the narrow cockpit."[95]

Sanderson was philosophical about his fate. He said the F2W-1 was "a great ship ... [but it] just won't stay in the air without gasoline."[96] The designers and engineers who installed the small 31.7-gallon gas tank in Sanderson's racer had figured fuel consumption exactly (or luckily). There had been sufficient fuel for the race. They hadn't allowed for Sanderson's postrace climb and circling.

A *Post-Dispatch* report indicates that the gas tank was not so small. The newspaper quoted Sanderson as having said, "I told them to put in more than 50 gallons of gas ... but the boys thought 50 gallons sufficient, so I let it go at that."[97] Whether 31.7 or 50 gallons, it was sufficient for the race, but no more. Sanderson owned the world speed record for only a few minutes. Williams, in Heat 2, finished 14 mph faster.

Sanderson won no races and set no records in three headlined air races in two years—the 1922 Curtiss Marine Trophy Race, the 1922 Pulitzer, and the 1923 Pulitzer. He had, however, set one unenviable record that may stand forever. All three of his races ended in forced landings or crashes. Out of gas, he alighted in the Curtiss 18-T seaplane on the Detroit River within sight of the finish of the Curtiss Race, and he and the airplane were towed to shore. His engine failed following an oil leak in the 1922 Pulitzer, and he had to ditch in shallow water near the shore of Lake St. Claire. He finished the 1923 Pulitzer before running out of gas and crashing into a haystack. Sanderson's worst injury was a sprained ankle in the 1923 Pulitzer.

Williams had still been in the air when Sanderson was being rescued. Puzzled spectators, ground crew, and officials wondered what Williams was doing when he crossed the finish line at the end of the fourth lap and headed into another one. Efforts to flag him down did nothing, and Williams raced around the course to complete five laps for a total of 250 km. Taxiing in from his race, Williams asked, "What did we do?" as soon as the engine was throttled down. He reacted enthusiastically and boyishly when he learned he had flown almost 244 mph: "Whoopee! Navy!"[98] But his victory wasn't certain yet. Other pilots were still in the air. Navy lt. Harold Brow in the other R2C had a good chance and navy lt. Steven Calloway in the other Wright F2W-1 had more of an outside chance to beat Williams' time. Only army lieutenants Walter Miller and Johnny Corkille in the R-6s had no chance to equal or beat Williams.

All the pilots flew the straightaways in a similar fashion. Knowing there were no bumps in the air over water, they dropped down when flying the first leg over the Missouri River, and they climbed to 300 feet on the other laps because of possible low-level turbulence. They flew turns differently. The army pilots banked tightly around the pylons. The navy pilots flew farther from the pylons, convinced that the higher speeds that could be maintained on wider turns more than compensated for the shorter distances from tight turns.[99] In 1923, the navy's wider, faster turns made little difference; the explanation for navy racers taking the first four places in the Pulitzer was simple. They were faster.

When all the racers had landed and the times were posted, Williams prevailed. Williams' time for the 200-km race was an astounding 244 mph. He had flown four laps of a 50-km triangle course, with 12 pylon turns and beaten the world's speed record, set over a straight course of three kilometers, by eight miles per hour. He had bested the 1922 Pulitzer-winning time by 37 mph. He set new world records for 100- and 200-km distances, although the FAI recognized only the record for the shorter distance. Whoopee, indeed.

Newspapers and magazine articles highlighted the fact that both Williams in first place and Brow in second had exceeded four miles a minute. Imagine! "At such a rate of speed in continuous flight one could travel the distance between New York and San Francisco in ten hours."[100] Today's traveler might have different reactions, most likely amusement at "ten hours" for a coast-to-coast flight. After all, one can now make the nonstop flight in six hours or less. Moreover, the 1923 traveler's airplane could cruise only 300 to 500 miles, making multiple takeoffs and landings a necessity; and at nighttime he or she probably got off the plane and took an overnight train ride. Coast to coast took a couple of days. The 1923 traveler's airplane flew at low, bumpy altitudes, making the availability of a stewardess who had been trained as a nurse a necessary luxury to comfort and treat the airsick.

On the other hand, the 1923 traveler could glory in the glamour of air travel, and was provided first class service and pampered to the limit of the airline and railroad companies' capabilities. There was no cramming into a cramped seat. "Economy" five-abreast seating had not yet been invented. Whatever today's traveler's reaction to the differences between 1923 and the present, he or she can marvel at the change the airplane wrought on concepts of speed and transportation in its first 20 years. The Wright brothers first flew at 20 or 30 mph. Two decades later, airplanes flew eight or ten times faster; 244 mph, four miles a minute.

The *Post-Dispatch* published a picture of an exhausted-looking Williams, still in his cockpit at the "instant his plane came to a rest after his winning flight," and published this description: "An extremely drawn young man in an extremely dirty white shirt wobbled on his athletic legs in the navy hangar later yesterday afternoon."[101]

Williams appears to have been in his "one and only greasy, tattered good-luck shirt"[102] when General Patrick shook his hand and congratulated him. The cut through the leather covering on the top and back of Williams' helmet clearly showed the effect of wind rushing

General Patrick (shaking hands, left) congratulates Lt. Williams as Lt. Maitland (center) looks on. The leather of Williams' helmet was cut through by the rush of air over the top of the Curtiss R2C's windshield (courtesy National Archives and Records Administration).

over the windshield of his racer. Evidently, Williams changed into a clean shirt and donned a surprisingly short tie when he received congratulatory flowers from Katherine Wright, the Wright brothers' sister.

For Orville Wright, who was at the race, the victory of the Curtiss racers may have been a bitter reminder of the Wright brothers' long legal and commercial struggles with Glenn Curtiss. More significantly, his presence at St. Louis and Katherine's presentation of flowers to Williams was emblematic of something far more important: The airplane was only 20 years old, so new that pioneers who had made it all possible in the first place were still alive and active.

Most probably neither Katherine nor any other woman saw Williams, along with Brow and Sanderson, who finished second and third in the Pulitzer, receive gold, silver, and bronze plaques on the evening after the race. The presentations were made at a "smoker," traditionally an all-male, by-invitation-only gathering where "boys could be boys."

Williams, like other Pulitzer and high-speed pilots, reported he had consciousness problems on the turns (the term "black out" had yet to be invented): "[S]omewhere in the third lap I went woozy. I felt just like I was asleep. It was those turns that did it. I couldn't see so I jerked off my goggles but that didn't help any, it seemed. I was all mixed up and lost track of the laps. Sure was woozy. I knew things were mixed up so I went around again to make sure."[103] The *Post-Dispatch* noted that Williams, a week before the race, had told a reporter he had been around the course 25 times and planned to go around another half dozen times so that he would always know where he was. After the race, the newspaper reported Williams "had piloted his blue, motor driven projectile at better than 243 miles

Katherine Wright presents congratulatory flowers to "Al" Williams before an admiring throng (courtesy National Archives and Records Administration).

an hour while in a semi-conscious state, [which] shows the wisdom of his intensive preparation for the test."[104] The wooziness could, at least in part, explain his flying the extra lap.

Williams had trouble with wooziness, but his R2C was free of the mechanical and other problems that compromised other racers' performances. Navy lt. Howard Brow in the other R2C flew almost as fast as Williams on the first lap and faster on the second, but his racer overheated on the third lap, and he had to throttle back: "But for a boiling water system, Lieut. Brow might today be wearing the crown of all speed kings."[105] He needed about 15 seconds more than Williams to finish the 200-km course (a two-mph difference).

Navy lt. Steven Calloway, flying the second F2W-1, which also overheated, finished fourth, less than one-tenth a mile per hour slower than Sanderson in the first F2W-1: "The water heated up before I was half way through the race, and had it not been for that, I am sure I could have tacked eight or ten miles an hour additional on to my speed."[106] An additional ten miles per hour would have, perhaps, earned Calloway second place. Williams finished at 244 mph; Brow at 242; Sanderson at 230.1; Calloway at 230.0.

Lt. Walter Miller, who finished fifth and next-to-last, reported that the 507-hp engine of his R-6 "kept missing and I knew I was out of it from the start. We did the best we could with the equipment we had and I guess that is all the explanation there is necessary."[107] Even so, he averaged nearly 219 mph, almost 13 mph faster than Maughan had flown in the same plane the year before. Miller's more powerful engine, 507 horsepower rather than Maughan's 450, contributed to, or accounted for the difference. The R-6 of the last-place finisher, Johnny Corkille, had a 450-hp engine. Corkille had trouble with his goggles that compromised his piloting, and he finished last at 216 mph, flying 10 mph faster than the winner's 206 mph of the year before.

The 1923 St. Louis air meet was the United States' largest-ever air meet in terms of attendance and its most successful. It had everything associated with an entertaining air show: exciting racing, parachute jumps, aerobatics, demonstrations of nighttime flying (a novelty in 1923), static and in-flight displays of the latest in aircraft, both lighter-than-air and heavier-than-air, good seating and adequate parking.

The decision to delay the air meet for three days to wait for better weather had paid off handsomely, with three fine days for flying. In all, more than 400 flights, logging 300,000 miles, had been flown, with Lt. Sanderson's sprained ankle the only injury beyond scrapes and bruises. Most surprising, perhaps, when the final accounting was done, the organizers had a profit of $13,717.[108] St. Louis' Second National Air Race was a rarity: A meet that finished in the black.

Attendance met expectations, and crowd-control problems emerged only when the racing was over. Some 100,000 people wanted to leave at the same time. Some fences were pushed down; great crowds swarmed onto the street cars; and some women reportedly fainted in the crush.[109] That hiccup at the end did not dim the enthusiasm and feelings of accomplishment that emerged from the success of the Second National Air Race. Even the traffic courts entered into the festivities. Newspapers and police had urged motorists headed for the air meet to drive as fast as reasonably possible to keep traffic moving. "Reasonably" is in the eye of the beholder, and some police took exception to some motorists' "reasonable." Tickets were written, but traffic court judges pronounced, "Forget it."

Of lasting importance to the city, St. Louis had a fine, new airport. More fleetingly, it earned praise for its hosting the meet. *Aviation* commented, "The arrangements for the air meet and the airport called for strong approval." General Patrick praised the city's efforts and added, "St. Louis will be rewarded by having one of the best airfields in the United States." Commander Mitscher, chief of the navy effort at St. Louis, thanked the city and officials for their hospitality: "For your efforts you are establishing yourself as an aviation center that shall become a national byword."[110] Admiral Moffett, who had briefly visited the St. Louis air meet on the ZR-1 (*Shenandoah*), returned for the prize presentation. His praise was unstinting: "St. Louis has done more for aviation than any other city to date."[111]

The Pulitzer family had not at first planned or envisioned that its trophy would reward outstanding performances in closed-course, around-pylons races or that military airplanes would dominate the contests. By the time of the first Pulitzer in 1920, pylon races were in place instead of the transcontinental contests originally envisioned because of safety concerns and, of lesser importance, the crowd-attracting potential of pylon races. Civilian airplanes competed in the first two Pulitzers, but individuals and companies without government contracts could not afford to race in the Pulitzer after 1921. However it differed from what the Pulitzer Family had intended, the Pulitzer had become the most important pylon race. The consistently increasing speeds in the four years of Pulitzer competition showed clearly that better airplanes and engines were being produced, and, gratifyingly, features of the airplanes and engines were being incorporated into other aircraft.

The army and navy were tied for Pulitzer wins after the 1923 race. An army racer won in 1920, a navy plane, flown by a civilian, won in 1921, the army won again in 1922, and the navy tied it up in 1923. The Curtiss Company had been the winner in the last three contests. What did 1924 promise? A new army race between navy racers? Would Curtiss, Wright, Verville, or some other company produce a new record breaker? No one knew, of course, but the heady excitement and satisfaction of the 1923 air meet surely convinced many that 1924 would be even better.

General Patrick was decidedly optimistic about the army's chances in 1924. After the Pulitzer, he told army pilots and ground crew, "We will do better next year."[112] Later that evening, at the prize-presentation smoker, he complimented the navy "but warned them that next year the Army Air Service intended to move up where it belonged."[113] The army would win in 1924, but it would be a very different contest from 1923, one marked by disappointment and disaster.

Chapter Six

1924

Dayton, Disappointment, and Death, Wilbur Wright Field, Dayton, Ohio, October 4

"As spectacular races, the Dayton events did not compare with those of preceding years...."[1]

On the day after the 1923 Pulitzer Race, 25,000 people showed up at Lambert Field in St. Louis to watch attempts to set new speed records. The disappointed thousands heard, and probably understood, Admiral Moffett's two reasons for postponing the trials— insufficient time to install electrical timing devices, and pilots worn out from flying the races— but they remained disappointed.[2] They left Lambert Field without having seen the speed spectacular they wanted, but in the months to come, thousands saw such flights, and millions read about them on the front pages of their daily newspapers less than a month later on November 2 and 4, 1923.

Navy lieutenants Harold Brow and "Al" Williams, whom the *New York Times* usually identified as "former pitcher of the Giants,"[3] set new world records flying the Curtiss R2Cs they had flown at St. Louis. The absence of the Wright F2W-1s—which had placed third and fourth in the 1923 Pulitzer—from the speed trials was confirmation, if confirmation was needed, of the superiority of the Curtiss racers over the Wrights.

In preparation for Brow and Williams' attempts on the world speed records, Air Service (USAS) officers and National Aeronautical Association (NAA) officials measured and supervised construction of a three-kilometer course at Mitchel Field. An extra half kilometer on each end of the three-kilometer course provided areas for the pilots to line up and level out their flight paths as they made required flights in both directions, and observation posts at the ends of the four kilometers enabled officials to verify that the flyers flew level at or below 50-meters altitude as required for an official record. The low altitude requirement had an unexpected benefit, if it *was* a benefit: spectators would hear as well as see the record attempts.

Harold Brow made the first attempt and averaged 257.4 mph over the 3-km course in four passes—two with the wind and two against.[4] His speed was a new world absolute speed record. Williams took off, flew the required four passes, and landed. A reporter drafting a story about the new world-record holder and speed would have had to scratch out "Brow" and "257.4," and insert "Williams" and "258.6." Brow made another attempt, and, "comet-like, increased his speed to 259.15 miles per hour."[5] That ended the day. Darkness was falling, and the timers, uncertain about being able to see the racers over the course, called off attempts for the day. Impatient Williams, eager to try again, had to wait until the next day. Given the differing reports of Williams' temperament, it's impossible to decide whether he was friendly, bantering, or taunting when he said to Brow, "You did fine work, but I'm going to beat you tomorrow." In any case, he exhibited the competitive spirit expected of "a former pitcher of the Giants."

High winds on November 3 canceled speed attempts on that day. Williams, first off, flew to a new record of 263.3 mph on the next afternoon, November 4. The record lasted a half hour. Brow clocked 263.5 mph. Planning to meet "20,000 of his friends and neighbors"[6] at Van Cortlandt Park in the Bronx, Williams told Brow, "I guess I'll have to take it away from you tomorrow.... I'm flying to Van Cortlandt to meet my friends as promised." He started to walk to the hangar to change out of his flying clothes—including, it is assumed, the blue trousers and "extremely dirty white shirt" he wore for luck in races and speed trials. He changed his mind. Determined to have another go at the record, he asked a reporter to call Van Cortlandt and explain to his friends "that he could not come without being the world's speed holder" and that he would be a bit late.[7]

The *New York Times* reported dramatically: "Twice escaping death, once in coming out of a dive only ten feet from the ground and again when he darted through a squadron of Martin bombers arriving at Mitchel Field, Lieutenant A.J. ("Al") Williams, former pitcher of the Giants ... won the international supremacy of the air by making an average speed of 266.68 miles an hour."[8] In making the diving start customarily used in starting air races and speed trials, Williams climbed higher than usual, to 9,000 feet, and dived down to zoom across the start line. In "the closest shave I have ever had in my flying days," Williams leveled out at 10 feet and made his first pass over the course. He averaged 266.7 mph in his four flights. As he sped across the finish line on the last of the speed runs, the horrified crowd of 2,500 saw a flight of two-engined Martin bombers headed directly toward the speeding blue racer. Seeing the bombers too late to turn away, Williams had no choice but to fly straight ahead, hoping to pass between them. A gasp went up from the crowd. With what seemed "only a few feet leeway," he streaked away, safe.

After Williams' 266.7 mph flight, Brow, who had "made the fastest time on one lap, attaining a speed of 274.2 miles as hour," said, "I'll be after that record in the morning."[9] He was not to have the chance. Admiral Moffett ruled out any more speed attempts with diving starts. Wholly or in part, Moffett was responding to concerns about the great strain placed on the racers and their pilots by the high speed diving starts and sharp pullouts.

Plans were made for Williams and Brow to "race each other with—and against the wind over the three kilometer course"[10] without diving starts for a total of 12 kilometers—at an aerial carnival to benefit the Army Relief Society on November 6, but the race was called off. So as not to disappoint 22,000 spectators—12,000 at Mitchel Field for the carnival and another 10,000 watching from around the field—the two "speed kings performed aerial antics that kept the spectators in a constant gasp." In addition, Williams climbed 5,000 feet in one minute in his Curtiss R2C, and "added what is believed to be another record to his credit."[11] If it was, indeed, another record, it was less news-worthy than the speed record. The *New York Times* had made the speed record page-one news. It put the climb record on page 19.

A month after the St. Louis air meet, "in point of attendance and gate receipts ... the greatest that has been ever held,"[12] the record-breaking flights kept airplanes and racing in the news. The *New York Times* published speculative articles about possible limits on airplane speed — 400 or 500 or even 1,000 mph — and pilots' and planes' survival at such speeds.[13] A major in the Air Service Medical Corps was convinced that flying at 1,000 mph would drive the pilot's brain against his skull with sufficient force to cause injury and likely kill him.[14]

At the end of the year, the *New York Times* took stock of the United States' 1923 accomplishments in aviation. For the second year in a row, the United States held records for absolute speed (set over a three-km course) and for 100 and 200 km (although only the 100-km record was officially recognized). A single U.S. Navy pilot, "Al" Williams, held those three speed records. The navy held a total of 22 world records; 19 were seaplane and flying boat records, the remaining three were Williams' land plane marks. The army's 11 records, for speeds over every distance between 500 and 5300 km and for duration of flight,[15] were "made mostly with obsolescent craft improved and operated with greater skill."[16] Underlining the maturation of the country's aviation industry, all the record-making airplanes and engines had been built in the United States.

Nonracing aspects of 1923 navy and army aviation generated news as well. The navy had designed, built, and was flying the largest dirigible. It continued encouraging the development of air-cooled engines, and it was perfecting aircraft carrier operations. What was more spectacular, on May 2 and 3, army lieutenants O.G. Kelly and J.A. Macready[17] flew a one-engined Fokker T-2 across the United States, nonstop, in just under 27 hours. With army-developed in-flight refueling techniques, Captain L.H. Smith and Lt. J.P. Richter remained airborne for 37¼ hours in a DH-4, having been refueled 15 times. The Barling bomber, underpowered and unable to climb over mountains, remained the largest airplane in the world and generated awe when it could be dispatched to a city not located beyond high hills.

On June 23, 1924, U.S. Army lieutenant Russell Maughan, who had won the 1922 Pulitzer in a Curtiss R-6, crossed the United States "between dawn and dusk," in 20 hours 48 minutes. He flew a new Curtiss PW-8 (Pursuit, Water-Cooled), which clearly traced its lineage back to the Curtiss 1922 R-6 racer, and he bettered by about a half-hour the previous record of 21 hours 19 minutes for a transcontinental flight, set by Lieutenant James Doolittle on September 4, 1922. Airmail and commercial operations also moved forward. The government and industry had developed and begun operation of the first night airway system in the world (to deliver mail), and on July 1, the first regular transcontinental airmail service began. During the year, commercial operators flew about 2,000,000 miles transporting passengers and freight.[18]

Expectations for aviation soared as 1924 began, and nowhere were they higher than in Dayton, Ohio. On December 28, 1923, The *Dayton Journal* had headlined the awarding of the 1924 Pulitzer Race to Dayton and predicted 250,000 people would attend the three-day-long air meet with 180,000 flocking to the Pulitzer Race on the third day. The newspaper predicted the Sperry-Verville R-3 would reach 275 mph in the Pulitzer, a terrific speed over a 200-km closed course; it further predicted that, for the first time, several foreign pilots and racers would compete in the Pulitzer.[19] Six weeks later, the newspaper estimated that the air meet would cost $150,000.[20]

At least one other Midwestern location, the "Twin Cities" of Minneapolis–St. Paul, Minnesota, had wanted to host the 1924 National Air Races and Pulitzer Race. A *Minneapolis*

Tribune editorial expressed regret that "the Contest Committee of the National Aeronautic Association could not see its way clear to schedule the Pulitzer International Air Races for the Twin Cities in 1924," and stated that the Twin Cities would try again in 1925. The editorial's comment that "there is, of course, no question about Dayton's ability to offer adequate facilities for the races"[21] somehow reads as damning with faint praise.

The success of the 1923 St. Louis air meet and the public's interest in airplane speeds stimulated rosy expectations for the 1924 Pulitzer Trophy and National Air Race. Fundraising was going well. In late spring, Dayton organizers announced they had raised $50,000 prize money, four times more than was raised for any other previous National Air Race.[22] And what better site for the Pulitzer than Dayton? The city dubbed itself the "birthplace of aviation." It was the home town of the Wright brothers, and Orville Wright still lived there. The Air Service Engineering Division at Dayton's McCook Field had become a critical component of aviation research, development, and testing during World War I and remained so. Dayton itself was an important manufacturing center for automobiles and office machinery.

There is little difference between the populations of Dayton in 1924 — about 173,000, and 2010 — about 166,000,[23] but population trends are different. In 1924, the population was increasing and would reach 262,000 in 1960. Population began declining that year and fell until 2000, when it stabilized at about 166,000.[24]

An aerial photograph of the city in the early 1920s shows well-maintained business buildings, hotels, and churches, cars parked at angles on wide streets and pedestrians thronging some sidewalks. A person walking in the city today bangs up against the reality of a 21st century rust-belt Midwestern city. Sturdy, sometimes stately, buildings that appear to have been built to be used for hundreds of years are without tenants or tenanted only as storefronts shopped only by people who live nearby and lack the means to reach suburban malls. Empty shops and office blocks with gaps where buildings formerly stood give the city a worn-out rather than a rusted-out look. Few cars move on the streets around the downtown library in mid-afternoon, and pedestrians are sparse except for near-stationary individuals loitering on corners. The downtown weariness is out of sync with upbeat reports about the growth of electronic, biotech, and other high-tech firms in and around Dayton. That growth, as in so many cities today, is taking place on the city's outskirts, not downtown. Wherever they are located, many of Dayton's firms — both well-established and start-up — have connections to the United States Air Force's huge Wright-Patterson Air Force Base ("Wright-Pat").

The National Museum of the United States Air Force, a spectacular collection of aircraft, missiles, aviation equipment and materials, and one of the nation's leading tourist attractions, is located at Wright-Pat. The Wright Brothers' Bicycle Shop, the brothers' home, Huffman Prairie, location of some of their early flights and factories, other Wright brothers–related sites, memorials and monuments remind everyone of the city's aviation history. Wright State University, a modern, attractive campus with a large collection of aviation-related material in its library, adjoins Wright-Pat. The main public library's local history room, much used by patrons interested in family trees, also holds many books, periodicals, and other records about Dayton aviation.

The economy of present-day Dayton, like that of much of the United States, depends on service firms. Following Wright-Pat, the area's largest employer, the second through fifth largest employers are service businesses: two health systems, Wright State University, and Lexis-Nexus, a legal publishing company. How different Dayton is today from the

1920s. A 1932 description, "Dayton has always been a manufacturing center, and it always will be,"[25] accurately captured the city's history from the late 1800s and predicted its future through the 1950s. The prophecy began going wrong in 1960.

Manufacturing boomed in the 1920s at the time of the Dayton Pulitzer, and the Dayton Chamber of Commerce map for visitors to the 1924 National Air Races called Dayton, "The City of a Thousand Factories." The "thousand" was an over count, but not by much. The number of factories in Dayton in 1924 (650) was impressive, and the chamber claimed the city led the world in manufacture of 55 products, some quaintly old-fashioned when viewed from 2012, including wooden steering wheels for automobiles and sealing wax for fruit jars, and others that are still important, including bicycles, automobile ignition systems, washing machines, and filing envelopes for court papers.[26]

One company was more important than all others in 1920s Dayton. Forty years earlier, in the 1880s, John H. Patterson had owned two grocery and general stores that had good sales but no profits. To understand why he was making no money, he observed his clerks at work. He inspected their bookkeeping and saw that they did not always charge the correct prices for goods, favored their friends with discounts or free goods, sometimes forgot to collect money (and sometimes simply decided not to), or deposited cash in their pockets instead of the till. A flyer describing a cash register, called the "Incorruptible Cashier," arrived in Patterson's mail as he was mulling over the missing cash in his store. He ordered two of the machines, invented by a Dayton bartender, and installed them in his stores. The stores turned a profit the next month.

Seeing commercial opportunity in selling cash registers, he purchased the rights to manufacture the machines, set up a company, the National Cash Register Company (NCR), to make and sell them, and initiated a vigorous research and development program to improve and perfect the cash register. He and the company were so successful that most other cash register manufacturers in the United States dropped out of the business before 1900, and Patterson claimed that NCR made 95 percent of all the cash registers in the world. He added other office machines to the company line and developed work methods for offices that used them. He developed a well-trained sales force (less charitably, he was "the 'inventor' of modern high-pressure salesmanship")[27] to market his machines and services. In the 1920s, NCR, which Daytonians "always called 'The Cash,'"[28] employed thousands in Dayton.

The importance of the cash register to the development of modern business methods was humorously underlined in a book about big league baseball (of all things): "The humorist Finley Peter Dunne, writing in an Irish accent as his famous character, Mr. Dooley, declared 'the crownin' wurruk iv our civilization — th' cash register.'"[29] Dunne/Dooley's comment was an overstatement, but the cash register, along with new market and office procedures introduced by Patterson and his company, revolutionized the conduct of business around the world.

The "beloved"[30] Mr. Patterson and his company were good citizens. They were instrumental in helping Dayton recover from deadly floods in 1913 and worked steadily to improve city services and parks. As befitting a man living in the birthplace of aviation with a businessman's eye for opportunities, Mr. Patterson's strong personal interest in aviation was paralleled by business considerations. During World War I, NCR, skilled in the manufacture of precision mechanical devices, had become "actively involved in the lucrative wartime manufacture of military aircraft components."[31]

Many other important manufacturing companies had headquarters and factories in

Dayton. The Maxwell Motor Company[32] produced automobiles from 1914 until 1925 when the Chrysler Corporation took it over. Thereafter the former Maxwell plant turned out Chrysler and Dodge parts. The Dayton Rubber Company ("Dayco") made automobile tires and many other rubber devices. To supply such major manufacturers, other Dayton companies built subunits, parts, and tools.

Edward A. Deeds and Charles F. Kettering, who had worked for NCR in the first decade of the 20th century, invented an "ignition, lighting, and starting system for automobiles," which was installed in the 1912 Cadillac. "Starting system" does not capture or project the revolutionary importance of the Deeds-Kettering innovation. Their "self-starter," turned on with a key, provided a new layer of theft protection and did away with hand-cranking the engine, which had never been a job for a woman, required a certain amount of strength, and sometimes broke wrists and arms when the crank jerked unexpectedly. From those beginnings, the two men formed the Dayton Engineering Laboratories Company, better known as "Delco," to manufacture auto components, and Delco subsequently spun off Delco Light Company, which manufactured generators and lighting systems for farms and homes beyond the reach of electric utilities in the 1920s and 1930s, and Frigidaire, which made refrigerators and other home appliances.

Keeting and, especially, Deeds were important to Dayton's developing as a center for military aviation. Writing in 1932, a Dayton booster praised Deeds' service in World War I:

> Mr. Deeds, at great personal and financial sacrifice, accepted a call to Washington.... He then became a member of the Aircraft Production Board.... It was largely through his efforts that aircraft production in the United States was built up as it was.[33]

Others saw Mr. Deeds' wartime service as not entirely selfless:

> [Deeds and Kettering had] formed the Dayton Airplane Company in 1917, which was soon reorganized as the Dayton-Wright Company.[34] When the war began, Deeds was commissioned [as an Army colonel] and put in charge of procurement for the Aircraft Production Board. He divested himself of his financial interest in Dayton-Wright but awarded the company two contracts to produce more than 4,000 DH-4 and Standard J-1 aircraft. Given the company's inexperience, the size of its contract led to charges of favoritism. A Senate committee corroborated these allegations, and President Woodrow Wilson appointed a commission headed by future Supreme Court Chief Justice Charles Evans Hughes to investigate. Although mismanagement and favoritism were documented, charges were not brought, and the company survived the scandal.[35]

Orville Wright also played a pivotal role in keeping his hometown at the forefront in aviation. When army officials asked him to recommend locations for training fields during World War I, he suggested an area east of Dayton that included the Huffman Prairie, where he and his brother made many flights and had constructed an early factory. Among other reasons, he favored the site because "the ground was soft and spongy and cushioned the shock of landing."[36] On behalf of the army, Deeds leased more than 2,000 acres of land, including Huffman Prairie, for use as a training facility and flying field, and the field was named Wright Field to honor Wilbur Wright, who had died in 1912. In keeping with his probable favoring of his old company, Deeds contributed to the decision to lease a parcel of land from the Dayton-Wright Company on the north side of Dayton. That parcel became McCook Field,[37] the home of the army's Aviation Engineering Division.

As early as 1920, the engineering division was outgrowing McCook's facilities, and *Aviation* magazine complained that it was unsuited "as a flying field for experimental types." The surroundings were "so dangerous to our splendid test flyers that they need a new field."³⁸ Having a McCook-like installation appealed to many cities.³⁹ It would provide high-tech, skilled, and unskilled jobs, attract high-tech firms that would provide more jobs, and anchor technical and economic development. Other communities, including Newport News, Virginia, site of many shipyards and related industries, and Garden City, New York, home to the Curtiss Aeroplane and Engine Company, were interested.

John Patterson, NCR's president, led the effort to keep the division in Dayton. To kick off the drive to keep McCook open or, if that proved impossible, to move the engineering division to another location in Dayton, he hosted a reception for General William Mitchell in March 1922. Upon Patterson's death two months later, his son, Frederick B. Patterson, took over the running of the company and the drive to keep the engineering division in Dayton. The younger Mr. Patterson established the Dayton Air Service Committee, an association of local businesses and individuals, which raised $400,000 by public subscription "within two days."⁴⁰ Some $263,000 of the total was used to purchase 5,000 acres of land adjoining Wright Field, to the east of Dayton, and the remainder was used to begin the construction of a Wright brothers memorial in Dayton.

When the legal hurdles of obtaining the land had been cleared, Mr. Patterson journeyed to Washington to transfer the land to the War Department on behalf of the citizens of Dayton in exchange for $1.00. President Calvin Coolidge invited Mr. Patterson to visit the White House and asked him to convey the president's thanks to Dayton's citizens for their gift. When McCook Field closed in 1927, the Air Services activities formerly housed there had been kept in Dayton and moved to the enlarged Wilbur Wright Field, which is now Wright-Pat's Patterson Field.

Anyone interested in identifying the most enthusiastic supporter of the Aviation Engineering Division, McCook and Wright Fields, army aviation, Dayton aviation, and the 1924 National Air Races would certainly consider and likely choose Fred F. Marshall. Marshall, who had edited the engineering division's in-house magazine, *The Slipstream*, became its publisher when the government no longer supported such in-house publications after World War I. Marshall edited and published and did everything else at *The Slipstream*. The army had trained Marshall as a photographer during the war, and his photographs appeared in *The Slipstream*. He recruited authors for articles, wrote and edited articles, drew cartoons, and wrote poetry. He may even have attended the softball games played by various army units; certainly *The Slipstream* reported the results of such local activities. At the time of the National Air Race, he produced a special, expanded issue of *The Slipstream* with a wealth of information about Dayton, McCook Field, military and naval aviation, and air racing.⁴¹ The Dayton Metro Library honors his memory in its Fred F. Marshall Collection, related to Dayton and military aviation.

The NAA did not publically announce that Dayton would host the 1924 Pulitzer until December 1923, but Frederick Patterson's election as NAA president during the association's meeting in St. Louis during the 1923 National Air Race was likely coordinated with an understanding that the 1924 Pulitzer would be held in Dayton. Patterson made a determined effort to bring foreign competitors to the 1924 Pulitzer and, in the December 1923 *Slipstream*, Fred Marshall wrote enthusiastically that foreign pilots were coming and praised Patterson:

> Through his [Patterson's] efforts ... Sadi Lecointe, famous French aviator, will compete in the Pulitzer.... There is a strong possibility of Belgium being represented by a team sent here by King Albert.... Italy will no doubt be represented by the famous Lieut. Brack-papa, foremost aviator of that country.... [R]ecent communications received from the [British] Royal Aero Club ... indicate that considerable interest in the event has been aroused in this body.⁴²

In February 1924, two months after Marshall's praise appeared in print, Patterson made a trip to Europe to solidify European participation and attract more European pilots and organizations to compete in the Pulitzer. On behalf of the NAA, he offered to cover the expenses of foreign entrants.[43]

Patterson, the son of the inventor of "high-pressure selling" and probably confident of his selling ability, apparently did not understand the great difference between funding of races and racers in Europe and the United States that had led to European decisions not to fly in the Pulitzers. European airplane manufacturers footed the cost of building racers, either from their own resources or from money provided by individuals or organizations such as flying clubs, and they employed pilots who, among their other duties, flew the racers. In 1920 and 1921, the same sort of private financing had paid for the Curtiss *Texas Wildcat* and Dayton-Wright RB-1 that raced in the 1920 Gordon Bennett, four nonmilitary racers at the 1920 Pulitzer, and the Curtiss-Cox *Cactus Kitten* and two Ansaldo racers in the 1921 Pulitzer.

The separation of United States financing from private sources had also begun with the 1920 Gordon Bennett. In that race the U.S. Army-built Verville VCP-R piloted by an army officer flew against the privately financed *Wildcat*, RB-1, and British and French entries, all flown by civilian pilots. The mixture of privately and government-financed racers continued for only the first two Pulitzer Races of 1920 and 21. A year later, private sponsorship was gone, unable to compete with the army and the navy. The disparity between resources available to European clubs and companies and United States government agencies had not boiled over into controversy because Europeans did not compete in the United States, and, until 1923, no Americans competed in Europe. In that year, boil over it did, when United States Navy racers took first and second in the Schneider Trophy Race.

One of two Curtiss CR-3s (CR-2s outfitted as seaplanes). The two CR-3s placed first and second in the 1923 Schneider Race, Cowes, England (courtesy Glenn H. Curtiss Museum).

The Schneider Races, first flown in 1913, had been international from the beginning, and privately financed American aviators and airplanes had flown in the 1913 and 1914 contests. The entry of two U.S. Navy seaplane racers—Curtiss CR-3s—in the 1923 Schneider changed everything from racer financing to racer design. Bert Acosta had flown a CR to victory in the 1921 Pulitzer, and navy pilots Harold Brow and Al Williams had flown updated CRs—the CR-2s—to third and fourth place behind the army's R-6s in the 1922 Pulitzer in Detroit. When the navy decided to enter the Schneider in February 1923, it pulled the CR-2s from storage, where they had been since after the 1922 Pulitzer, had Curtiss fit them with twin pontoons as long as the racers' fuselages, and dubbed them "CR-3s." The racers, pilots, mechanics, and other crew members sailed across the Atlantic to Cowes, England, site of the Schneider, in late summer 1923.

To say the CR-3s won is an understatement: "The Schneider Trophy Race became a one-two romp for the Curtiss machines."[44] Understanding precisely what Thomas Foxworth meant when he wrote that the CR-3s transformed the Schneider from a race of "ladylike flying boats doing minuets of navigability into a brew smoking with demons and banshees in a fire dance of speed"[45] is difficult, but his meaning is clear.

The pilots of the CR-3s, navy lieutenants David Rittenhouse and Rutledge Irvine, who had cut their racing teeth in the 1922 Curtiss Marine Trophy Race at Detroit (see Chapter Four), were competitive with each other at Cowes, but the rest of the field puttered along in a slow-motion quest for third. Rittenhouse's winning speed of 177 mph was a full 32 mph faster than the 1922 winning speed. Some Europeans were outraged by the incursion of government-owned aircraft in the Schneider, and, correctly, forecast that Europeans, to stay competitive, would have to turn to governments for financing. The Schneiders were on their way from being sporting contests among aero clubs to struggles for national prestige.

Under the Schneider Trophy Race rules, the United States would host the 1924 race, and enticing foreign entrants to Baltimore, chosen as the race site, was going to be a tough sell. European competitors faced the costs of building competitive airplanes and shipping racers, pilots, ground crew, tools, and supplies to the United States, but signs were good in early summer 1924. At least, signs were good to *The Slipstream* reporter who wrote, "Great Britain, France, and Italy will enter seaplanes in the Schneider Cup Races."[46]

The Slipstream and other newspapers and magazines that reported foreign pilots and racers would enter a 1924 Schneider either overly optimistically interpreted Europeans' comments, gave into wishful thinking, or accurately reported intentions that were rendered incorrect by subsequent events. There's little evidence that anyone in France had serious plans to enter the Schneider. The Air Ministry in Great Britain ordered racers from two firms. One was built, but it crashed with damage so severe that it could not be made ready for the Baltimore race. A month before the scheduled race, Italy's air attaché in Washington informed the Flying Club of Baltimore, the host for the 1924 race, that he had received a cable saying that Italy was withdrawing its three entries. No reason was given.[47]

The Baltimore Flying Club and the NAA could have held an all-American event to secure a second straight United States' Schneider Race victory. An actual race was not necessary. A single American plane flying over the course would have given the United States its second "victory," and a subsequent third victory would give the United States permanent possession of the Schneider Cup. The Baltimore Flying Club decided against that course of action, and postponed the race a year to 1925 to enable other countries to prepare racers. At the same time, the club arranged an event to capitalize on the enthusiasm from the 1923

Schneider victory. The club and the navy planned a one-day event on Chesapeake Bay on October 25, 1924, at which U.S. Navy seaplanes would attempt to set new speed records over various distances.

The navy had contracted with Curtiss to convert the Curtiss C2R that had won the 1923 Pulitzer, and that Al Williams had flown to a world speed record, into a seaplane. Like the CR-2s that had been modified for the Schneider, the seagoing C2R, Williams' racer, was equipped with two large floats and designated "C2R-2." The other C2R, which had been flown by Brow, remained a land plane and was called "C2R-1." The C2R-2 was impressively fast, flying 228 mph on a time trial over Long Island Sound, more than 40 mph faster than the winning CR-3's 177 mph in the 1923 Schneider. Fast it was, but the navy decided to save it for the 1925 Schneider and not fly it in the 1924 time trials over the Chesapeake. The navy flew its CR-3s, veterans of the 1923 Schneider, now three years old and elderly for racers, in the 1924 speed trials. They did well; the faster one set a new absolute speed record for seaplanes of 189 mph, and, in all, navy seaplanes set 17 new world records.[48]

Well before the cancellation of the planned 1924 Schneider that took place in late summer, Mr. Patterson had returned from his early spring European trip to report that British, French, and Italian racers were expected to fly in the Pulitzer. A memo from an official in the U.S. Embassy in London painted a different picture. According to the memo, neither the Air Ministry nor the RAF was interested in sending a racer to Dayton. The Gloucestershire ("Gloster") Aircraft Co., Ltd., was interested, but it was said, "holding them back is the enormous cost of building an entry and sending it to Dayton." More specifically the company knew it could not successfully compete against "the light Curtiss engine, the Reid [sic] Propeller and the Wing Radiator." Possibly encouraged by Patterson's offer to cover expenses of foreign entrants, "they [the Gloucestershire Company] will expect the Aeronautic Association [NAA] to assist them in obtaining wing radiators and Reid Propellers."[49] No British entry was forthcoming.

Nothing definite was heard from Italy, but an entry from France promised international competition of the highest order. Joesph Sadi Lecointe, an eminently skilled, popular, gregarious French pilot, winner of the 1920 Gordon Bennett and holder of 19 world records at various times, entered the race and would fly a new racer being built by the Dewoitine Company. Whatever disappointments he had met elsewhere, Mr. Patterson's visit to France had paid off. Sadi Lecointe was coming. Already-high expectations for success at Dayton climbed higher. High expectations crashed down two weeks before the date for the Pulitzer. The September 13 *New York Times* published a two-sentence news article with headline and sub-headline:

Lecointe Won't Race Here.
No Foreign Competition Now for Pulitzer Plane Speed Prize.

Paris, Sept. 12. Sadi Lecointe, French aviator, has abandoned his plan of going to the United States to participate in the races for the Pulitzer Cup. The sporting paper L'Auto says Lecointe found that his airplane would not be ready in time to make the trip."[50]

Foxworth cites a contract dispute as the cause of the withdrawal. The Dewoitine Company had built the racer, but Sadi Lecointe was under contract to Nieuport-Asta, which would not allow him to fly another manufacturer's racer. In any case, Foxworth considered it unlikely that any French racer could have met the Pulitzer requirement to land at 75 mph or less.[51]

The disappointment of no foreign competitors had been nine months in the future when pleased United States readers saw the article "The Development of High-Speed Aircraft" in a December 1923 issue of the British magazine *Flight*. In a speech to the Royal Aeronautical Society, Royal Air Force (RAF) Major R.H. Mayo praised the United States and, in particular, the Curtiss Company for progress in high speed flight and predicted that clipping the wings and making small improvements to Curtiss racers would enable them to reach 300 mph. New racers, he said, would achieve 350 mph within two or three years. Primarily, Mayo praised the United States policy that allotted part of army and navy aviation appropriations to purchase and fly racers: "The result of this policy was that America had become the leading air power for an insignificant expenditure.... America has placed herself well ahead of any other nation in the design of high-speed aeroplanes and the development of suitable engines, and her position as the leading Air Power is secure for some time to come."[52]

In contrast to the United States, he said, was the British government's failure to spend money on racers. Speed was not only important to racing; its importance extended to commercial and military aviation. Raising the speed of commercial airplanes by 20 mph without reducing load carrying or increasing engine size would, Mayo said, revolutionize the finances of commercial aviation. The *Flight* reader who enjoyed Mayo's praise of the United States' progress and its policies toward racing airplanes could have been depressed and worried by the remarks of USAS major William N. Hensley Jr., commander of Mitchel Field, on April 23, 1924: "Unless Congress comes to the immediate assistance of the country's air force, three years will find us with absolutely no aerial defense."[53] Common as it is, such "the sky is falling" rhetoric among those seeking greater government funding seldom fails to attract attention.

How to reconcile the views of Majors Mayo and Hensley? They were discussing different subjects. Major Mayo discussed speed increases and generalized from that to general progress in aviation. Major Hensley talked about a clearly impossible future to question the adequacy of funding for aerial defense. Nevertheless, the remarks of Majors Mayo and Hensley pointed in far different directions. The RAF officer thought that U.S. aviation — including military aviation — was in good shape; the U.S. officer concluded that U.S. military aviation was in deadly peril.

Gloom and pessimism apparently filled the room where Major Hensley spoke. NAA president Frederick Patterson, speaking on the same program, lamented that the United States was far behind Europe in commercial aviation, especially in comparison to the promotion of passenger services that was ongoing in Britain, France, and Germany. Somewhat surprisingly for the man who was in charge of the Pulitzer, he said, "France expects to win the coming Pulitzer and ["I" omitted from text?] hope she will.... The stimulation will be the best thing in the world for American flying."[54] Who else, besides Mr. Patterson, wished for a foreign victory in the Pulitzer or would say it, even in frustration or hyperbole? Perhaps it was a facet of NCR Company high-pressure selling technique.

While Patterson wrestled with problems of attracting foreign racers and Marshall's *Slipstream* "talked up" the races, others faced up to the stark decline in expectations for the 1924 Pulitzer. Reductions in federal spending, beginning in 1921, had continued downward, and the army and navy did not contract for new racers for 1924. Furthermore, the navy announced it did not have funds to race in both the Pulitzer and the Schneider and would not enter the Pulitzer. To assure speeds at least equivalent to those seen in the 1923 Pulitzer, the navy sold the Curtiss R2C that finished second in the 1923 Pulitzer to the army for $1.00. Both services now had an R2C: The navy's R2C-2 seaplane for the Schneider; the army's R2C-1 land plane for the Pulitzer. On May 12, the army's R2C-1 arrived at McCook

Army Curtiss R-8, originally a navy R2C. In preparation for the 1924 Pulitzer, it was painted silver and given an army serial number (courtesy National Archives and Records Administration).

Field, where it was given an army serial number, painted silver and designated "R-8." The army assigned Lt. Alexander Pearson, Jr., to fly it.

A month before the Pulitzer, Pearson, in the R-8, and Burt Skeel and Wendell Brookley, flying the army's R-6s, were in the air near McCook to make speed runs over a three-kilometer course. The army had not followed Admiral Moffett's lead to ban diving starts. Pearson climbed, leveled out, and pushed the nose down to begin his speed run. With a loud crack, the left wings sagged down, and the racer rolled several times. Pearson throttled back or stopped the engine, appeared to get out of the cockpit, and either fell or jumped before the racer slammed into the ground. His body was found 75 yards from the aircraft.[55] A joint army-navy board investigated the crash and traced the cause to a laminated interwing strut, which had been hollowed out to save weight. The navy immediately substituted solid struts on the R2C-2 that it intended to race in the Schneider Race, and the army replaced hollow struts on its R-6s.

Unlike its forebears, the CRs and the R-6s, which competed in multiple races, the R2Cs had short racing careers. Pearson's crash destroyed the R-8 (former R2C-1) before the 1924 Pulitzer. Almost two years later, the navy was using the seaplane version, the R2C-2, as a trainer for navy pilots preparing for the 1926 Schneider. On a routine flight, its pilot lost control and crashed into the Potomac River near the navy's Anacostia airfield. The crash killed the pilot and destroyed the R2C-2. Both R2Cs flew in only a single race—the 1923 Pulitzer—but they made more news in Brow's and Williams' repeated flights to set a new absolute speed record. Their history was touched with glory, raising the Pulitzer speed to 244 mph and the absolute speed record to 267 mph, but distressingly short and deadly, as two pilots died.

Optimism for exciting new speeds in the 1924 Pulitzer withered with the R-8's crash. There was no new racer, and the fastest available racer had crashed. What was left? It wasn't encouraging. A Verville R-3 and the two Curtiss R-6s, all veterans of both the 1922 and

1923 Pulitzers, would fly their third race. No one could fool himself that the any of those racers would break the record set by Williams in the C2R the year before. The R-3 at Dayton was the only one remaining of the three that had been built for the 1922 Pulitzer. One R-3 was tested to destruction in early fall of 1923, and the second was scrapped in February 1924. The third, which army lieutenant Fonda Johnson had flown to 7th place in the 1922 Pulitzer, had been updated with a Curtiss D-12 engine and wing radiators for the 1923 Pulitzer. Alex Pearson, designated to fly the R-3 in the 1924 Pulitzer, had flown the same airplane in the 1923 Pulitzer, where its unbalanced propeller (or some such problem) forced the racer out near the start of the race.

Following the 1923 Pulitzer, the R-3 was trucked to McCook Field where it was used to test a variety of wood and metal propellers through December 1923, and Alex Pearson flew it to an official world record speed for 500 km. On January 1, 1924, it went into storage. In August 1924, it was pulled from storage, painted over-all silver like the R-6s and the R-8, and reconditioned for the October Pulitzer. Lt. Harry Mills, assigned to fly the R-3, made his first flight on September 13 and flew it almost every day until October 4, race day. A week before the race, the D12 engine was removed and replaced with a 507-hp D12A engine. Finally, the R-3 had what Fred Verville had wanted in 1922 — a high-powered, smooth-running V-12 engine and wing radiators. Mills, its pilot, said that the R-3 "handled well and the engine functioned perfectly."[56] It was ready for the race.

Both R-6s had spent much of the time after the 1923 Pulitzer in storage at McCook. Consideration of costs caused cancellation of plans to replace their engines with larger Packards and to upgrade their control systems to R2C standards. In late summer 1924, the decision not to upgrade the control systems was reversed, and new controls and 507-hp Curtiss D12As were installed. With those modifications, the same engine powered all three racers, two R-6s and the R-3, in the 1924 Pulitzer.

By 1924, metal propellers were recognized as superior to wood ones in strength, reliability, and resistance to rain, hail, rocks and other foreign objects. Nevertheless, the Air

Verville-Sperry R-3, prepared for the 1924 Pulitzer (courtesy National Archives and Records Administration).

Service declared that suitable metal propellers were "unavailable" for its three racers. To make certain that the racers were "evenly matched," all would be equipped with wood propellers manufactured at McCook. The army's assertion of its inability to procure metal propellers for its racers must be met with incredulity. Metal propellers were entering routine Air Service use, and Curtiss-Reed metal propellers were standard equipment on the army's PW-8 pursuits.[57]

In early 1923, Reed and Curtiss had entered into an agreement under which Curtiss would manufacture and sell metal propellers. Curtiss "equipped all its 1923 special aeroplanes with this airscrew [propeller], viz., the Navy-Curtiss seaplane racers for the Schneider Cup, Navy-Curtiss racers for the Pulitzer Trophy, Army-Curtiss Pursuit 'planes and the Curtiss Night Mail 'plane.'"[58] Curtiss-Reeds were not the only metal propellers. For example, three-bladed metal propellers had been used on the Wright F2W racers in 1923. For racing, the great advantage of a metal propeller was that it could be safely spun faster than a wood one, and higher propeller speeds increased airplane speeds.

The army's decision to manufacture wood propellers at McCook Field may have influenced its requiring such propellers on its 1924 racers. The army was certain to win the Pulitzer, and its own wood propellers would be on its racers. What better publicity could be imagined for the army propellers? Wood propellers, like the Curtiss interplane strut that failed on the R-8, were laminates made of glued-together strips of wood. The glue, not so strong as the wood, limited the strength of the propeller. Some wooden propellers had been found faulty when insufficient glue had been applied and when glue had been squeezed out between laminates, leaving "starved joints." Various tests were devised to ferret out such problems, and McCook Field wood propellers passed the tests with no reported difficulty.

On race day, an army pursuit, the Curtiss PW-8B, joined the three racers. The PW-8B was the second modification of the army's PW-8. The PW-8, like the racers, was cooled with wing radiators, which had been perfected on racers and had "contributed so much to the racers' speed."[59] Wing radiators required constant attention and maintenance to make certain air bubbles did not form and block water flow that caused steam to form and blow holes in the less-than-paper-thick brass of the radiators. Ordinary events on an airfield, a bit of gravel thrown up by prop wash, a mechanic's dropped tool hitting a wing radiator, or any kind of collision involving a wing could cause damage that required hours of labor to repair. The diligent attention of the large number of mechanics that tended racers could take care of such mishaps. When installed on service airplanes, the radiators improved speeds, but contributed "even more to maintenance officers' nightmares."[60] No one could ignore, either, that the radiators' large areas would made them vulnerable to combat damage.

In recognition of the radiator problem and to try out other modifications, the army contracted with Curtiss to rebuild one of the prototype PW-8s with strengthened, single-bay wings and a conventional radiator mounted flat in the upper wing with cooling fins projecting from the top and bottom of the wing. That radiator configuration on the PW-8A eliminated the service and maintenance problems associated with the wing radiators, but the army had been impressed with the "tunnel radiator" installed on the Boeing PW-9, and it had Curtiss modify the PW-8A. The upper wing radiator was removed, and a tunnel radiator was installed below the nose. This airplane, the PW-8B, was prepared for, and raced in, the 1924 Pulitzer. It had no chance against the specialized racers, but its appearance in the Pulitzer combined with the appearance of PW-8s in the John L. Mitchell Race gave the Air Service opportunities to show off its speedy successors to the Thomas-Morse MB-3s. Later in 1924, *Flight* magazine praised Russell Maughan's "great Dawn-to-

Top: Curtiss PW-8A (first modification). Note single-bay wings rather than double-bay wings on PW-8s and radiator mounted flat in the upper wing, replacing wing radiators. The radiator cooling fins were exposed on the bottom as well as the top of the wing (courtesy Glenn H. Curtiss Museum). *Bottom:* Curtiss PW-8B. Note "tunnel" radiator under nose. This is the airplane that raced in the 1924 Pulitzer. In the collection of the National Archives and Records Administration, this airplane is identified as a PW-8A, but it is not an 8A. The single 8A with the conventional radiator in the upper wing [upper photograph] was rebuilt with a tunnel radiator, and, in that configuration, it is variously identified as the "XPW-8A, second modification," "XPW-8B" and "PW-813." It was the prototype for the Curtiss P-1 service pursuit (courtesy National Archives and Records Administration).

Dusk" transcontinental flight in a modified PW-8 and called the PW-8 "a Successful American Fighter."[61]

Anyone who looked at the PW-8s with wing radiators would see resemblances between the pursuits and Curtiss racers and could answer for himself questions about whether the military's racing expenditures had influenced military aircraft design. More knowledgeable observers, familiar with engines, knew that the PW-8s' engines were derived from the Curtiss V-12s that first roared into the public's eye and ear in the Curtiss 18-T triplanes at the 1920 Pulitzer. Today, the military's racing program would be cited as an example of successful technology transfer. The military invested a little of its budget to purchase racers, the companies used that money to design, test, and develop features that proved useful in racers, and some of those features were then incorporated into higher-performing airplanes.

In Dayton in the summer of 1924, the army prepared its R-3 and two R-6s, its pilots practiced, and the National Air Race organizers worked hard to advertise the October races and persuade people to attend. Dayton was "the smallest city yet to handle the races,"[62] but its "systematic and businesslike advertising and publicity campaign" was expected to produce a good crowd "between 225,000 and 250,000" for the three days of racing.[63] The following is an example of Dayton's advertising, beginning on September 22, 1924: "Some five million people passing Times Square ... will see material ... telling of Dayton and the plans for the race meet." Literature about the air races was distributed in 600 cities in the central states; and in a two-week-long campaign, nine trucks fanned out from Dayton to put up posters advertising the races, with the reported result that "interest in the races in every city they visited is at high pitch."[64]

Roads from Dayton to Wilbur Wright Field were improved to handle the many automobiles expected at the races, and Dayton's streetcar company laid tracks to the field. Box seats for 40,000 people were arranged along the fence that marked the boundary of the airfield, and space for 25,000 parked cars "upon a natural elevation" would provide views of the races for the automobiles' occupants. Behind those spaces, cars parked in thousands more "dead" places provided no view of the races, and their "occupants will either find seats in boxes or watch the races from the promenade."[65]

The NAA arranged special train fares for its members. A member could buy a round-trip ticket from his city to Dayton and exchange his return ticket for a half-price ticket and a refund of half the fare for his return.[66] Dayton and race officials worried that the city had insufficient housing for the expected crowds, and the NAA urged its members to make reservations early: "It is anticipated that every hotel, rooming house, and private dwelling will be needed.... Tourists camps and Pullman cars will also be utilized."[67] Wilbur Wright Field would provide housing for visiting military officers, although accommodations might be tight: "[E]ach officer has agreed to share his quarters, ordinarily large enough for two men, with ten officers, if this is found necessary."[68]

The writers and publishers of the *National Aeronautic Association Review*[69] and *The Slipstream*[70] had vested interest in the success of the air meet and published upbeat articles in the weeks before the races. In the summer, in a turnabout from earlier stories, neither publication printed quantitative estimates of the expected crowd. Neither repeated the earlier estimates that ranged up to 250,000. Writers might have tired of making quantitative estimates or, more ominously, they might have sensed a slacking of air race enthusiasm. Reasons to worry about the success of the 1924 National Air Races abounded as time to the races shortened. Railroad bookings for Dayton did not surge; Dayton hotels' reservation books did not fill up. The expected crowd was either putting off making reservations until the last minute or, worse, attendance at the Dayton races was about to flop.

How to explain the diminished interest? Dayton is a bit off the beaten track. It offered few of the attractions found in larger cities like Detroit and St. Louis. The absence of new racers meant no attention-attracting news about prerace claims and publicity about new high speeds. The absence also meant no chance of record-breaking speeds in the Pulitzer, further dampening the interest of potential attendees. Far more general events, including the 1924 economic downturn, could have convinced potential race-goers to pinch their pennies and stay at home. Whatever the reasons, expectations were in decline in mid–September when Sadi Lecointe's bombshell announcement that he would not race eliminated the attraction of foreign competition.

General Mitchell may have noticed the declining interest. Rather than tout the races, he focused on obtaining a remarkably faster world speed record, and, at his request, Fred Verville designed an R-3-like airplane with two V-12 engines mounted in tandem driving a single propeller through a gearbox. Wind tunnel tests predicted a top speed of 290 mph, but funds were not forthcoming; the plane was not built.[71]

Race officials at Dayton laid out near-rectangle racecourses with distances of five and fifteen miles for all the events except the Mitchell and Pulitzer races. The 50-km Mitchell and Pulitzer course were hexagonal in shape extending to the northeast of Wilbur Wright Field. The last leg of the hexagonal course brought the racers in from the west and past the grandstand before turning northeast to start the next lap.

More than 120 private and commercial airplanes flew into Dayton and parked in the "Civilian Camp." Two of the planes belonged to unknown pilots who would be headline news in a few years—Charles Lindbergh and Roscoe Turner. The addition of more than 200 military and government aircraft brought the total of visiting aircraft to over 300, "the largest number [of aircraft] ever assembled on one spot."[72]

Public interest in racing might have been waning, but aviation achievements were attracting national attention. A week before the 1924 National Air Race, on September 28, two of the four U.S. Army Douglas World Cruisers that had set out on the first round-the-world airplane flights successfully completed their trips. The other two other World Cruisers crashed on the flight, but without fatalities. As the World Cruisers made their ways around the world, stories about the struggles of the crew members whose airplanes went down may have outnumbered those about the successful flights, but there were plenty of both kinds.

Barring reports of a fatality, it is difficult to imagine a more discouraging headline than the one chosen by the Associated Press to open its reporting of the opening day of the 1924 National Air Races: "Few attend the air races opening day."[73] Anyone in Dayton on that day probably sensed as much. Dayton hotels had not burst with guests. No crowds overtaxed public transportation, ticket sellers, or ushers. Automobiles were not turned away from already-filled parking lots. About 12,000 people attended, maybe a third of the 35,000 that had been predicted by H.W. Karr, the NAA's general manager.[74] A month before opening day, a writer in the *National Aeronautic Association Review* had revised the then-current estimate of 250,000 attendees for the three-day meet down to 200,000.[75] After day one, the revised figure looked wildly optimistic. In the fall of 1923, St. Louis' National Air Race, the navy's R2Cs' speed records, and the navy's victory in the Schneider had been front page news in the *New York Times*. The same newspaper ran its story of Dayton's opening day on page 10.[76]

TABLE 5
1924: Closed-Course Competitive Events at the Third National Air Race, Fifth Pulitzer Race, Dayton, Ohio, October 2, 3, and 4[a]

Day One, Thursday, October 2

Event	Entrants	Course	Pilot, Airplane (hp) Speed
National Cash Register Company Trophy Race. Comparable to *FlyingClub of St. Louis Trophy Race, 1923*	Civilian free-for-all race for two-seat planes, engines less than 520 cu. in. displacement, 340 lb. load. 1st, 2nd, and 3rd place finishers had Curtiss OX-5 engines.	6 laps of 15-mile course for total of 90 miles.	1. Walter E. Lees, Hartzell VC-1, (90) 98 mph. *Same pilot/plane won in 1923 at 89 mph over ~93-mile course.* 2. Perry Hutton, Laird Commercial, (90) 93 mph. 3. Randolph Page, Yackey Sport, (90) 87 mph
Central Labor Union Trophy. Comparable to *Aviation Club of Detroit Trophy Race, 1923*	Civilian free-for-all race for two-, three-, or four-seat planes, engines less than 800 cu. in. displacement (~220 hp), 340 lb. load	8 laps of 15-mile course for total of 120 miles	1. C.S. "Casey" Jones, modified Curtiss Oriole, Curtiss C-6 engine (160), 125 mph 2. J.C. Ray, Curtiss Oriole, Curtiss C-6 engine (160) 107 mph 3. C.C. Caldwell, Glenn Martin Commercial-70, Wright E-4 (~200) 102 mph
Liberty Engine[b] *Builders' Race*	Two-seat military observation aircraft. All 11 entrants, DH4s powered by 400-hp Liberty engines.	12 laps of 15-mile course for total of 180 miles	1. Lt. D.G. Duke, 130 mph 2. Lt. A. Simmonin, 128 mph 3. Lt. C.A. Cover, 124 mph

Day Two, Friday, October 3

Event	Entrants	Course	Pilot, Airplane (hp) Speed
Aviation Town and Country Club of Detroit Trophy Race	Civilian pilots/engine of less than 800 cu. in. displacement (~ 220 hp), 340 lb. load	8 laps of 15-mile course, 120 miles. Two Events. *Speed* and efficiency. *Efficiency* calculated: *payload in lbs × mph / hp*	*Speed*. 1. B.L. Rowe, S.V.A., Curtiss C-6 (160) 111 mph 2. J.G. Ray, Curtiss Oriole, Curtiss C-6 (160) 107.5 mph 3. W.L. Stulz, Atlantic S-3, Wright E engine (~200), 106.9 *Efficiency*: 1. W.H. Beech, *New Swallow*, Curtiss OXX-6 (100) 320 points 2. C.C. Caldwell, Glenn L. Martin 70, Wright E-4 (~200) 291 points 3. B.L. Rowe, S.V.A., Curtiss C-6 (160) 265 points
Dayton Chamber of Commerce Trophy Race. Comparable to *Mercantile Exchange of St. Louis Trophy Race, 1923*	Planned for large capacity civilian and military airplanes. Martin bombers only entrants.	10 laps of 15-mile course for total of 150 miles	1. Lt. D.M. Myers, same Martin Bomber (2 × 440) that won in 1923, 110 mph *1923 winner flew 114 mph over ~ 186-mile course* 2. Lt C.F. Woolsey (2 × 440) 108 mph 3. Lt H. McClellan (2 × 440) 105 mph

Event	Entrants	Course	Pilot, Airplane (hp) Speed
Dayton Daily News Light Airplane Contest	Speed contest for light airplanes, powered by motorcycle engines of 20 or fewer horsepower and carrying 150 lb loads.	5 laps of 5-mile course for 25 miles	1. J.M. Johnson, Driggs-Johnson "parasol" monoplane, (20-hp, Henderson engine) 64 mph 2. E. Dormoy, Dormoy "flying bathtub" (20-hp, Henderson engine) 54 mph. Eight starters; two finishers
Detroit News Air Mail Trophy Race	Cancelled: Planned for U.S. Air Mail pilots/ U.S. Air Mail airplanes.		

Day Three, Saturday, October 4

Event	Entrants	Course	Pilot, Airplane (hp) Speed
Dayton Bicycle Club and Engineers' Club Trophy Race	Speed and efficiency race for light airplanes, powered by motorcycle engines of 20 or fewer horsepower. Winner based on computation of speed divided by fuel consumption.	Ten laps of 5-mile course for 50 miles	1. H.C. Mummert, Mummert "Sport" (18-hp Harley-Davidson) 38.2 mph 2. J.M. Johnson, Driggs-Johnson (20-hp Henderson) 22.8 mph 3. E. Dormoy, Dormoy design (20-hp Henderson engine) DNF but awarded 3rd because "mock bombing" destroyed a course pylon
John L. Mitchell Trophy Race	Pilots of 1st Pursuit Group flying Curtiss PW-8 pursuits, all powered by 333-hp Curtiss D-12s.	4 laps of 50-km course for 200 km (~124 miles)	1. Lt. Cyrus Bettis, 175 mph 2. Lt. D.F. Stace, 174 mph 3. Lt. T.K. Matthews, 173 mph. *1923 winner, B.E. Skeel, 148 mph*
Pulitzer Trophy Race	Civilian[c] and military pilots/airplanes w/ landing speed 75 mph or less and top speed greater than 170	4 laps of 50-km course for 200 km (~124 miles)	1. Lt. H. Mills, Verville R-3, USAS, 507 hp, 217 mph 2. Lt. W. Brookley, USAS, Curtiss R-6, 507 hp 214 mph 3. Lt. R. Stoner, USAS, Curtiss PW-8B, 375 hp, 168 mph *1923 winner, A.J. Williams, USN, Curtiss R2C, 244 mph*

[a]Primary sources: "The Dayton Air Races," *Aircraft Year Book 1925* (New York: Aeronautical Chamber of Commerce of America, 1925) 189–193 and from same book, "Race Results, Dayton Air Meet, Oct. 2, 3, 4, 1924," pp. 299–302. Other sources: "Largest Gathering in History of Sport Expected to Witness Three Days' Program," *National Aeronautic Association Review*, September 2, 1924, p. 1; "Air Racers Open Meet at Dayton," *New York Times*, October 3, 1923, p. 10; "The International Air Races, Dayton," *Flight*, October 30, 1924, pp. 693–695.
[b]"Contests" in italics shared same or similar name with contests held in 1923.
[c]Had Joseph Sadi Lecointe flown, he would have flown a civilian airplane.

Eleven now-antiquated DH-4s flew in the Liberty Engine Builders' Trophy Race on Day one. The DH-4s, evenly matched for speed, provided opportunities for army pilots to show their mettle at flying big, slow, single-engined airplanes; the race was a crowd-pleaser, but the DH-4s were at their limit for speed. The DH-4 that won in 1922 had flown 129 mph; the 1924 winner flew one mile per hour faster, and the second and third place planes flew slower than the 1922 winner. In 1923, the more modern Fokker CO-4, powered by the same 400-hp Liberty that powered the DH-4s, had introduced variety and speed—139 mph, 10 mph faster than the fastest DH-4—into the race. No CO-4 nor any other plane different from the DH-4s flew in 1924, and the winner came in at 130 mph, the same as the 1922 winner's. The other events of day one were similar to events in 1923, with no remarkable new airplanes and no new records.

The *New York Times* ran its story about day two on page one, but the headline, "Airship Launches Plane While Aloft,"[77] indicates the writer found an exhibition or stunt more newsworthy than any race that day. For the exhibition, an army dirigible flew over Wilbur Wright Field with a small Sperry Messenger biplane (20-foot wingspan and less that 19-foot length, powered by a 60-hp air-cooled engine) slung below the cabin. The airplane was released, engine off, and glided down a short distance. The pilot turned on the engine, the plane climbed, and then returned to earth after a short flight. General Patrick said that the airship was equipped with a hook to which the airplane could fasten with a catch on its upper wing, but a midair hookup was not attempted.

Sponsorship of the annual race for "heavy lifters" or "heavy-weights" had changed as the site of the National Air Races changed. The Detroit Chamber of Commerce Trophy Race of 1922 became the Merchants Exchange of St. Louis Trophy Race in 1923, and, a year later, the Dayton Chamber of Commerce Trophy Race in 1924. Each race had been open to military and civilian aircraft, but only military planes had flown in 1922 and 1923, and such would again be the case in 1924. The "heavy-weights" race, a contest among Martin two-engined bombers, shared an unfortunate characteristic with the Liberty Engine Builders' Race: all its racers had raced before. A new airplane or a variety of planes would have generated more interest. As it was, the event was crowd-pleasing but old hat: "Although the skill displayed by the Army pilots ... left nothing to be desired — the steep banking in many cases being very thrilling — it was felt by many of those present, who had witnessed the same performance on previous occasions, that this event was becoming decidedly out of date."[78] Judgments that "thrilling" races could be flown by airplanes "becoming decidedly out-of-date" were hardly ringing endorsements of the 1924 National Air Race. They must also have had a chilling effect on discussions about future National Air Races.

The Dayton Air Meet, like all air meets, had exhibition flights and static displays of aircraft and equipment to complement the races. The monstrous Barling bomber, built at McCook, awed the crowd with its size, and, for good measure, set a record. On October 3, day two of the air races, with all six of its 400-hp Liberty engines throbbing and its triplane wings getting a grip on the air, the Barling trundled over the grass at Wilbur Wright Field, its tail lifted, and it slowly left the earth. A shallow climb took it and its 4,000-kg useful load to 4,000 feet, where it flew for an hour and forty-seven minutes to set a record for longest flight with a 4,000-kg payload. The Barling did not bear a charmed life. Even its record-setting flight did not end well. Landing, it "ran out of gas and the Barling landed in a three-foot ditch."[79]

Watching the behemoth trundling through its takeoff and laboring upward might have given knowledgeable spectators pause. What would happen to such an airplane, struggling to reach 90 mph at a mere 4,000 feet if it were attacking a target in wartime? Surely, fighters such as the Air Service PW-8 pursuits buzzing around Wilbur Wright Field, at more than 140 mph, would make short work of shooting it down.

"[S]ome excitement — not to say amusement" greeted the flight of a 1910-type Wright biplane. Lt. John Macready, who had an enviable flying record — a world altitude record in 1921, third place in the 1921 Pulitzer, and pilot of the Fokker T-2 on the first nonstop transcontinental flight in 1923 — flew the Wright biplane past the grandstand to "made its bow to Orville and Katherine Wright."[80] Orville Wright, interviewed during the air meet, said, "[T]he principles of flight we used twenty-one years ago are still being used. I am extremely proud.... The airplane has undergone amazing developments during the past years but not the principle of flight."[81] In his thoughts, did Wright include in the "amazing developments," the many features of the Wright aircraft displaced by other inventors, especially Glenn Curtiss? Ailerons had replaced wing warping, wheels had replaced skids, tail-mounted elevators had replaced elevators on poles in front of the wings, and stick and rudder bar had replaced the Wright's control system. Few particulars of the early Wrights remained in 1923, but the Wrights' "principle of flight" remained. In any case, the fly-by of the "1910 type Wright" provided a pointed contrast with the airplanes competing at Dayton 14 years later.

Two contests at Dayton — the *Dayton Daily News* Light Airplane Contest on day two and the Dayton Bicycle Club and Engineers' Club Trophy Race on day three — introduced new, small airplanes powered by motorcycle engines of 20 or fewer horsepower. While far from identical to today's "home-built" or "ultra-light" aircraft, they are, in some ways, comparable.

The *New York Times* stated that the races for "baby planes" demonstrated that "engines of less than twenty horsepower are able to sustain a plane in safe as well as efficient flight,"[82] but only two of the eight aircraft that started in the 25-mile-long *Dayton Daily News* speed race finished. The race, flown around a triangular five-mile course, required rounding a pylon at 60 feet, climbing to pass a moored balloon at 500 feet that marked the end of a two-mile straightaway, then easing back down to turn another 60-foot-high pylon.[83] The course was designed to show that the airplanes were fully controllable. J. Johnson of Dayton won the contest in a 20-hp Driggs-Johnson airplane, built in Dayton.

The other small airplane race, the Bicycle and Engineers' Clubs' Race on day three, was an efficiency contest with the order of finishers determined by dividing speed by the volume of gasoline used in the race. Mechanical problems forced all five airplanes down before the third lap, but two were able to complete the course before the race was ended because of the "mock bombing" that was the final event (after the Mitchell and Pulitzer races). A third plane was awarded third place because it had been closest to the finish when the bombing curtailed the race.

At least some of the "baby airplanes" were not much smaller than racers. For instance, the Driggs-Johnson had a wingspan of 27 ft., 0 in. compared to the R-3's 30-ft., 7-in. wingspan, but it was much lighter, 511 pounds compared to 2,006 pounds (empty weights). The Driggs-Johnson was far less powerful. The nominal horsepower of the R-3's Curtiss engine was 507; the maximum horsepower of the Driggs-Johnson's motorcycle engine was 20. The large wingspans, light weights, and low power of the lightweight airplanes must have made them appear to flutter more than fly, but their appearance at the National Air Race seemed to bring nearer the possibility of "everyman" being able to afford an airplane.

Later in the month Alexander Klemin, associate professor of aeronautics at New York University, estimated a young man could build one of the light airplanes, a "flivver," as he called it, for $500 or purchase one already built for $1,500 (about $22,000 in 2013 dollars). Those estimates seemed high to Henry Ford. After examining a Sperry Messenger at the 1922 Detroit air races, he had pointed out changes in the manufacture of the engine that, he said, would reduce the cost from $1,500 to $50. If sufficient demand justified mass production of the little airplanes, the prices were bound to fall — to "cheaper than Ford?" as Klemin asked. Two-seat flivvers with bigger engines — 30- or 35-hp rather than the 20-hp seen at Dayton — would make the small airplanes even more popular. Klemin

Driggs-Johnson "baby" airplane (courtesy National Archives and Records Administration).

imagined doctors visiting patients and hospitals and salesmen making their calls in the little airplanes, flying back home, folding the aircraft's wings, parking it in the garage, and taking it out next morning. The goal (or dream) of an airplane in every garage that has popped into the minds of many visionaries (and little boys and girls) over the 100 years of flight may never have seemed closer to realization than in 1924.[84]

Any hope of attendance approaching expectations had expired by day three. There was no surge of spectators. The predicted 180,000 spectators for day three seemed a hurtful, cruel memory or bad joke. At most, 50,000 people entered the race grounds. Whether the empty seats dismayed the average spectator or made him thankful for extra room and knowing that getting home after the day's racing would be easier, is impossible to know, but surely he noticed.

The Mitchell and Pulitzer races, the biggest attractions, were the penultimate and final events. According to *Aviation* magazine:

> [T]he John L. Mitchell was by far the most interesting. In the first place, there were eleven entries — as against four in the Pulitzer race — and the greater number of competing ships, all evenly matched because they were strictly service types made the race an exciting thing to watch. Second, the public was treated for the first time to a race between eleven PW8 Curtiss Pursuit ships, and the way these formidable fighters whizzed around the course must have been an eye opener even to many aeronautical people.[85]

Foxworth wrote simply that the Mitchell Race was the "true highlight" of the air races.[86] The *Mt. Clemens Daily Leader*, the hometown newspaper for Selfridge Field, characterized the Mitchell as "probably the hardest fought event of the entire three-day program."[87]

The competitors in the Mitchell took off and circled the field to gain altitude before diving down, flattening out, and crossing the start line directly in front of the grandstand. The air shrieked around the PW-8s' wires, the Curtiss D12 engines roared, engulfing the spectators in "an inferno of noise,"[88] and the aircraft flew off. The race satisfied everyone who wanted to see the Air Service's latest pursuits in close competition. The PW-8s "flew

... in such close order that it seemed at times as though some of the craft would collide."[89] Lt. Blackburn, who did not place, flew the most spectacular race. On his second lap, he banked sharply around the pylon in front of the grandstand and his wingtip severed a guy wire, leaving the pylon momentarily aquiver. He flew the homestretch at 25-foot altitude. The flying of Blackburn and the other ten closely bunched pilots gave the spectators their money's worth.

Lt. Cyrus "Cy" Bettis led the eleven PW-8s across the start line, stayed in front, and finished first at 175.4 mph. Less than two mph behind Bettis, at 173.7 mph, Lt. Donald Stace, winner of the 1922 contest, crossed the finish line in second place, barely beating out Lt. Thomas Matthews, who finished third at 172.3. Bettis finished 26 mph faster than Bert Skeel, who had won the 1923 Mitchell in a Thomas Morse MB-3 and answered the often-asked question, "What is the Army (or Navy) getting for all the money it's spending on racers?" The PW-8 was a leap forward from the MB-3, and the PW-8 was clearly derived from the R-6 racer.

The Pulitzer Race was the only contest remaining. Three racers—the R-3 and two R-6s—had identical engines, and identical propellers, and were two-year-old designs. The fourth competitor, the modified PW-8B, looked much like the PW-8s that had just completed the closely contested Mitchell except that a tunnel radiator under the nose had been substituted for the standard PW-8s' wing radiators. The PW-8B was expected to finish a distant fourth. The four pilots took off in the sequence of their race numbers. Capt. Burt Skeel in R-6, race number 68 was first, followed by Lt. Wendell Brookley in R-6 number 69, Lt. Harry Mills in the R-3, number 70, and last, Lt. Rex Stoner in the PW-8B, number 71.[90] They climbed into the cloudy skies, west of Wilbur Wright Field, circled upward, and prepared to dive across the start line. Mills, in the R-3, started his dive from a lower altitude while the two R-6s were climbing to altitude and Stoner was getting into the air.

A tragedy, not a new record or brilliant race, put the Pulitzer on front pages. Decisions about how high to climb before diving for the start line were "entirely up to the pilot's judgment."[91] In a stiff southwest wind, Captain Burt Skeel, the popular commander of the 1st Pursuit Group at Selfridge Field and the winner of the 1923 Mitchell Race, led Lieutenant Brookley upward through a thick cloud layer and burst into sunlight. Brookley flattened his climb. His R-6 had oscillated—its nose pitching up and down—during practice dives, and, cautiously, he had decided to begin his dive from only 1,000 feet, level out at 350, and cross the start line from that altitude.

As Brookley watched, Skeel's racer climbed higher and higher, growing smaller and smaller. Seeing Skeel peel off to begin his dive, Brookley held his altitude at about 1,000 feet, from where he could see the ground through a skim of clouds. Skeel, traveling nearly vertically at terrific speed, had dived below Brookley before Brookley put the nose of his R-6 down and opened the throttle. Brookley's worries about his R-6 were borne out. The racer oscillated violently. Brookley eased off the power, regained control, and flattened his dive.

Spectators saw Skeel's R-6 hurtling downward in a 60-degree dive from far higher than Brookley's airplane. The loudspeakers blared, "Here comes Skeel. Note his speed." About "a mile and a half from the starting point ... a sudden swerve of his plane showed something was wrong."[92] Sunlight flashed on what looked like "a bursting shell" as the cooling water burst from the ruptured wing radiators, sprayed out, and refracted the sunlight. Skeel's R-6 collapsed so rapidly that no coherent picture of the sequence of events ever emerged. Many observers reported that the upper wing tore loose from the fuselage first, but others reported a lower wing came off first. The fuselage, part of the lower wing still attached, hit the earth nearly vertically at the tremendous speed built from the dive and buried itself in swampy earth, disappearing from view. Several hours passed before the fuselage was found, and even more before Captain Skeel's body was located at a depth variously reported as 10 and 14 feet. The upper wing fluttered down, largely intact, at some distance from the fuselage's burial place.

The apparent explosion, the plunging airplane, the inevitability of the crash, the geyser of earth and debris thrown up at impact stilled the 50,000 spectators except for gasps and moans. Captain Skeel's wife was in the stands. Some reports say she was looking toward the start line and not at the descending airplane, but others report she saw the crash. She was taken into Dayton, and later that evening she returned to her mother and two small children at Selfridge Field. One of their children, Burt E. Skeel II, commissioned a plaque that now stands in the Memorial Garden of the National Museum of the United States Air Force.

How many people in the stunned crowd paid attention to the three other racers continuing methodically around the hexagonal racecourse? Even for those who did, it was a boring exhibition. Mills, in the R-3, first across the start line, led all the way. Brookley finished second in the R-6, and Lt. Rex Stoner flew the PW-8B to a distant third, 40 mph behind Mills. The *New York Times* reported the three pilots "flew a careful race, taking no chances in making sharp banks in rounding the pylons."[93] Foxworth describes the end of the race in different terms: "As a tribute to his dead comrade, and a mark of confidence in his racer, Brookley stunted briefly before landing."[94]

Forty-five minutes after Skeel crashed a hastily assembled Air Service board of inquiry was at the site and established that the cause of the crash was either a collapse of the wings or the splitting apart of the propeller. The death of Lt. Pearson in the R-8 because of the faulty interplane strut added to the plausibility of the wing collapse explanation, and it fit with many reported observations. Finding that the propeller had split along the laminations rather than having been broken and splintered by the impact of the crash bolstered the propeller failure theory. Equally important for the propeller failure theory, parts of the engine were scattered 100 yards from the point of impact, which could have been caused by a damaged, unbalanced propeller flailing the air and tearing apart the engine.

The scattering of engine parts over a wide area and several observers' reports that the top wing came off first were consistent with the propeller delaminating. The wildly spinning unbalanced propeller would have torn the engine apart, flinging pieces ripping through the cowling and the central carbane strut that held the wing above the

Captain Bert Skeel's memorial plaque at the National Museum of the United States Air Force (author's photograph).

Top: Site of the crash of Burt Skeel's R-6. The fuselage of the racer buried itself in the soft damp earth, leaving a few shreds on the surface. The depth and size of the crater can be judged from the height of the man standing in it (courtesy National Archives and Records Administration). *Bottom:* The torn-apart upper wing of Bert Skeel's racer, which landed hundreds of feet from where the fuselage crashed (courtesy National Archives and Records Administration).

Verville-Sperry R-3 in flight. The 1924 Pulitzer winner, despite its poor speed, was a glimpse of the future (courtesy National Archives and Records Administration).

fuselage. With the carbane strut gone, turbulence could rip off the upper wing. Curtiss engineers stated that the wood propeller had been a mistake, and the editor of *Aeroplane* weighed in to say, "Later it was proved fairly conclusively that the airscrew burst," and concluded it had been "singularly imbecilic to fit a wooden airscrew to an engine of such power designed for such speed."[95]

The Air Service, which had ordered the use of wood propellers on the racers, would have been interested in shifting the blame elsewhere, and it tried: The official conclusion for the cause of the crash was collapse of the lower wing and its ripping away. Many knowledgeable people rejected that conclusion. The *Mt. Clemens Daily Leader* expended little effort in trying to understand or explicate the cause of the crash. It interpreted the crash simply. The blame lay in Washington:

> Aviation authorities who witnessed the crash ... are loud in the condemnation of the plane which he used in the race. It was the same plane in which Lieut. Russell Maughan won the Pulitzer at Selfridge, two years ago....
> Because of a foolishly economical administration, the flower of the land are ordered to go to their death in old broken down planes, when a few thousand dollars would save their lives.
> Lieut. Brookley ... was forced to cut his speed because of the fact that his plane nearly collapsed in the same manner that Skeel's did, and Lieut. Mills, winner of the race, came down with the fuselage of his ancient plane torn to bits, barely landing before collapsing.[96]

Blaming Washington absolved everyone lower down in the Air Service hierarchy of responsibility. The importance of Selfridge Field to the economy of Mt. Clemens might have influenced the newspaper's deciding all blame rested in the capital.

The tragic death of Alexander Pearson when testing the R-8 and of Burt Skeel in an R-6 at the start of the Pulitzer surged up as an especially ugly part of the 1925 court-martial of General William Mitchell. In line with Mitchell's arguments that high-ranking political and military officials were guilty of criminal neglect of the Air Service, his attorney charged that General Mason Patrick and Theodore Roosevelt, Jr., the oldest son of former president Teddy Roosevelt and an official in the Navy Department in 1923, had hatched a sleazy deal to save money and fix the Pulitzer Races. During the court-martial, Mitchell's lawyer called an Air Corps officer to the stand and introduced letters that he interpreted to show that Patrick and Roosevelt had arranged that the army and navy would win the Pulitzer Race in alternating years.

Mitchell charged that because the navy didn't enter a racer, the army knew it could win with "older-model planes," and Pearson and Skeel died when "the wings broke off 'the two old crates.'" The charge, which amounted to an attack on General Patrick, backfired. "Patrick, who was still steamed at Mitchell for insinuating that he conspired with the navy to fix the competition, set the record straight."[97] He insisted that Roosevelt and he made plans to divide up racers for the air meets, not the outcomes. The fact that the navy sold one of its R2C racers to the army for $1.00 in 1924 supports that interpretation. At the same time, the agreement, barring accident or breakdown in 1924, assured that the army would win the Pulitzer and the navy would win the Schneider. It's possible to construe the agreement to share the R2Cs as amounting to support for Mitchell's charge that Roosevelt and Patrick had arranged for the army to win, but that argument was not made in the trial. Mitchell's criticism and Patrick's defense of policies affecting the races seem unlikely to have influenced the court-martial's outcome.

Spectacular speeds in the Pulitzer might have partially lifted the pall that Skeel's death caused to settle on the 1924 National Air Race. However, there was no spectacular speed. Mills' winning speed in the R-3—a disappointing 216.6 mph—was fully 27 mph below Al Williams' winning speed the year before and Brookley's R-6, in second place at 214.4 mph, finished about five mph slower than its speed of 219 mph at St. Louis the year before. Stoner in the PW-8A, third at 168 mph, flew eight mph slower than Bettis had in winning the Mitchell. Its speed fell far short of any record, but the 1924 R-3 was much superior to the original 1922 version. The 1924 racer's Curtiss engine did not vibrate and shake the airplane as the Wright engine in the 1922 airplane had, and the 1924 model flew the Pulitzer 36 mph faster than the 1922 model's 181 mph. Had the army made a Curtiss engine available to Verville in 1922 and had he been able to use wing radiators as he wanted, the speed of the 1922 R-3, if it had been close to the 217 mph of the 1924 R-3, would have been better than the Curtiss R-6's 207 mph winning speed in 1922. A win by the R-3 in 1922 might have hastened the introduction of monoplane pursuits into the Air Service. In part because the biplane R-6 won, the Air Service equipped its pursuit squadrons with Curtiss and other companies' biplanes, and biplanes were "to dominate American military aviation for a full decade."[98]

All the speeds in the 1924 Pulitzer were disappointing. The R-3's was far less than the 1923 winners; Brookely's R-6 was slower than an R-6 in 1923; Stoner's PW-8B was slower than five of the PW-8s in the Mitchell Race. An engine that "sputtered on hitting heavy bumps"[99]—held down the speed of Brookley's R-6. Blustery, errant winds buffeted him and the other pilots, and all pilots likely flew more conservatively after seeing Skeel's crash. The hexagonal course with more turns than the 1923 triangle course would have further reduced speeds. Stoner, stationed at Langley Field, not at Selfridge, may have had fewer opportunities to practice for racing, and Langley ground crew may not have reached the heights of tuning up racers seen at Selfridge.

The slow speeds could be explained. They could not be made exciting or satisfying. On December 28, 1923, in its article announcing the award of the Pulitzer Race to Dayton, the *Dayton Journal* had predicted that the R-3 would fly 275 mph.[100] No one who remembered that article could find any joy in the 1924 Pulitzer results or speeds.

Whether or not Skeel's R-6 was "worn out," the safety of the 1924 racers can be questioned. The R-6s were not "old" in terms of flying time. Thomas Foxworth says that each of the two R-6s had fewer than 50 hours' flying time and maybe fewer than 25 and quotes someone that both had been put "in the best possible condition."[101] Nevertheless, almost all the flights of the two-year-old R-6s had been made at or near all-out speeds with high-speed dives and sharp pullouts to start races or speed trials and tightly banked turns in races putting great stresses on the racers. The "best possible condition" did not mean the racers were good as new. The army's crash investigation committee declared that examination revealed some deterioration in the wings of the surviving R-6. No special examination was needed to see that the R-3 was the worse for wear. After its 1924 victory, it "showed obvious signs of degeneration — joints beginning to open out, dope and fabric stripped in places — and obviously its days were over."[102]

The deaths definitely made one point spectacularly clear. Wooden structures were not strong enough to bear reliably the stress and strains of racing airplanes in the 1920s. Wood airframes and propellers lingered on in lower-powered airplanes, but metal structures became increasingly common in higher-performance and passenger-carrying airplanes. On the other hand, Curtiss racers were tough. The two CRs/CR-2s/CR-3s had been flown very hard over three years. One of the two had flown three major races, the 1921 and 1922 Pulitzers and the 1923 Schneider; and the other had flown the 1922 Pulitzer and 1923 Schneider. Both flew in the 1924 seaplane speed trials on Chesapeake Bay in 1924. There are no reports that they had been weakened, but because neither was involved in a serious crash there may simply have been no report about the condition of the racers.

The dramatic and violent end of Bert Skeel's life silenced the crowd at Dayton. Many left the airfield, but some remained to watch the "bombing" of a (small-scale) mock New York City, the air meet's final exhibition. How many who stayed and watched thought about the shock and sadness of an individual's death and wondered how it was possible to enjoy the spectacle of a bombing, which, if real, would have killed hundreds or thousands? On a more peaceful and happier note, the army fliers who had just completed the first round-the-world flight flew into Wilbur Wright in three airplanes. General Mitchell awarded them metals on behalf of President Coolidge.

General Patrick canceled the "smoker" scheduled for Saturday evening. The smoker, with attendance limited to 1,000, was to have been held in Dayton's Memorial Hall, with trophies and prizes awarded to victorious fliers. Entertainment had not been neglected on the planned program. Two six-round boxing matches would serve to warm up the crowd for the finale, "a battle royal between numbers of husky colored athletes."[103]

In depressed after-race Dayton, individuals and companies that had underwritten the costs of the air show were left to scramble for an additional $75,000 to make up for the revenue shortfall when less than one-third of the much ballyhooed quarter-million spectators attended. Two publications, heretofore indefatigable boosters of the Dayton International Air Meet, faced up to explaining what went wrong to their readers. The *National Aeronautical Association Review* dealt with disappointment by apparently not noticing it. Its article with the Panglossian headline "Dayton's Big Race Meet Scores Unusual Success" splashed in very large red letters across the front page of the broadsheet ended with the statement that "every aim and object [had been] accomplished." There is not a word about how many people attended or the speeds attained.[104] A second article in the *National Aeronautical Association Review* is intriguingly entitled "Four World Records Were Set in Races." Indeed they were, but the records were for loads carried to altitude, not pulse-quickening speed records.[105] Only on page 6 of the *Review* was there reference to tragedy, "The One Sad Note in Successful Air Meet,"[106] with a picture of Burt Skeel and a short article about his death. Ignoring the realities of the disappointments might have rubbed some salve on the NAA's wounds, but no official was unaware of the grievous damage done to the National Air Races by poor attendance, lackluster races, and the death of a popular pilot who left a widow and two small fatherless children.

Fred F. Marshall's headline for his article/editorial in *Slipstream* was more honest and straightforward. The headline was a question: "The Dayton Air Race Meet: Was It a Failure and Why?" Inclusion of "why?" acknowledges it was a failure. Marshall singled out those who had expected to make money from the races as among the most disappointed (and why should they not have been?). He himself was disappointed in the spectators who, because of ignorance or other reasons, were "so engrossed in talking crops and politics that poor Burt Skeel was buried in twelve feet of muck before they were even aware that the Pulitzer Race — the feature drawing card on the annual Air Race meet was in progress."[107] The sentence is filled with understandable bitterness, but how well did it describe the crowd? After all, people not interested in racing had stayed at home, accounting for the poor attendance.

Responsibility and blame there were aplenty, much of it directed at Air Race management. Management's prerace warnings about expected crowded conditions may have convinced some potential spectators to stay at home. Somewhat off-beat posters might have left potential spectators uncertain of the seriousness of the race organizers. For instance, a poster featuring a drawing of a monoplane racer advertised the "1924 International Air Races, Dayton, Ohio," which was straightforward enough. But beneath the drawing, the poster read, "Kentucky is famed for it's [sic] beautiful women, good liquor [this during Prohibition], and fast horses. Dayton cordially invites all Kentucky to the great International Air Races.... P.S. We can't take care of the horses."[108] Was the poster likely to lure an additional spectator or to convince someone who was wavering about attending to decide that the organizers were not serious people?

It was impossible not to look back one year to St. Louis in 1923 and compare its spectacular crowds to Dayton's smaller numbers. There was no shortage of possible and partial explanations. The cost of tickets, the absence of foreign competitors, the refusal to shower visiting newsmen with gifts, or to endorse the putting up of thousands of road posters were all cited as reasons for low attendance. A "less rabid group" suggested that economic conditions far beyond the control of anyone connected with the races — fear about bad times to come and a resulting tendency to tighten belts — kept people at home. A most likely, and perhaps most important, reason was that for three years — 1921, 1922, and 1923 — the United States military had bought new, faster racers and publicized their capabilities and promise well before the races. Those airplanes and that publicity were missing in 1924, and their absence must have affected interest and attendance.

In addition, something special in St. Louis was missing from Dayton. St. Louis had a sparkling new airfield, which General Patrick said would be one of the best in the country, and Admiral Moffett said St. Louis had done more for aviation than any other city (see Chapter Five). Residents in and around St. Louis knew about Mr. Lambert, who had sold the land for the airport to the city for the same price he'd paid for it years before, and they knew the airport had been improved for the National Air Race. There might have been a strong feeling of "I'd like to see our new airport" to complement interests in going to the races, which promised new records because of the new Curtiss and Wright racers.

Dayton was going to have an expanded airport thanks to the businessmen who had contributed the money necessary to buy land and present it to the federal government to expand McCook and Wright fields, but for Dayton's citizens, it was not "our" airport, it was the army's. Furthermore, the land had been purchased, but the new facilities had not been built. Dayton citizens who wished to could drive over to McCook and Wright and see military airplanes whenever they wanted. They knew too that there would be no new records in 1924 because there were no new racers.

Attendance plummeted between 1923 and 1924; but, Marshall wrote, without convincing support, "Aeronautically, at least, the 1924 Air Race was in every way successful." He reported that business executives, government officials, and military officers had praised Dayton for being well prepared, "its facilities for holding the meet were demonstrated as being unsurpassed." The praise, well-merited as it was, could have provided scant comfort to those in charge, who realized the races had fallen well below their expectations for attendance and performances and who were scurrying around to find money to pay off the race's debt. Marshall may have interviewed too few NAA members when forming his conclusion that the races had been well run. It is difficult to reconcile it with the NAA's electing a new president "[i]n a riotous election described as 'more fun than an acre of billy-goats.'"[109]

A year later, the *Aircraft Year Book, 1925* published a qualified favorable evaluation of the 1924 air meet.[110] Not so spectacular as in previous years, the 1924 National Air Race had instituted an emphasis on "practical utility," referring to the popularity of "On to Dayton Race," the races for two- and three-place civilian airplanes, and the large number of civilian aircraft and pilots in attendance. Practical utility may be good for selling an airplane to a customer, but it doesn't draw spectators to air meets. Spectaculars do, and there was none but a tragic one at Dayton. Marshall wrote about changes in the public's attitude toward air shows:

> The public can no longer be depended upon to attend ordinary air races as they are now conducted. The average spectator wants excitement, thrills, unusual events and a crash or two ... to give one his money's worth at a dollar general admission fee....
>
> In the future it must be an air circus throughout with possibly a goodly assortment of country fair side shows.... Air Races can never again be expected to attract huge throngs from the masses, at least, in North America and any city, no matter how large it is, would be foolish to make the preparation instituted by Dayton in anticipation of this happy thought.... The International Air Race if carried through in the usual way in 1925 is bound to fail again in attendance. For the benefit of aeronautics it will be well to continue them, but as a business proposition the task is hopeless.[111]

Those like Marshall who faced up to the unpleasant fact of diminished popularity, must have realized that the Pulitzer Trophy Race, the National Air Races' main attraction, would wither and die for want of new racers if the army and navy did not pay for them. What would happen in 1925? With no new airplanes, any 1925 Pulitzer would be a rerun, with even older aircraft.

How many spectators would pay to see more "bomber races" or DH-4 races or, even, Pulitzer races flown by years-old airplanes? The small motorcycle-engined "baby" airplanes were a novelty, but why would crowds pay money to watch them putter and splutter around a racecourse until a few survivors crossed the finish line at speeds achievable by most cars? The year 1924 had come in on a wave of enthusiasm because of speed and other records made in 1923. It was going out in a slough of discouragement, if not depression, because airplane development, especially speed-plane development had stalled during 1924.

Probably few in the NAA or anywhere else expected both the army and the navy to purchase even faster Curtiss racers, and the navy to invite Lt. Al Williams back to fly its racer in 1925. Had anyone seen that future—faster airplanes and a well-known speed pilot—would he have predicted a continued, glowing future for the most important closed-course races for land planes in the world? Or was the popularity of air races spiraling downward, likely to be unaffected even by a better effort in 1925?

Predictions about the future are always uncertain, and predicting the future of air races might be even more difficult. As Fred Marshall looked back at the 1924 race, he may have remembered a paragraph from *Slipstream* published six months before the air meet: "Dayton will this fall add itself to the list of these cities whose names are synonymous with successful aviation meets, and the dividends from the investment of time and money will be valuable even to the present generation; their value to our children and our children's children will be incalculable."[112]

The Dayton air meet had fallen short of expectations. It had been deadly, its speeds were poorer than in years past, the crowd had been tiny compared to the size predicted. What was the future of the Pulitzer?

CHAPTER SEVEN

1925

Back to the Origin, Mitchel Field, Long Island, New York, October 12

> *The public can no longer be depended upon to attend ordinary air races as they are now conducted. The average spectator wants excitement, thrills, unusual events and a crash or two ... to give one his money's worth at a dollar general admission fee.... Air Races can never again be expected to attract huge throngs from the masses, at least, in North America and any city, no matter how large it is, would be foolish to make the preparation instituted by Dayton in anticipation of this happy thought....*[1]

The Dayton International Air Meet, especially the 1924 Pulitzer Race and preparation for it, had not gone well. Two army pilots, Lieutenant Alex Pearson and Captain Burt Skeel, their airplanes disintegrating, had crashed and died. No new racer had been built for the Pulitzer. The old ones were slow, their speeds disappointing. Featuring military airplanes "of standard types, old if not absolutely obsolete," the other races at Dayton had seemed "old hat."[2] Attendance was a third of the expected number. The results of the 1924 National Air Race plunged Fred Marshall, editor/publisher of *The Slipstream* and unfailing champion of Dayton, army aviation, and the National Air Race, into dismay and despair. Were the days of popular and record-setting races a thing of the past? Would 1925 see a continued slide in interest and attendance? Or was the Dayton disappointment a one-off event? Would locating the 1925 race in a more populous area, and, above all, would building new, faster racers bring back the crowds?

Aviation experts pondered more technical questions than air race attendance. Did the two benefits of air racing — increasing public awareness of, interest in, and, with luck, support of aviation, and providing a dramatic venue for the full-throttle testing of airframes and engines — justify its costs and risks? Some experts said "no." A year after Pearson's and Skeel's deaths, an *Aviation* magazine editorial writer acknowledged that racing had been aviation's "great drawing card" but argued that times had changed. The "publicity and technical advancement attained may not be worth the cost and risk."[3]

Whatever doubts and questions existed elsewhere, the National Aeronautical Association (NAA) was determined to hold a 1925 Pulitzer Race. In March, the NAA was "looking for a city to sponsor the International Pulitzer Trophy land plane speed contest" and issued "a statement deploring the lack of "American sporting pride.... [O]nly New York and

Wichita, Kan., have unofficially requested the competition."[4] In addition, the Twin Cities, turned down to host the 1924 race, had announced its intention to bid for the 1925 Pulitzer.[5] Despite the nonappearance of a foreign pilot in any race, Pulitzer promoters continued to refer to an "international" race.

New York was the preferred site for the 1925 Pulitzer. The city's huge population provided an enormous number of people from which to draw spectators and there was plenty of public transportation to move people to and from the races. NAA headquarters were in the city, making it easy for NAA officials and staff to pitch in with race preparations and for national newspapers and magazines to carry news, announcements, and publicity about the races. In addition, advertising agencies were centered in New York.

In late April, the NAA issued a sequence of press releases, surely intended to dramatize what may have been a straightforward arrangement between the NAA and the Air Service to hold the race at Mitchel Field. On April 25, a delegation of Air Service officers flew from Washington to the Long Island airfield "to inspect a proposed course for the Pulitzer air race.... There has been a good deal of sentiment for the selection of this course and the opinion has been expressed here [Washington] that it may be decided upon, if the army experts make a favorable report."[6] After the visiting officers' inspection, the *New York Times* reported, "rumors were current in aviation circles that the chances of the races being held here this Fall were extremely favorable."[7] The next day's *Times*, April 26, confirmed that the rumors, which the NAA may have originated or fueled, had panned out: "Application is about to be made by the National Aeronautical Association for authority to hold the Pulitzer air races on Mitchel Field, Mineola. A representative of the association today sought such authority from General Patrick ... who informed him that the permission could be granted only by the Secretary of War. It is expected that the Secretary of War will grant the permit."[8] As expected, the secretary granted permission, and on May 1, the NAA, meeting at St. Joseph, Missouri, announced the races would be flown "at New York late in October."[9]

Ten weeks later, on July 12, the president of New York 1925 Air Races, Inc., fixed October 8, 9, and 10 as the dates. Those early fall days seemed to provide a buffer against chilly, windy, wintry weather such as had cloaked the 1920 Pulitzer at Mitchel, held on Thanksgiving Day, November 25. The Air Race president expected there would be new records in the races, that the winning speed in the Pulitzer would be raised from 244 to 265 mph, and that the racers would fly over 300 mph on the straightaways. As always, foreign interest was claimed, and England and Canada were mentioned.[10] Most important of all for the success of the 1925 Pulitzer, the army and navy were buying new racers. The new aircraft would generate publicity during their prerace testing, keeping the race in front of the public and almost guaranteeing higher race speeds and new world records.

The postcards prepared by the air race committee to advertise the Pulitzer Race indicated a lack of planning. The postcards feature a drawing of a De Havilland DH-4, an antiquated design and certainly no racer. Were there no drawings of Curtiss racers or other, more modern airplanes to illustrate the cards? The DH-4 drawing suggests either a last-minute effort, with no time to find an appropriate drawing, or unfamiliarity with racers. Either possibility denotes a lack of professionalism among those promoting the races.

No DH-4 would fly in the 1925 Pulitzer, but early in that year, no one could be certain about which airplanes would fly. On the other hand, anyone picking winners in the 1925 Pulitzer would have considered a Curtiss Company product and would expect either the army or the navy to contract with Curtiss for racers. One or both the services might contract with another company or even other companies, but both would want to go with winners and Curtiss racers were proven winners. Curtiss racers had won in 1921, '22, and '23, and Curtiss engines had powered all those racers as well as the Verville-Sperry R-3 that won the 1924 race.

Curtiss prospered, partly because of its racing success. In 1925, it was paying dividends to its investors; English[11] and Italian firms were buying its D-12 engines, which had been developed and proven in racing, Curtiss-Reed metal propellers were selling well in foreign and domestic markets and generating steady profits, Curtiss commercial airplanes were beginning to sell, and the Air Service was buying Curtiss fighter planes clearly derived from Curtiss racers. Even so, Curtiss Company management was turning away from racing. The highly skilled and expensive design and engineering team that produced, at best, two, three, or four racers a year could more profitably work on military, commercial, or private aircraft to be made in greater numbers.

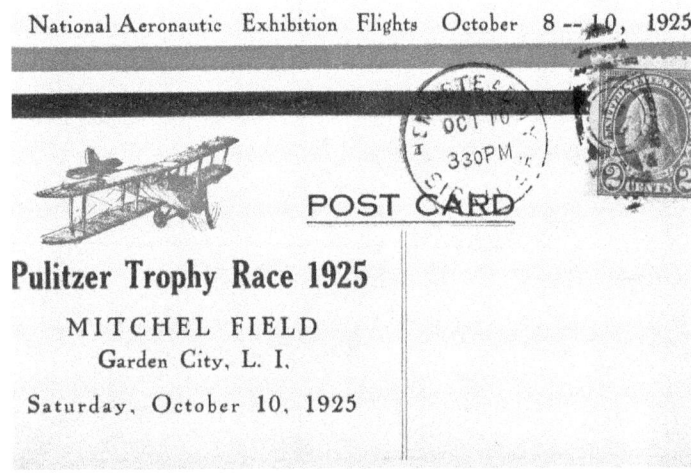

Postcard advertising the 1925 Pulitzer (courtesy Cradle of Aviation Museum, Garden City, Long Island, New York).

Moreover, racers were an uncertain business. After 1921 there had been only one customer for Pulitzer Racers—the United States government—and 1924 demonstrated the customer did not always buy. The government did pay in 1925, and pooling of army and navy money purchased three complete racers and a partial one from Curtiss. The company's racer business remained, but no one could be confident that it would continue. Criticism of air racing because of its danger and expense was evident in the comments of an *Aviation* writer reporting the government's decision to buy new racers: "[I]n spite of the better judgment of the Air Services, political and other influences induced the Secretaries of War and Navy to order the building of three new racers for this year, and participate in speed racing with experimental machines."[12]

The navy had most of the money for racers. The army had enough to comingle with the navy to purchase one airframe, only partially completed, intended for testing to destruction. navy funds purchased three complete racers, two for the navy and one for the army.[13] Pilots of both services would test the first flying version; the second would go to the navy; the third to the army. Once the army had its racer, the first one would be turned over to the navy to be fitted with pontoons and prepared for the 1925 Schneider. The other navy racer and the army racer would be flown in the Pulitzer. All four racers had navy serial numbers and received naval designations R3C-1 when flown as land planes, and R3C-2 when flown as seaplanes.

The R3Cs featured detail refinements over the R2Cs, which had set records in 1923 and '24, but the most significant change was the engine. Arthur Nutt, long associated with Curtiss engines, had designed the new engine, which fitted a larger combustion volume into a slightly lighter and slightly smaller engine. The new engine, the V-1400, weighed 660 pounds,

Army Curtiss R3C-1 and its pilot, "Cy" Bettis, Mitchel Field, Long Island, 1925. Note the small size of the R3C-1 and Bettis' "tennis shoes" and Alpine-patterned socks he wore when flying (courtesy National Archives and Records Administration).

some 10 to 40 pounds lighter than the Curtiss D-12A (reports differ). Nutt had larger cylinders bored into the engine block and increased the stroke to bring the new engine's combustion volume to almost 1,400 cubic inches (hence the name), about 200 cubic inches more than the D-12A (a 17 percent increase). Twelve V-1400s were built; the first was a "freak."

Fitted with an improved Curtiss-Reed propeller, drop-forged to the correct pitch (or twist) rather than twisted from a straight forging, the first V-1400 performed phenomenally in a bench test. It produced 619 hp, leading to press reports that the V-1400s were "620-hp engines." The 620 figure was 55 horsepower higher than the engine's nominal power rating; none of the other engines matched it, and the freak engine may never have generated such power again.[14] It also differed from the others in that it did not consume several quarts of oil in each hour's running.

The R3C's dimensions differed little from 1923's R2C (see appendix). Both had upper wings of 22-foot spans; the R3C's lower wing, at 20 feet, was seven inches greater than the R2C's. The R3C was two and a half feet shorter; 20 feet as compared to 22 feet, seven inches. Its wings had a new, very thin airfoil, called the C-80, jointly developed by the Curtiss Company and the Massachusetts Institute of Technology (MIT). The R3C's control system was an advance over any found in earlier Curtiss racers, and, likely, racers from any company. For the first time, a racer could be flown "hands off."[15] In straight and level flight, the pilot could release the controls, and the racer would continue to fly straight and level.

The R3C-1 weighed 1,792 pounds, 100 pounds more than the R2C-1; loaded, at 2181 pounds, it was 80 pounds heavier. The seaplane versions of both racers, the R2C-2s and R3C-2s, were about 600 pounds heavier than the land planes. The R3C-1's heavier weight resulted in an increase of wing loading to 15.15 pounds per square foot compared to the R2C-1's 14.64. Its more powerful engine reduced the power loading from 4.22 pounds per horsepower to 3.64.[16] *Aviation* magazine calculated the power loading at 3.75 pounds per horsepower for the R3C-1 with pilot and fuel for two-hours and compared it to the power loading of the widely used Curtiss OX engine *by itself*, with no airplane attached. Weighing 400 pounds and generating 90 horsepower, the OX engine, by itself, had a power loading of 4.5 pounds per horsepower,[17] or ¾ pound per horsepower heavier than the complete R3C, engine and airframe combined. The *Aircraft Year Book, 1926* observed that the V-1400 generated three times the horsepower from each pound of engine as the best engine of eight years earlier.[18]

Construction of the R3C followed the methods developed for the R2C, but no explanation is available for the inattention to finish, which left the R3Cs noticeably rougher than earlier Curtiss racers. The absence of a silky-smooth finish must have affected performance, but the racers exceeded their guaranteed straightaway speeds of 250 mph.[19]

Racer pilots in 1925 were required to wear parachutes. The navy R3Cs had "breakaways," metal panels in the sides of the fuselages that could be jettisoned by moving a lever on the control panel, to better the pilot's chance of being able to leave the plane during an in-flight emergency. So long as the racer was flying at a good speed, even at low altitude, it was expected that the pilot could zoom up to about 1,000 feet, spring open the breakaways, and parachute to safety. The army's racer had no breakaways. The army might have considered its pilots smaller or more agile, better able to wiggle out of the cockpit.

Not everyone was happy with the all-military domination of the Pulitzer. In 1925, "Shorty" Schroeder, the winner of the first Pulitzer Race in 1920 and the current NAA vice president, proposed that the NAA change the Pulitzer Race to an all-civilian contest. Such a move would have reduced speeds in the races, but it would have increased the number of companies preparing racers and probably increased the number of racers. No action was taken.

The NAA and armed services added a "Heat B," to the 1925 Pulitzer for operational aircraft from the army and the navy. They could not compete with the speed of the Pulitzer racers, but their addition would increase the number of airplanes involved, and head-to-head competition between army and navy aircraft would generate additional excitement, as the John L. Mitchell race for army pilots had done. More than one plane in a Heat B was a new idea, but it had a precursor in the Curtiss PW-8B (second modification) that had flown in the 1924 Pulitzer.

The army and navy used different procedures to select race pilots. The army purposely limited its pilots to one opportunity to fly in the Pulitzer and designated the winner of the previous year's Mitchell Race as the first Pulitzer pilot. In 1925, Lt. Cyrus "Cy" Bettis, winner of the 1924 Mitchell Race, flew the army's R3C-1. For his backup, the army selected Lt. James "Jimmy" H. Doolittle. The army assigned the same pilots to fly the Schneider but designated Doolittle as first pilot and Bettis as backup. The navy relied on race-tested veterans when it could. Lt. "Al" Williams, victor at St. Louis in 1923, would fly the navy's R3C-1 in the Pulitzer. Lieutenants George Cuddihy, who had set the absolute speed record for seaplanes, and Ralph A. Ofstie, who had set records for 100-, 200-, and 300-km distances in speed trials on Chesapeake Bay in 1924, would fly in the Schneider.

Cyrus Bettis is not as well-known as Doolittle and Williams (see Chapter Three). Doolittle earned a Ph.D. in aeronautical engineering at MIT in two years, flew and won many races during the 1920s and '30s, and earned glory by leading the famous "Doolittle Raid" on Japan early in World War II. He subsequently led the 8th Air Force in Europe and remained active in aviation until his death at age 95 in 1993. Williams was a widely read writer on aviation subjects before World War II and an executive in Gulf Oil Company before and after the war and flew aerobatic flights in Grumman fighters painted in Gulf's bright orange with blue and white trim colors. He died in 1958.

Bettis was born in Iowa, completed high school and some college in that state and went to work for a Michigan telephone company before volunteering as an aviation cadet in February 1918.[20] Following his commission, he flew at army fields in the United States and in the Philippines. In 1924, Bettis was stationed at Selfridge Field, Michigan, with the 1st Pursuit Group and won the John L. Mitchell Race to earn the army's starting position in the 1925 Pulitzer. Two weeks after winning the 1925 Pulitzer, he was in Baltimore as backup to Doolittle, where the two pilots won *Aviation* magazine's praise for their "businesslike, modest bearing" and mutual support.[21]

Less than a year after his Pulitzer victory on August 23, 1926, Bettis led a flight of three pursuits from Philadelphia, where they had flown exhibitions at the Sesquicentennial Celebration, back to Selfridge. They became separated when flying through heavy fog over the eastern slope of the Allegheny Mountains. His companions flew onto Selfridge, but Bettis did not show up. His companions flew back over their course to search, joined by two other airplanes from Philadelphia.[22] The searches proved fruitless. In the fog, Bettis had crashed into a tree on a mountainside. After a period of unconsciousness, he awoke to find his left leg broken and his jaw broken in two places. He remained with his plane overnight, and the next day painfully crawled more than two miles (some accounts say six miles) to a road, where he collapsed. Road workers found him the next morning and took him to the Bellefonte, Pennsylvania, hospital, where he was treated. Ira Eaker, an Air Service pilot who was to go on to fame in World War II, flew him in an army ambulance plane from Bellefonte to Walter Reed Hospital in Washington. After a period of improvement, Bettis developed meningitis and died on September 1, 1926.[23]

Bettis's short, event-filled life provides a story for those interested in good luck charms and talismans. After his victory in the Pulitzer, someone asked him about a small, quarter-size, blue and gold pin on his flying helmet. "[H]e laughed and said it had won the race for him. It was a pin he found in the sand at Long Beach a few days ago, he said, and which he pinned on his cap to bring him luck."[24] Bill Wait remembered that Bettis had "attached it to the front of his leather flying helmet and wore it all during the meet and every time he flew after that."[25] After Bettis crashed, Wait asked one of Betttis's companions what had happened to the pin. "His friend said he didn't know, because somebody had stolen it off Cy's cap the night before they began their return flight."[26]

Doolittle and Williams were on hand at Curtiss' Garden City, Long Island, factory when the company rolled out the first R3C, powered by the freak engine, on September 11. Defects arose. On Williams' first flight, the streamline on one wheel cracked; the other blew off. Quickly repaired, the racer was ready for Doolittle. He climbed to about 200 feet and sped across the measured one-km course at nearly 300 mph. As he eased off on the speed, he felt the left wings grow heavy and saw the upper left wingtip collapsing. He landed, and repairs were made.

A week after the first flights, on September 18, Doolittle was clocked at 254 mph on a circuit of the 50-km Pulitzer Race course from Mitchel Field, about 10 mph faster than the R2C's winning speed at St. Louis. On the same day, Williams and Doolittle both posted unofficial world records. Williams flew 302.3 mph over a one-km course.[27] Doolittle averaged 285 mph on upwind and downwind flights. Two days later, on September 20, Bettis flew the first R3C-1 around the Pulitzer course at more than 260 mph,[28] at least 16 mph faster than Williams' 244 mph in the 1923 Pulitzer.

On the same day, the Dutch Fokker company, which was beginning to sell passenger airplanes in the United States, offered a different attraction at Mitchel. Nearby motorists stopped and joined everyone on or near the field to watch "eight chorus girls from a Broadway show" climb down from a single-engine Fokker passenger airplane, look at the new racer, climb back on the Fokker, and fly back to nearby Curtiss Field from where they had come.[29]

Bettis was flying the first R3C on September 24, and a routine flight became dramatic when the spinner broke loose from the propeller, flew back along the fuselage, cut through two sections of the wing radiator, bounced off the interwing strut, and smashed into the stabilizer. Bettis landed the racer without further damage. It was repaired and shipped to Baltimore, put on pontoons, and readied for racing as a navy entry in the Schneider.[30]

The 278-mph absolute speed record held by France seemed within reach of the United States. Buoyed with predictions of 300 mph on the straightaways, "Army, Navy, and civilian aviators are confident that by the evening of October 10 [the Pulitzer's scheduled date] the speed record of airplanes will be wrested from France and once more be held by America."[31]

As race week approached, newspapers used "Pulitzer races" to refer to all the competitions of the air meet. The change in nomenclature made no practical difference, but it underlined the significance of the Pulitzer. Race organizers predicted 500,000 spectators, and arrangements were made to park 50,000 cars near, but outside, the Long Island airfields. The Long Island Railroad planned special trains and added extra cars to its regular trains. Parking places for visiting airplanes were arranged at neighboring airfields. In the days before the 1925 National Air Race, the army staged daily flights and exhibitions over New York City to stimulate interest in the races. On the afternoon of October 7, Bettis and Doolittle in PW-8s, "which can do nearly 200 miles an hour," dived and looped, side-slipped and climbed over thousands of upturned faces in Manhattan. One of the pilots "swooped" below the tops of buildings around Times Square, "then with a rush he swept up again."[32] The two pilots raced to the Statue of Liberty, turned, and flew back to Mitchel.[33]

That night, three army airplanes, rigged with flashing colored electric lights, flew below a cloud layer on a hazy evening, dropping fireworks that burst in sparks and colors as they drifted to the ground. Lt. Eugene Barksdale climbed upward in a much-faster pursuit until the navigation lights were almost invisible. From 12,000 feet, he dived to about 1,000 feet, dropped magnesium flares and flew several stunts.[34] Surely by the time Barksdale departed to return to Mitchel few people in Manhattan had failed to look upward to see airplanes that day or night. How many of them then made up their minds to go to the "Pulitzer races"? Thursday and Friday were workdays, and people with jobs might have a hard time getting away. But Saturday, the day of the Pulitzer would be easier.

Would the soft fall weather persist? Would it be good for flying all three days of the meet? The answer on Thursday morning was yes. Day one started well with "brilliant blue sky dusted by a few fleecy clouds," but, balmy weather aside, Mitchel Field was not an inviting, pleasant place. "[W]herever one turned there was a sentry with fixed bayonet to repel those not dignified by an official badge,"[35] and infantrymen's bayoneted rifles and their lieutenants' swords rigorously controlled people's movements around the race site. Spectators without precious armbands that showed they could be permitted to enter particular places on the field were herded into areas where the soldiers could contain them.

Members of the British Schneider Race Team had journeyed to Mitchel Field from Baltimore, where they were preparing for the upcoming Schneider. *Flight* reported, "The United States Army—complete with fixed bayonets was obdurate, and for a time it looked as if the British had made a fruitless journey."[36] But the British visitors found an army officer who made everything "plain sailing." Not everyone was so fortunate in finding a helpful officer. The NAA blamed the army for unneeded security, the army blamed the NAA. Race officials announced that security would loosen up on days two and three sounded hollow. "Cy" Caldwell, an amusing, sometimes caustic writer for *Aviation*, summed up the security arrangements:

> Soldiers on guard at Mitchel Field hadn't been given an opportunity to stick bayonets into anyone since 1918 and ... would be bitterly disappointed if the crowd didn't come round to get stuck. So, to make everyone happy, the events were held at Mitchel amid glistening bayonets and harsh commands.
>
> Admiral Moffett and Lieutenant Maughan had no arm bands and were driven away by guards when they tried to look at the planes.... I'll say this for the soldiers, they certainly played no favorites, and before the day was half over had us so cowed that whenever a solider approached we lay flat and uttered low moaning sounds to placate him.[37]

Day one at Mitchel, like day three at Dayton, had its disaster. A soon-to-be-famous pilot crashed and his passenger was killed in the Glenn H. Curtiss Trophy Race [Table 6] for light commercial aircraft. In that race, the "low-speed civilian machines ... drifted by in what seemed to be a perfectly safe contest"; but within minutes after the race started, the judges learned that a Bellanca, piloted by Clarence Chamberlain, had crashed, injuring him and killing his passenger, Lawrence Burnelli.[38]

Chamberlain, who had flown in the Air Service during World War I, purchased his airplane, built as a biplane in 1918 and left outside for long periods of time, in the spring of 1925. He converted it into a low-wing monoplane, braced with struts from the fuselage to the tops of the wings, and installed a more powerful engine. Race officials had asked the pilots to carry no passengers but to carry ballast equivalent to a passenger's weight. Late to start the race, Chamberlain was ready to begin taxiing to the start line when Lawrence Burnelli rushed out and climbed into the passenger's seat. (The *Times* prints Burnelli's name as "Buranelli" and provides no information about whether Chamberlain carried ballast as well.)

At a height of about 200 feet, Chamberlain's airplane "began to wabble ... the machine nosed down sideways, and a wing hit a telegraph wire.... After a short wavering course, it crashed.... Burranelli was thrown fifty feet[39] and his head crushed."[40] Burnelli's brother, Vincent Burnelli, had designed the "widebody" Remington-Burnelli RB-2 cargo carrier with two engines mounted at the front of its spacious fuselage, which could accommodate a small truck. It was a favorite exhibit and elicited many comments at Mitchel Field, but it turned out to be a design dead-end.

In its commentary about day one, the *New York Times* said, "Aside from this accident the day was all that those who watched could have wished for," which rivals the apocryphal question "Other than that, Mrs. Lincoln, how did you enjoy the play?" The *Times* writer, despite having all he could have wished for, dismissed the Curtiss Trophy Race (see Table 6) because it "produced little of interest."[41]

Clarence Chamberlain recovered from his injuries. He and Bert Acosta set an endurance record of 56 hours, 11 minutes circling over New York City in 1927, and later that year, soon after Lindbergh's flight, he flew across the Atlantic Ocean headed for Rome. Charles Levine, sometimes called a passenger, sometimes a navigator despite knowing little about navigation, and sometimes taking the controls for short periods of time despite having no piloting experience, accompanied him. Over the English Channel, winds blew them northward, and they decided to fly to Berlin. Lost in fog and running out of fuel, they landed safely in Germany and returned to America as celebrities.

TABLE 6
1925: Closed-Course Competitive Events at the Fourth National Air Race, Sixth Pulitzer Race, Mitchel Field, Long Island, New York, October 8–13[a]

(Weather forced postponements. "Planned Flying Days" were October 8, 9, 10, Thursday, Friday, and Saturday. Thursday, October 8, went as planned. One event only was flown on Friday, October 9. No races were held on October 10 and 11. On Monday, October 12, some events originally scheduled for Friday, October 9, and the John L. Mitchell and Pulitzer races, originally scheduled for Saturday, October 10, were flown. The remaining events from October 9 and 10 were flown on Tuesday, October 14.)

Flying Day One: Thursday, October 8

Event	Entrants	Course	Pilot, Airplane (hp) Speed[b]
Glenn H. Curtiss Trophy Race. Similar to *National*[d] *Cash Register Company Trophy Race, 1924.*	Two-seat, low-horsepower airplanes (engines no larger than 520 cu. in. displacement, ca. 220 hp)	8 laps of 12-mile course (96 miles)[c]	1. Basil Rowe, Thomas-Morse S4E, 103 mph 2. Bert Lott, Thomas-Morse S4E 3. Clyde Emrick, Hartzell FC2
Merchants Association of New York Race. Similar to *Central Labor Union Trophy Race, 1924.*	Two-, three-, and four-seat civilian airplanes with engines less than 800 cu. in. displacement	8 laps of 12-mile course	1. Casey Jones, Curtiss Oriole, 134 mph 2. Fred Becker, Wright-Bellanca WB-1. 121.8 mph 3. Basil Rowe, S.V.A. 121.5 mph
Liberty Engine Builders' Trophy Race	Two-seat military observation aircraft. 11 of 16 entrants were DH4s powered by 400-hp Liberty engines.	12 laps of 15-mile course for total of 180 miles	1. Capt. Henri Lemaitre, French Air Force, Breguet 19, (464), 129 mph 2. Lt. Gene Bayley, USAS, Douglas XO-2, (400), 128 mph 3. Lt. George Henderson, USN, Boeing O2B-1, (400), Navy version of DH4

Flying Day Two: Friday, October 9

Three events scheduled. Only one completed because of bad weather. Other two rescheduled to Monday, October 12.

Event	Entrants	Course	Pilot, Airplane (hp) Speed[b]
Aviation Town and Country Club of Detroit Trophy Race similar to *1924 race of same name.*	Civilian pilots/engine of less than 800 cubic in. displacement, 340 lb. load, and a minimum speed of 80 mph	8 laps of 12-mile course for total of 96 miles. Two contests, speed and efficiency. Efficiency = payload lbs × mph / hp	Speed. 1. Casey Jones, Curtiss Oriole, 128 mph 2. Basil Rowe, C.6-SVA, 120 mph 3. Fred Becker, Wright-Bellanca WB-1, 112 mph Efficiency: 1. Fred Becker, Wright-Bellanca WB-1, 602 points 2. Casey Jones, Curtiss Oriole, 394.3 points 3. Dick Depew, Sikorsky S-31A

Misty, lowering skies followed by steady rain caused judges to suspend Friday events. Near-hurricane force winds caused postponement of Saturday program.

Scheduled Flying Day Three, Saturday, October 10. Bad weather forced postponement of all events. Saturday's planned events moved to Monday, October 13. Events that had previously been rescheduled from Friday to Monday were moved again to Tuesday.

Flying Day Three: Monday, October 12, Events Originally Scheduled for Friday and Saturday, October 9 and 10

Event	Entrants	Course	Pilot, Airplane (hp) Speed[b]
Aero Digest Trophy Race. Similar to *Dayton Bicycle Club and Engineers' Club Trophy Race*, 1924.	Light planes. All but Powell had motorcycle engines of 18 or 19 hp. Powell, Bristol "Cherub" of 17 (or 22) hp (sources differ) 150 lb. load	Speed contest of 10 laps of 5-mile course.	1. Jerry Dack, Powell Racer 2. Clyde Emrick, Driggs-Johnson DJ-1 3. Harvery Mummert, Mummert Sportplane
Scientific American Trophy Race. Similar to *Dayton Bicycle Club and Engineers' Club Trophy Race*, 1924	Light planes. All but one had motorcycle engines of 18 or 19 hp. 150 lb. load. Powell Racer had Bristol "Cherub" engine.	Efficiency contest of 10 laps of 5-mile course. To be considered for a prize, the airplane had to finish in top four in Aero Digest Trophy Race. Efficiency calculated: *payload lbs × mph / hp*	1. Jerry Dack, Powell Racer. 76.1 mph 2. Clyde Emrick, Driggs-Johnson DJ-1. 62.6 mph 3. E. Dormoy, Dormoy, 57.2 mph (No record of efficiency points found)
John L. Mitchell Trophy Race	Ten pilots of 1st Pursuit Group flying Curtiss PW-8 pursuits	10 laps of 12-mile course for total of 120 miles.	1. Lt. Tom Matthews, 161.1 mph 2. Lt. George Schulgen, 158.7 mph 3. Lt Al Lyon. *1923 Winner*, "*Cy*" *Bettis, USAS, PW-8,* 175 mph
Pulitzer Trophy Race, Heat 1	All 470 hp airplanes w/ landing speed 75 mph or less and top speed greater than 170.	4 laps of 50-km course. Both pilots flew Curtiss R3C-1s, 565 hp.	1. Lt. Cyrus Bettis, USAS, Curtiss R3C-1, 249 mph 2. Lt. A.J. "Al" Williams, USN, Curtiss R3C-1, 242 mph *1924 Winner, H. Mills, USAS, Rerville R-3,* 217 mph
Pulitzer Trophy Race, Heat 2	Operational USAS single-seaters. USAS loaned pursuits to Navy for race. All 470 hp.	4 laps of 50-km course	1. (3rd overall in Pulitzer). Lt. L.H. Dawson, USAS, Curtiss P-1, 170 mph 2. (4th overall in Pulitzer) Lt. H.J. Norton, USMC, Curtiss PW-8, 169 mph[e] 3. (5th overall in Pulitzer). Capt. H.W. Cook, USAS. Curtiss PW-8, 167 mph

Flying Day Four: Tuesday, October 13, Events Originally Scheduled for Friday, October 9.

Event	Entrants	Contest	Pilot, Airplane, Speed
Dayton Daily News Trophy Race. Similar to race of same name, 1924.	Light planes. All but one had motorcycle engines of 18 or 19 hp. Powell racer had Bristol "Cherub" engine.	10 laps of 5-mile course. All other races required flight below 400 feet. This contest required passing pylons 1 and 3 at 60 ft and climbing to 500 ft to round balloon at pylon 2.	1. John Faucher, Powell Racer, 71.2 mph 2. Clyde Emrick, Johnson "Bumble Bee," 64.8 3. E. Dormoy, Dormoy airplane, 52.9 mph

Event	Entrants	Contest	Pilot, Airplane, Speed
Detroit News Trophy Race. Similar to *Dayton Chamber of Commerce Trophy Race*, 1924.	Fourteen large capacity airplanes, capable of 85 mph and lifting 2,000 lbs.	10 laps of 12-mile course for total of 120 miles	1. Lt. E.E. Harmon, USAS, Huff-Daland XLB-1, 119.9 mph 2. Lt. C.F. Schilt, USMC, Douglas DT-4, 118.2 mph 3. Lt. Kenneth Wolfe, USAS, Epicyclic C-1 (modified Douglas C-1), 113.4 mph

[a]Sources: The 1926 *Aircraft Year Book* did not, as had earlier year books, publish detailed results from the 1925 National Air Race. In its place, data were obtained from "The Events of the New York Air Races," *Aviation*, October 19, 1925, pp. 533–543; T.G. Foxworth, *The Speed Seekers* (New York: Doubleday, 1974] p. 459; "Instructions for Judges and Scorers, October 8th–9th–10th 1925," photocopied typescript from Air Force Museum; "Pulitzer Race Won at 249-Mile Speed; Disappoints Fliers," *New York Times*, October 13, 1925, p. 1; "Results of Yesterday's Air Races at Mitchel Field," *New York Times*, October 13, 1925, p. 2.

[b]Horsepower ratings and speeds are missing for some entrants because no source was found for that information.

[c]Some sources report that this race and others were 100 miles long. There was no 10-mile course, making a 100-mile race impossible unless it was flown as 20 laps of the 5-mile course. There is no mention of any 20-lap races.

[d]Races in italics were similar to or same as indicated 1924 races.

[e]This airplane had won the Mitchell Race. It was loaned to the navy for the Pulitzer Heat 2.

Remington-Burnelli RB-1, 1925 (1:72 scale model) (built by Gabriel Stern and photographed by him at http://www.internetmodeler.com/2008/february/aviation/burn.php, reproduced with Mr. Stern's permission).

Remington-Burnelli RB-1 in flight, 1925 (courtesy National Archives and Records Administration).

National Air Race organizers, who had made strenuous efforts every year to attract foreign entrants, achieved their goal in 1925, but not for the Pulitzer. The French airplane builder Jacques Breguet had informed the organizers that his company would like to enter two airplanes in the Liberty Engine Builders Race for observation planes, but his entry ran afoul because of objections to civilian entries in what was a military contest. The French War Department circumvented that problem and entered two Breguet XIXs.[42]

A total of sixteen airplanes were entered in the Liberty Engine Builders Trophy Race. The U.S. Army entered nine DHs, a CO6, which was a modified DH-4 powered by an inverted 400-hp Liberty V-12 engine, and a Douglas XO2 powered by a Liberty engine and credited with a speed of 151 mph. The U.S. Navy entered three DH-type airplanes, two being O2B1s built by the Boeing Company. The 14 United States entries and the two French Breguet XIXs made the Liberty Race "the most interesting event of the day."[43] The airplanes flew in three heats "running simultaneously with a few moments interval between their starting times."[44]

Captain Pelletier d'Oisy, piloting the Breguet XIX powered by a 325-hp engine, Lt. George Henderson, USN, in a Boeing O2B1 (400-hp Liberty engine), and Capt. Henri Lemaitre, flying the Breguet powered by a 464-hp engine, led the field in that order for eight laps of the ten-lap race. Then Lemaitre opened the throttle, and "the sensation of the meet"[45] won.

But the speeds disappointed. Lemaitre won at 129.1 mph, far below the Douglas O2's reported speed of 151 mph. Also puzzling, Henderson's Boeing-built O2B1 posted the best time for any DH-type aircraft, 127.9 mph, which was about 3 mph slower than the winning DH at Dayton in 1924. In the same issue where it called the Liberty Engine Builders Race the "most interesting event of the day," *Aviation* concluded, "The general slipping back for

Breguet XIXs, Mitchel Field, Long Island, 1925 (courtesy National Archives and Records Administration).

two successive years should be given the careful attention it deserves."⁴⁶ No suggestion was made about what the "attention" could be. Cy Caldwell discussed the race more chauvinistically and straightforwardly. "How it comes that a French bomber is faster than an American Observation plane is a very technical question that is too deep for me."⁴⁷

Few people saw the day one races. Attendance was "not more than 5,000 ... but as it was the first day, those in charge of the races were not disappointed." It is difficult to imagine that there was no disappointment or, at least, concern. How likely did it seem that 495,000 people, necessary to reach the projected 500,000, would stream into the races on days two and three when only 5,000 trickled in on day one? Furthermore, few of the invited dignitaries and high foreign and government officials "were in the [VIP] boxes ... apparently deferring their visit to the field until today [Friday] and Saturday."⁴⁸

Fred Marshall, writing about the disappointing Dayton race, concluded that "the average spectator wants excitement, thrills, unusual stunts."⁴⁹ A *Times* reporter, writing about day one, agreed: "There was nothing at the meet which held attention as did these circus tricks [parachute jumps, formation flying and aerobatics by the 1st Pursuit Group] although they have nothing to do with racing." In part, the reporter's comments reflected the fact that "what went on during the races was somewhat difficult to ascertain." Airplanes were sometimes out of sight or hard to see, and the spectators were informed only by "information coming spasmodically from an announcer on the judges' stand."⁵⁰

Day one had had its problems, but Admiral Moffett, looking at rows of parked planes while a few flew overhead, said, "This is a wonderful sight.... It should show the people just how efficient the service is ... and make them proud of their aviators."⁵¹ Even so, surely the admiral and everyone else involved in the air meet hoped for good weather, better attendance, and fewer bayonets on days two and three. Day two started poorly and ended prematurely. The Detroit Town and Country Club Race had two prizes, one for speed and the other for efficiency in load-carrying at speed. Blustery winds caused the judges to delay its start, but the race, once started, was flown without incident. Casey Jones finished first. His mechanic questioned how his yellow Curtiss Oriole could manage the 128 mph necessary for him to have completed the eight laps of the course in the time he took. He wondered if Jones had miscounted and flown only seven instead of eight laps. Jones checked the chalk marks he made for each lap he flew. There were eight, the right number, but he remained uncertain. He took off and flew another lap (which, it turned out, was unnecessary). Edward Becker, in a Wright-Bellanca, won the efficiency competition, as had the Bellanca in 1923. His margin of victory, resembling 1923 again, was surprisingly large. He earned 602 points. Jones, second, managed less than half that, 295 points.

After the Detroit Town and Country Club Race, the weather worsened, intermittent rain turned steady, and the judges postponed the remaining day two events. The judges moved the postponed events to Monday, October 12, and planned to fly the day three events, including the Pulitzer and Mitchell races, on Saturday as planned. It all made sense: the Pulitzer was the biggest drawing card, many people had already written the 10th on their calendars and attending on a Saturday was easier for most people than getting there on a workday. Sunday was not a possibility as a race day. Neither the army nor the navy would risk the public censure it had earlier borne after staging public contests on Sunday.

The weather worsened. Friday night's was worse than Friday's during the day. Saturday's was worse still: "City's Coldest Oct. 10 Brings 72-Mile Gale; Plane Races Put Off."[52]

C.S. "Casey" Jones of the Curtiss Aeroplane and Motor Corporation rounds a pylon in his Curtiss Oriole (courtesy National Archives and Records Administration).

Clouds of dust and paper whirled over the field, scoreboards were blown down, one plane in landing ... was smashed, and the few women who ventured out to the field were frequently blown before the wind, desperately clutching their hats. One was bowled over.

The 65 mile per hour gale whipped pebbles off the ground, which stung ankles and filled every one's eyes with grit.

One pylon near Hicksville was blown over....

The Ford all-metal plane brought here for the John Wanamaker store, which is being dismantled preparatory to delivery on a Ford truck which it carried here from Detroit, was damaged also. When mechanics arrived at the field yesterday [Saturday] morning ... they could not find the wings. After an hour's search they discovered them 200 yards away where they had been carried by the wind. They were undamaged.[53]

Saturday, mechanics, pilots, soldiers ordered to the job, and willing hangers-on and would-be spectators fought the wind and packed airplanes that fit into hangars and pegged down others, tail to the wind, in the lee of buildings.

The weather sometimes got the upper hand. "Gale Damages 10 Planes"[54] headlined an article that detailed the overturning of eight DH-4s with damage to their wings. Another airplane was tossed into the air and hung up on a ten-foot-high sign. Others, not overturned, also suffered wing damage. Four Martin bombers, flying from Baltimore, reported flying around and through snowstorms between Philadelphia and Mitchel Field. The Martins and all the other airplanes en route to Mitchel Field landed safely throughout the windy, wintry day except one. That one, an army DH-4, turned turtle but the crew escaped unharmed. The Racing Committee had no choice but to postpone and reschedule the races for the second time. Weather had made Saturday impossible and public relations problems did the same for Sunday, but Monday was fine in terms of possible attendance. It was Columbus Day, a holiday for many people.

The Saturday schedule, with the Pulitzer and Mitchell, was moved to Monday, along with some day two events. The move to Monday, because of its being a holiday, must have been accompanied with hope for a big crowd. Competitors in events originally scheduled for day two, Friday — which had been rescheduled once to Monday and were moved again to Tuesday, October 13, which was not a holiday — saw any hope for decent attendance disappearing. "It was officially explained at Mitchel Field that the day's [Saturday] program of air racing was called off because of the possible injury to civilian planes and not because of any fear over the winged bullets of the army and navy."[55] The official explainers had not talked to Al Williams, who said, "To run the Pulitzer racers [in Saturday's wind] would have resulted in a very unsatisfactory performance ... and would have been hazardous." He also wrote that "'Bumps' [in the air] are bad enough when one is in an ordinary ship, but in a racing ship — buffering is tremendous ... extremely grueling ... and hazardous."[56] Williams made his concerns personal when, pointing to the turned-turtle DH-4, he asked what chance a light, small wingspan, fast-landing racer had to land safely if the DH-4, a stable, sturdy service airplane could not do it. Not much, was the answer, "'And I don't want to crack up this baby,' he [Williams] added, tapping his chest."[57] Al Williams was a skilled, courageous pilot. If he deemed conditions unsatisfactory, they surely were.

On Sunday afternoon, the weather turned fine, and Bettis and Williams took their racers up. "Thousands of people were thrilled as the two diminutive airplanes, almost resembling toy airplanes when viewed at 1,000 feet or more, turned, dived, side-slipped and then whizzed off."[58] Observant spectators noted that Williams flew aggressively, with wide-open throttle. Bettis babied his engine in its first flight. On the ground and close-up, the two racers were identifiable by their paint schemes and race numbers, 43 on Bettis' machine and 40 on Williams'. The fuselage of the army R3C-1 and its fin/rudder were black; its wings and stabilizer/elevators gold. The navy's had light blue fuselage and vertical tail surfaces, silver wings and horizontal tail surfaces. At a distance and against bright skies, the two planes must have been impossible to tell apart.

Little remained of the snowy, rainy, windy weather on Monday, October 12, but what lingered influenced racing; "Bettis had no complaints to make about anything except the haze, which he said made it difficult to keep the course at high speed."[59] Some negative aspects of the races, unlike the weather, were under human control, and much to everyone's relief, Lt. Col. Benjamin Foulois[60] had ordered the soldiers to put away their bayonets and swords. "[T]here was much greater courtesy than before from everyone connected with the affair."[61]

The projected half-million spectators proved more illusionary than Dayton's quarter-million the year before. Only 25,000 people paid admission on Monday, Pulitzer Race day. The *New York Times* estimated another 10,000 people looked on from cars around the airfield. Discounting the price of box seats from $6.00 to $2.00 after the gates were opened to the airfield produced few additional sales. Later in the day, at the start of the Pulitzer Race, the public address system urged spectators to spend a dollar and fill the box seats. Most box seats remained unfilled. About 5,000 spectators had attended day one on Thursday and perhaps that number paid admission on Friday only to be sent home after one event. Those 10,000 plus Monday's 25,000 brought the total to 35,000. *Aviation* magazine estimated that fewer than 10,000 people in all saw the civilian races. Even if 10,000 paid on Tuesday, the total attendance of 45,000 was less than one-tenth the hoped-for half-million.

Day two events, which had been moved to Monday, October 12, began with two contests for light planes. University of Detroit professor of aeronautics C.H. Powell's tiny Powell Racer—wingspan, 15 feet, 9 inches; length 15 feet; empty weight 310 pounds—powered by a 16.7-hp Bristol Cherub engine, looked exactly like a regular biplane except for its small size. *Flight* magazine commented on "the almost ridiculously small size of the machine.... The machine could stand comfortably in a fairly small living room.... The question of garaging is a simple one."[62] With a top speed of 85 mph and landing speed of 50, it won all three contests for light planes—the *Aero Digest* Trophy, the *Scientific American* Trophy, and the *Dayton Daily News* Trophy. In total, Dr. Powell's airplane won $2,000, a good haul, equal to the $2,000 awarded to the winner of the Pulitzer.

In his novel *Pylon*, about air racing, William Faulkner described an air race seen at a distance: "The noise was faint now and disseminated; the drowsy afternoon was domed with it and the four machines seemed to hover like dragonflies."[63] Hovering "like dragonflies" is more applicable to the low-powered Powell Racer and its competitors, the airplanes in the 1925 Glenn H. Curtiss Race, and other civilian racers than the snarling Curtiss PW-8s in the Mitchell Race and the R3Cs in the Pulitzer.

Whatever the speed of a racer, flying it involved more than simply climbing into the plane, running up the engine, taking off, and flying as fast as possible around the course. Successful pilots made decisions about how to "turn the pylons" and how high to fly before they climbed off the airfield, and they were ready to alter decisions as they flew the course. Many photos and, probably, most drawings and paintings of 1920s and '30s air races feature near-vertically banked racers turning the pylons, and movies catch the visually and audibly exciting turns as representative of air races. The ten equally matched airplanes in the 1925 Mitchell Race turning pylons in a closely packed gaggle has to have been among the most thrilling sights in all of racing. Surprisingly, perhaps, not all race pilots turned pylons the

Powell Racer, Mitchel Field, Long Island, 1925. The small size of this airplane is apparent (courtesy National Archives and Records Administration).

same way. Some pilots made their turns in near-level flight, maintaining their altitude as they went around the pylons. Others climbed as they turned, and some dived around.

Most navy pilots maintained near level flight, but the army pilots banked sharply, staying close to the pylons but losing speed, while the navy pilots swept wider arcs and maintained higher speeds. Casey Jones climbed through his turns "swooping up as he cuts close to the pylon and then sliding down again toward the ground for speed, a maneuver that catches the eye."[64] Italian race pilots who would fly in the 1925 Schneider favored a nearly reciprocal approach, as did Curtiss engineer "Bill" Wait: "Our [Curtiss] planes tended to climb on the straightaway and you had to hold them down [by holding the elevator slightly down], which caused drag. I thought that if you let the plane rise to the maximum height permissible under the rules and then dove at the pylons, you could probably save some time."[65] Wait decided to discuss his ideas with the Pulitzer pilots. Al Williams, whom he had met before, brushed aside his suggestion, assuring Wait that he knew how to fly races. Wait had not met Bettis or Doolittle but knew that Doolittle had been at MIT and had "quite a reputation as an engineer." Bettis and Doolittle listened.[66] Two weeks later, Doolittle[67] won the Schneider, diving around the pylons.[68]

The straight lines that connect pylons on diagrams of racecourses do not at all represent the courses pilots flew. A racer's momentum coming off a turn carried it far outside a straight line between pylons: "If anybody wants to be directly underneath us ... he'd better get about 700 feet outside the line of the pylons."[69] Pulitzer race rules required pilots to fly no higher than 300 feet. Wind conditions could vary between 300 feet and "on the deck," and race pilots often flew downwind and upwind legs at different altitudes. When close together, prop wash and slipstreams from nearby racers caused the racers of the 1920s to fly in the bouncing fashion visible in many movies. Spectators had plenty to watch.

The John L. Mitchell Race, "by far the most spectacular feature of the day" and "a thoroughly exciting contest,"[70] provided opportunities for pilots of the 1st Pursuit Group to demonstrate their racing skills and competitive spirits. Because they are mentioned only

Top: Curtiss PW-8s from Selfridge Field, Michigan, at Camp Skeel, mounted on skis for winter maneuvers, winter of 1924-25 (courtesy National Archives and Records Administration). *Bottom:* Curtiss PW-8s and two P-1s (fifth, sixth and seventh from right) at Mitchel Field, 1925. Their "tunnel" radiators beneath their noses make the P-1s identifiable (courtesy National Archives and Records Administration).

in the context of the Mitchell Race in this book, it is possible to forget or ignore that the 1st Pursuit Group's pilots and pursuits participated in other Air Service duties. The only PW-8-equipped unit in the Air Service, it was frequently involved in the development of new operational procedures. For example, in the winter of 1924-25, some of the 1st Pursuit Group's PW-8s participated in winter maneuvers at Camp Skeel, where they were mounted on skis. It can be assumed that one or more of those ski-mounted PW-8s, with its wheeled landing gear reinstalled, flew in the 1925 Mitchell.

At 11:30 on Monday, Columbus Day, October 12, delayed by a half hour because of wind, ten pilots of the 1st Pursuit Group in PW-8s,[71] took to the air to fly the John L. Mitchell Race climbed, and circled to be certain that engines were ticking and the airplanes were ready. At 15-second intervals, they flashed across the start line. The closely staggered starts, almost identical speeds, and small differences in piloting abilities bunched the pursuits together, wings level on the straightaways and banked nearly vertically around the pylons. Excellent piloting was on display as slightly faster racers snaked among, and passed close to, the outside of the slower airplanes.

Race numbers—48 through 55—painted on the sides of the PW-8s made it possible for spectators to keep track of which pilots were gaining and which were falling behind. They saw Lt. T.K. Matthews start from fifth place, post the best time, and win. Equally dramatic, Lt. G.F. Schulgen moved from last to second place weaving from close on the outside of the bunched-together, slightly slower airplanes to pass between some planes that were flying a bit faster. (As in all air races, passing on the inside, between overtaken racer and the pylons, was forbidden.) The winning speed, 161.7 mph, was about fourteen miles per hour slower than Bettis' 175.4 in the 1924 Mitchell.[72] Three factors may have contributed to the slower speeds. The 1925 Mitchell course was 12 miles around rather than 50 km (about 31 miles) as in 1924, reducing the lengths of the straightaways and increasing the number of speed-robbing turns. The prohibition against diving starts reduced initial speed and average speeds. And finally, there was the wind.

No contest between two airplanes—such as the Pulitzer would be—could match the excitement of the Mitchell Race with ten planes jockeying for position, maneuvering among and around each other, passing and being passed, banking for the fastest, closest pylon turns.

As it turned out, the Pulitzer was a poorer show than expected. It would not look like a race. With flame bursting and noise blurting from their exhausts, the R3C-1s taxied to the takeoff line. Much was expected from the racers and the pilots. Williams' 302.3 mph had been widely published, and Bettis had predicted a Pulitzer-winning speed of more than 270 mph, 25 mph faster than the 1923 record.

Williams took off first, and Bettis followed about two minutes later. Both circled to warm up their engines, and Williams flew to a corner of the field, accelerated, and got a running start across the start line at 3:15 P.M. He sped away at over 240 mph, visibly faster than the PW-8s had flown in the Mitchell. Bettis followed Williams across the start line but far enough behind him that there was no chance for him to overtake Williams or to fly a neck-and-neck race. Instead the Pulitzer looked like two time trials as the black and gold racer followed the blue and silver one to disappear to in the southeast. Pilots, ground crews, race officials, and spectators who had seen and heard earlier Reed-propeller-equipped racers fly were aware of an unexpected dividend of the switch from twisted to forged aluminum propellers. The R3C-1s were noticeably quieter.

When Bettis followed Williams out of sight to the southeast, spectators may have looked elsewhere for other sights to see, but they continually glanced back to look to the northeast to spot the racers' reappearance. Hundreds shouted, "There he comes" or "There he is" when the speck that became recognizable as Williams' plane appeared. His racer flashed in front of the grandstand and "took a long sweeping turn around the pylon, a turn far different from the quick bank made by the pursuit planes of lower speed, for it was not possible to snap around a turn at the rate he was going." Bettis, turning closer to the pylon, following what might have been Bill Wait's suggestion, dived around the pylons, "dropping toward the pylon with a hum that rose in swift crescendo to a shriek as the flier rounded the turn."[73] After the race, Bettis said, "The turns are important, but to win you have to have a combination of everything in your favor."[74] Bettis had the combination.

Despite the racers' being out of the crowd's sight for much of the race and the separation between the two racers, keen-eyed observers saw the distance between Williams' airplane and Bettis' shorten on each lap. The lap times, posted on a scoreboard after each lap, confirmed Bettis' better speed. His speed increased between laps one and two, fell back a bit on lap three, and was fastest on four. Williams' speed, on the other hand, dropped with each lap; his fourth was 16 mph slower than that of Bettis. Fingers were pointed at Williams for having run his engine at full speed on the day before the race, while Bettis had kept his engine throttled back. Whether this happened and whether it made any difference in speeds is unknown. The V-1400 engines were not without fault. V-1400s powered all three R3C-2 racers in the Schneider Race, held two weeks after the Pulitzer. Only the army's completed the race; both navy engines, including the 619-hp, non–oil-using freak, broke down. It is entirely possible that Williams' engine had defects not of his making.

TABLE 7
1925. Lap Speeds in Sixth Pulitzer Race

Heat One

Pilot	airplane	1st lap[a]	2nd lap	3rd lap	4th lap	200-km avg.
Lt. Cy Bettis, USAS	R3C-1	247.8	249.4	248.7	249.97	248.99
Lt. Al Williams, USN	R3C-1	243.7	242.1	241.4	234.66	241.71

Heat Two

Lt. Leo Dawson, USAS	P-1	170.3	170.0	169.5	169.8	169.9
Lt. Harmon Norton, USMC	PW-8	169.3	168.9	163.4	173.8	168.8
Capt. Harvey Cook, USAS	PW-8	167.0	167.3	167.4	167.8	167.4
Lt. George Cuddihy, USN	PW-8	165.2	165.8	165.0	out	DNF[b]

[a]All speeds in mph.
[b]Did Not Finish.
Source: Table "Results of Yesterday's Air Races at Mitchel Field," *New York Times*, October 13, 1925, p. 2.

Drama may have been missing from the race, but an interpersonal drama played out after the race. Bettis, "quiet and retiring on the ground, a bachelor subject to fits of lonesomeness,"[75] was all smiles and grins when he landed. He stood in the cockpit and shouted, "Where's Al?" Before he could find out, Admiral Moffett rushed up to congratulate him. Lt. Maughan, 1922 Pulitzer winner and coast-to-coast flyer, and lieutenants Eric Nelson and Lowell Smith, who had flown in the army's round-the-world flight, clustered around the army racer. Bettis happily batted questions and answers back and forth with the excited, delighted crowd around him.

The "acerbic"[76] Williams, an excellent flier, was a different personality. He was "used to glamour; he strutted and swaggered, he occasionally blustered, especially to non-fliers, often giving a petulant, immature impression, and certainly that of a poor loser."[77] Those close enough to see saw a world of difference between the Bettis/army group and the Williams/navy group. Williams barely spoke and clearly didn't want to talk. His sister Frances tried to console him, telling him he'd flown a beautiful race and reminding him

Lt. Cyrus "Cy" Bettis, USAS, standing in the army R3C-1 being congratulated after his victory in the 1925 Pulitzer by Major T.G. Langier, commander of Mitchel Field. (Had Bettis already yelled, "Where's Al?") (courtesy National Archives and Records Administration).

that he would be leaving the navy soon.[78] She did not cheer him up. Shrugging off the attention of friends and others who tried to talk to him, Williams started to walk away without shaking Bettis' hand. His friends prevailed, and he reluctantly walked to the black and gold racer to congratulate Bettis. His reluctance to congratulate Bettis followed him as a black mark.

At some time on Pulitzer Race Day, Williams and Bettis got together with Air Service general Mason Patrick for a photograph in front of an R3C-1. Bettis looks as if he's enjoying himself, Patrick's mind may be miles away, and Williams looks like he wants to get miles away as soon as possible. Bill Wait tells a different story about the postrace events, which seems unlikely and was undoubtedly burnished by the passage of time Wait says Williams was ignorant of the race's outcome when he had landed: "Later [after the Pulitzer racers had landed], I was out talking with him [Bettis] and a group of reporters. Al Williams came up to me and asked, 'How much did I win by?' I said, 'You didn't win, Cy did.' Al muttered, 'Son of a bitch,' and stomped off."[79] Williams, of course, had known from before he touched down that he had lost.

The plan for Heat Two of the Pulitzer Race — that the army and navy would enter three racers each — proved impossible. No navy fighter approached the performance of the army's Curtiss pursuits, and the army loaned two of its fighters to the navy. The race was among four army racers, two piloted by army officers and two by navy officers. The navy

Lt. Alford "Al" Williams, USN, General Mason Patrick, USAS, and Lt. Cyrus "Cy" Bettis, USAS, in front of a Curtiss R3C-1 (probably the navy racer, based on the light color of the fuselage) after the 1925 Pulitzer, Mitchel Field, Long Island. Williams and Bettis are in their racing clothes — Williams in white shirt and blue trousers, Bettis in army officer's uniform with non-uniform Alpine-patterned socks pulled over the bottoms of his uniform trousers and almost certainly tennis shoes (out of the picture) (courtesy National Archives and Records Administration).

pilots flew two PW-8s that had flown in the Mitchell Race, and the army pilots flew a PW-8 and a Curtiss P-1. Curtiss 470-hp D-12s powered all Heat Two racers.

The Heat Two racers were expected to fly an anticlimactic 90 mph slower than the R3C-1s, but it promised to be a crowd-satisfying race. The closely matched airplanes in Heat Two would take off within seconds of each other, making that heat a race instead of two speed trials as flown by Bettis and Williams. The P-1 was slightly faster than the other Heat Two airplanes, and Lt. Leo Dawson, USAS,[80] won and placed third in the Pulitzer at 169.9 mph (see Table 7). Lt. Harmon Norton, USMC, and Capt. Harvey Cook, USAS, followed Dawson across the finish line, and Lt. George Cuddihy, USN, dropped out of the race because of engine failure on the last lap. Except for Norton's speed on his fourth lap, which was remarkably faster than his other laps, all four pilots flew lap speeds that varied by less than two miles per hour. The speed of the second-place PW-8 in the Pulitzer's Heat Two, 169.8 mph, was eight miles per hour faster than the winning PW-8 in the Mitchell, 161.7 mph. The longer Pulitzer racecourse, 50 km with longer straightaways, as compared to the 12-mile Mitchel course, likely accounted for the faster speeds in the Pulitzer.

In contrast to Foxworth's comment that the Mitchell had been "more of a crowd pleaser that [*sic*] the Bettis-Williams trial had been,"[81] the *New York Times* reported "there was little interest in it."[82] It may have been too much to expect that any race could displace the disappointing Pulitzer speeds. Disappointing as Bettis' speed was, it was a new Pulitzer record. He finished at 249.99 mph, which was 3.5 mph faster than Williams had flown in 1923 and exceeded world records for 100 and 200 km. New records, but far short of predictions. For instance, Bettis' own prediction of 270+ mph in the Pulitzer, probably based on Doolittle's 260+ mph over 100 km and Williams' 300+ mph in the R3C-1 powered by the "freak" V-1400 engine, proved too high.[83] Perhaps to put the best face on the Pulitzer, the *Aircraft Year Book, 1926* de-emphasized speed as an objective of the race: "The principal object of the Pulitzer ... this year was to test the latest improvements made in the design of pursuit type planes and engines."[84] Oddly, this objective had not been mentioned in the pre-race and immediate postrace comments on the Pulitzer.

On Tuesday, October 13, a true anticlimax closed the air meet. A tiny crowd, estimated at fewer than 2,500,[85] showed up to watch the two events originally scheduled for, and postponed from, the previous Thursday. The few spectators saw Professor Powell's *Racer* complete its sweep of all light-plane events when it won the *Dayton Daily News* Trophy race. The 1925 National Air Races ended with the small crowd breaking up, some walking to nearby airfields to look at the visiting planes that remained on view, some heading for the Long Island Railroad Station to catch a train home, and others driving away. Before the crowd dispersed, NAA officials must have mulled over and begun talking about the challenges they faced in selecting a host city for the 1926 Pulitzer and managing to increase attendance and interest.

More immediately, aviation enthusiasts and participants' eyes turned toward Baltimore and the Schneider Trophy Race, two weeks away. The unexpectedly low speeds in the Pulitzer worried experts and newspapers alike that the R3C-2s being prepared for the Schneider would fall short of the speed predicted for a British "mystery plane." The worries were misplaced. The "mystery plane," the Supermarine S-4 monoplane, progenitor of faster racers and World War II's Spitfire, had set a world's seaplane speed record before being shipped to Baltimore for the Schneider,[86] but it crashed before the race. Flying the same plane that Bettis had flown to victory in the Pulitzer, Doolittle flew away from the other British entry (a biplane) and an Italian monoplane, comfortably winning the Schneider.[87] Neither navy R3C-2 — not even the one powered by the "freak" V-1400 — finished the race; both dropped out because of engine problems. Even if the Supermarine S-4 had been in the race, Doolittle would likely have won. His winning speed, 233 mph over the 350-km Schneider course, was six miles per hour faster than the 227 mph that the S-4 had established as the absolute seaplane speed record over a straight three-kilometer course six weeks earlier. The day after the race, Doolittle settled any lingering doubts. He and the R3C-2 flew the three-kilometer course at 246 mph, fully 19 mph faster than the S-4 had managed.

The Schneider results buoyed up Americans disappointed by speeds in the Pulitzer, and American readers must have been cheered when they read a report of C.R. Fairey's speech to the Royal Aero Club on November 26, 1925.[88] Fairey, head of the Fairey Aviation Company and president of the Society of British Aircraft Constructors (1922–24), summarized United States achievements— domination of airspeed records in the 1920s, first round-the-world flight, development of "scout type of machine [probably a reference to PW-8s] ... superior to any other," and, clearly rankling the British, victory in the Schneider Race. Fairey praised wing radiators, metal propellers, multi-spar wing construction with resulting stronger wings than in European racers, and the Curtiss engines, which "enabled the fuselage cross-section to be reduced to the minimum which would accommodate the pilot and the fuel." Fairey saw the increase in the difference between top speeds and landing speeds, which had been encouraged by the Pulitzer requirement that racers land at under 75 miles per hour, as being as significant as higher speeds.

The army had now won four Pulitzers—1920, '22, '24, and '25—and the navy two, in 1921, when a navy racer loaned back to Curtiss won, and 1923. The rivalry had not diminished over time, and fierce competitors kept it at a boil. Al Williams exclaimed, "Whoopee, Navy!" after the 1923 Pulitzer, and Bettis yelled "Where's Al?" after the 1925 race. With a bit of luck, the government would buy new racers in 1926, and the army and navy would duel with faster airplanes. Probably no one looked forward more enthusiastically to the 1926 contest than Lieutenant T.K. Matthews, who had won the Mitchell and would "be the army's pilot in next year's Pulitzer race."[89] Thousands of people thrilled by air racing and speed joined him in thinking about the next contest for the Pulitzer Trophy, the "perpetual prize for annual closed circuit air races."[90] Until then, quiet and unassuming Lt. Cy Bettis would be the Pulitzer Race champion.

Chapter Eight

1926
"Perpetual" Dies with a Whimper

>*No Pulitzer Race This Year*
>*There will be no airplane races this year for the Pulitzer prize, given for speed, the National Aeronautical Association announced today. A principal reason was the lack of entrants. The prize was won last year by Lieutenant Cyrus K. Bettis.*[1]

"(L)ack of entrants" was more than a "principal reason." By itself, it was sufficient to end, after six contests, the "perpetual" annual air races that had begun with high hopes, heady excitement and expectations in 1920.

Opportunities may have existed to arrange a different sort of race, without high-performance, government-funded racers, but there was no or little willingness to try. The NAA had seen the 1925 National Air Race featuring the Pulitzer Trophy Race attract little attention. It had cost the association $70,000. Evidently the NAA was tired of the whole thing.

There would be no more Pulitzers.

CHAPTER NINE

Pulitzer Legacies

[W]hat a time it had been! In the decades between the wars, the public had passionately followed races [and the] planes that could win races, set records, and go faster, faster, and faster.[1]

The six Pulitzer Trophy Airplane Races, flown over the six-year period of 1920 through 1925 were the most important events in American aviation between the end of World War I and Lindberg's flight in 1927. They provided the public with dashing, sometimes romantic pilots, faster and faster airplanes, and world records. They gave the aviation industry incentives to build better airframes and engines and produced race results that were compelling evidence for the excellence or failure of their products. An important development was that the Pulitzers established the United States at the apex of speed airplane and engine design, influenced decisions about aircraft design in this and other countries, and generated United States sales in overseas markets.

Without the Pulitzer Races, United States airplanes would not have flown at world record speeds in the 1920s. Neither would they have won the most prestigious international air race, the Schneider Trophy Races for seaplanes, two times. Without them, United States aviation would have developed differently — perhaps better, perhaps worse, but differently. In 1920, no expert would have ranked United States aviation as competitive with Europe's. United States manufacturers had built thousands of British, French, and Italian-designed warplanes in World War I. The Wright Airplane Company had manufactured thousands of Hispano-Suiza high-performance engines under license. Wright had not slavishly copied the "Hisso" design. It had strengthened and improved it, and by 1920, Wright "H" engines, derived, but different, from the Hissos, which had generated 150 to 180 hp, were churning out up to 400 hp. The Curtiss Aeroplane and Motor Company had made thousands of its "Jenny" trainers powered by Curtiss OX-5 engines of 90 or so horsepower and a couple hundred flying boats used by the British for antisubmarine patrol.

By war's end, American airplane manufacturing productivity was a marvel after a slow start. In contrast, American design of high-performance airplanes was still in the start-up phase. No American-designed warplane had flown in combat over France or Italy. No American design had been ready for combat by war's end. Early postwar American aviation endeavors did little to raise the United States' status in aviation. A mixture of American- and foreign-built aircraft flew in the army's 1919 Transcontinental Air Race. The majority of starters and finishers, including the winner, were prosaic De Havilland DH-4 light bomber/

observation planes, designed and first produced in England in 1916. Underlining the difference between manufacturing and design in the United States, almost all the competitors had been built in the United States. None had been designed here (see Chapter One).

What was expected to be a glorious exhibition of U.S. designed and built speed planes at the September 1920 Gordon Bennett Race ended in humiliation. Accompanied with great ballyhoo, including preposterous claims that an American racer had flown 220 mph, when the world speed record was about 175 mph, three United States racing teams sailed to France. No U.S. racer completed the 300-km race. None completed even the first 100-km lap (see Chapter One).

In 1916 the sons of Joseph Pulitzer had announced their intention to award an annual trophy and cash prize for the fastest transcontinental flight. Waiting until the end of World War I, they intended to initiate their contest in 1919 but deferred because the army was farther along in preparing its own transcontinental contest. Several deaths in long-distance races convinced the Pulitzer sons and the Aero Club of America to change the Pulitzer to a closed-course, "around-the-pylons" race.

The first Pulitzer Trophy Race, two months after the 1920 Gordon Bennett, brought a welcome burst of glory, even if the most spectacular glory lasted only a day or two. Based upon its time around a measured course, the winning racer, the army's VCP-R, had apparently flown a closed course race faster than any other airplane, including the winner of the Gordon Bennett Race. At least equally satisfying to United States advocates, the second-place finisher, a Thomas Morse MB-3, a standard army pursuit had made a closed-course flight faster than any other standard fighter in the world. The United States had designed, built, and flown the fastest airplane in the world and the fastest production fighter. How disappointing and embarrassing that an error in marking a turning point on the triangular Pulitzer Race course had made the course shorter than planned. When the error was discovered and the course correctly measured, recalculation of the speeds reduced them below world records (see Chapter Two).

The first Pulitzer Race was different from the rest. It was flown by aircraft that the army, navy, and a few private companies had on hand, and there were a big bunch of them. More than 60 airplanes were entered; about three dozen took off, and two dozen completed the 116 mile course. The racers were divided into seven flights so that airplanes with expected comparable speeds flew against each other. From then on (1921 through 1925), races were flown by racers specially built for the army and navy that were powered by increasingly powerful engines. There were progressively fewer racers; the number of starters ranged from a high of 15 in 1922 to two in 1925.

Curtiss Aeroplane and Motor Company airplanes and engines dominated all the races from 1921 on. Walter Boyne, former Air Force pilot, former director of the National Air and Space Museum, and aviation historian, wrote that Curtiss "made a fortune stretching basic designs [for biplane pursuits] far beyond normal life expectancy."[2] His observation is equally applicable to Curtiss racers. From 1921 through 1925, the basic Curtiss racer design was modified, smoothed, and fitted with streamlined radiators and more powerful engines, but the racers' family resemblances remained clear.

Gentlemen-balloonists had organized the Aero Club of American (ACA), a nongovernmental organization and the United States' affiliate of the International Aeronautics Federation, the Fédération Aéronautique Internationale (FAI), in the first years of the 20th century.[3] The ACA, which was responsible for sanctioning aviation contests and certifying records, hosted the first Pulitzer Trophy Race at Mitchel Field, a United States Air Service airfield, at Garden City, New York, on Long Island, about an hour from New York City. No admission was charged for the race on Thanksgiving Day 1920, and somewhere between 25,000 and 40,000 spectators attended. More thousands saw the race from outside Mitchel Field or parts of it as racers buzzed over their heads.

The ACA expected its local affiliates to host the remaining races, and cities with adequately sized airports, sufficient parking spaces, and other needed accommodations or access to Air Service airfields hosted the 1921 through 1924 Pulitzers: Omaha in 1921; Detroit and the Air Service Selfridge Field in 1922, St. Louis in 1923, and Dayton, Ohio, and Air Service Wilbur Wright Field in 1924. After disappointing attendance at Dayton, the National Aeronautic Association (NAA), successor to the ACA, hosted the 1925 contest at Mitchel Field, the site of the first Pulitzer, where the race could draw spectators from New York City's huge population.

The Curtiss-dominated races saw winning speeds rise 93 mph, from 156 mph in 1920 to 249 mph in 1925 (see figure 6 and table 8). Curtiss racers set new closed-course ("around-the-pylons") speed records in four of the last five Pulitzer Trophy Races—an unofficial record in 1921 and FAI-sanctioned records in 1922, '23, and '25. The 1922 and '23 Pulitzer winners also set new world absolute speed records for flights over a measured, straight-line distance (see table 8).

The army and navy contracted with six different aircraft manufacturers for new racers in 1921, '22, and '23, and speeds increased each year. Growing crowds came to see the increasing speeds. At Omaha in 1921, at least 16,000 people paid admission to the airfield on one of the three days of the air meet and more than that many watched from outside the field. (See Table 8 for information about pilots and airplanes that finished the Pulitzer Races and the appendix for specifications for all aircraft that started the races.) A year later, seventy-five thousand people swarmed onto Selfridge Field on Pulitzer Race Day, and 200,000 or so watched the Curtiss Marine Trophy Race over the Detroit River in downtown Detroit. The 1922 air meet, which featured six closed-course races for civilian and military airplanes, was the first air meet to be called the "National Air Race."

For the Pulitzer Races, St. Louis in 1923 was the apex of attendance and performance. A half-million people attended the three-day-long second National Air Race, and many saw navy Lt. Alford "Al" Williams fly a new Curtiss R2C racer to the new world record of 245 mph. They were also entertained by visits of the world's largest dirigible, the U.S. Navy's ZR-1 (being renamed *Shenandoah* on October 10, 1923, and crashing under that name); the world's largest airplane, the army's six-engine Barling Bomber; and Orville and his sister Katherine Wright. At the air race's end, St. Louis had one of the best civilian airports in the country, and "Lambert Field" is still the site of the St. Louis International Airport (see Chapter Five).

Performances had improved and attendance had risen through 1923. Both collapsed in 1924. Dayton, Ohio, the host of the 1924 Pulitzer, riding the swell of enthusiasm from the St. Louis meet in late 1923, expected great things for its National Air Race, but it was not to be. Neither the army nor the navy purchased a new racer, making it impossible to expect a new record speed. Dayton, where airplanes were old hat because of its Air Service fields, was the smallest city to host a Pulitzer. How many residents of Dayton who could go to the airfields on almost any day and see modern, even experimental, aircraft in the air, would pay to see a race featuring two-year-old racers? Initial predictions of a quarter million attendees fell to 200,000 before quantitative estimates disappeared from press releases. A disappointing 75,000 people, at best, attended the Dayton National Air Race.

Worse, the winning speed was 216 mph, 28 mph slower than the 1923 winner's speed. Worse yet, Capt. Burt Skeel, a popular Air Service pilot died when his racer came apart in the air as he dived to cross the Pulitzer start line (see Chapter Six).

TABLE 8
First, Second, and Third Place Racer and Pilot Finishers in the 1920 Pulitzer Race, All Racer and Pilot Finishers, 1921 through 1925, and Records Set by the Racers[a]

First Pulitzer, November 25, 1920, Mitchel Field, Long Island.
Four laps of 29.02-mile triangle course. Total distance, 116.08 miles.

Place in Race/ Racer Name	Pilot	Engine (nominal hp)[b]	Speed (mph)	FAI-Sanctioned[c] World Records
1. Verville (army-built) VCP-R	Lt. Corliss Moseley, USAS[d]	Packard A-2025 (578)	156.5	
2. Thomas-Morse MB-3	Capt. Harold Hartney, USAS	Wright H-2 (326)	148.2	
3. Ansaldo A.1 Balilla	Bert Acosta	S.P.A. 6A (210)	134.0	

Second Pulitzer, November 3, 1921, Omaha, Nebraska.
Six laps of 30.72-mile triangle course. Total distance, 153.60 miles.

Place in Race/ Racer Name	Pilot	Engine (nominal hp)	Speed (mph)	FAI-Sanctioned World Records
1. Curtiss CR	Bert Acosta	Curtiss CD-12 (372)	176.8	
2. Curtiss *Cactus Kitten*	Clarence Coombs	Curtiss C-12 (338)	170.3	
3. Thomas-Morse MB-6 (R-2)	Capt. John Macready, USAS	Wright H-2 (326)	160.7	
4. Ansaldo A.1 Balilla	Lloyd Bertaud	Curtiss K-12 (415)	149.7	

Third Pulitzer, October 14, 1922, Selfridge Field, Mt. Clemens, Michigan.
Five laps of 50-km triangle course. Total distance, 250 km (155.4 miles).

Place in Race/ Racer Name	Pilot	Engine (nominal hp)	Speed (mph)	FAI-Sanctioned World Records
1. Curtiss R-6	Lt. Russell Maughan, USAS	Curtiss D-12 (450)	205.9	October 14, 1922, 100 km, 205.3 mph; 200 km, 206 mph. Absolute world speed record. October 18, 1922, William Mitchell, 223 mph. March 29, 1923, Russell Maughan, 236.6 mph.
2. Curtiss R-6	Lt. Lester Maitland, USAS	Curtiss D-12 (450)	198.8	
3. Curtiss CR-2	Lt. Harold Brow, USN[e]	Curtiss D-12 (450)	193.7	
4. Curtiss CR-2	Ens. Alford Williams, USN	Curtiss D-12 (450)	188.0	

Place in Race/ Racer Name	Pilot	Engine (nominal hp)	Speed (mph)	FAI-Sanctioned World Records
5. Verville-Sperry R-3	Lt. Eugene Barksdale, USAS	Wright H-3 (333)	180.7	
6. Verville R-1	Lt. Corliss Moseley, USAS	Packard 1A-2025 (578)	178.8	
7. Verville-Sperry R-3	Lt. Fonda Johnson, USAS	Wright H-3 (333)	178.0	
8. Loening R-4	Lt. Ennis Whitehead, USAS	Packard 1A-2025 (578)	170.1	
9. Loening R-4	Lt. Lester Schulze, USAS	Packard 1A-2025 (578)	160.7	
10. Thomas Morse R-5	Lt. Clayton Bissell, USAS	Packard 1A-2025 (578)	155.4	
11. Thomas Morse R-5	Capt. Frank Hunter, USAS	Packard 1A-2025 (578)	149.2	

Fourth Pulitzer, October 6, 1923, Lambert Field, St. Louis, Missouri (weather forced postponement from October 3).

Four laps of 50-km triangle course. Total distance, 200 km (124.3 miles).

Place in Race/ Racer Name	Pilot	Engine (nominal hp)	Speed (mph)	FAI-Sanctioned World Records
1. Curtiss R2C-1	Lt. Alford (507) Williams, USN	Curtiss D-12A (507)	243.7	October 6, 1923, 100 km, 243.8 mph; 200 km, 243.7 mph. Absolute world speed record. November 4, 1923, Alford Williams, 266.6 mph.
2. Curtiss R2C-1	Lt. Harold Brow, USN	Curtiss D-12A (507)	241.8	
3. Wright F2W-1	Lt. Lawson Sanderson, USMC[f]	Wright T-3 (680)	230.1	
4. Wright F2W-1	Lt. Steven Calloway, USN	Wright T-3 (680)	230.0	
5. Curtiss R-6	Lt. Walter Miller, USAS	Curtiss D-12A (507)	218.9	
6. Curtiss R-6	Lt. Johnny Corkille, USAS	Curtiss D-12 (470)	216.5	

Fifth Pulitzer, October 4, 1924, Wilbur Wright Field, Dayton, Ohio.

Four laps of 50-km hexagonal course. Total distance, 200 km.

Place in Race/ Racer Name	Pilot	Engine (nominal hp)	Speed (mph)	FAI-Sanctioned World Records
1. Verville-Sperry R-3	Lt. Harry Mills, USAS	Curtiss D-12A (507)	216.6	March 29, 1923, Alexander Pearson, 500 km, 167.8 mph.
2. Curtiss R-6	Lt. Wendell H. Brookley, USAS	Curtiss D-12A (507)	214.4	

Place in Race/ Racer Name	Pilot	Engine (nominal hp)	Speed (mph)	FAI-Sanctioned World Records
Non-racer[g]				
3. Curtiss XPW-8A	Lt. Rex Stoner, USAS	Curtiss D-12 (470)	167.9	

Sixth Pulitzer, October 12, 1925, Mitchel Field, Long Island, New York (weather forced postponement from October 10).
Four laps of 50-km triangle course. Total distance, 200 km.

Place in Race/ Racer Name	Pilot	Engine (nominal hp)	Speed (mph)	FAI-Sanctioned World Records
1. Curtiss R3C-1	Lt. Cyrus Bettis, USAS	Curtiss V-1400 (565)	248.9	Oct. 12, 1925, 100 km, 249.3 mph.
2. Curtiss R3C-1	Lt. Alford Williams, USN	Curtiss V-1400 (565)	241.7	
Non-racers				
3. Curtiss P-1	Lt. Leo Dawson, USAS	Curtiss D-12 (470)	169.9	
4. Curtiss PW-8	Lt. Harmon Norton, USMC	Curtiss D-12 (470)	168.8	
5. Curtiss PW-8	Capt. Harvey Cook, USAS	Curtiss D-12 (470)	167.4	

[a]Data from T.G. Foxworth, *The Speed Seekers* (New York: Doubleday, 1974] pp. 423–546.
[b]"Horsepower" refers to "nominal horsepower" as characterized by T.G. Foxworth, *The Speed Seekers* (New York: Doubleday, 1974] p. 543). Foxworth discusses rated, nominal, and maximum horsepower. "Rated" horsepower is the manufacturers' guaranteed power output. "Nominal" horsepower was actually delivered in conditions of stress, as in a race. "Maximum" horsepower was achieved at least once, maybe only once, and for an unknown length of time, probably in a bench test.
[c]Foxworth reported records in km/hr; all were converted to mph by multiplying by 0.6214.
[d]United States Air Service, predecessor of United States Army Air Force and United States Air Force.
[e]United States Navy.
[f]United States Marine Corps.
[g]Probably to provide more interest, race organizers introduced service airplanes into the 1924 and 25 Pulitzer Races. In 1924, the experimental XPW-8B (called here "PW-8B") and also referred to as XPW-8A, second modification, flew in the same heat as the racers. In 1925, the P-1 and three PW-8s (one DNF) flew in a separate second heat.

The NAA, sensing that interest in both the Pulitzer and air racing was falling and also facing criticism of racing's dangers because of Skeel's death, was unsuccessful in arranging a new host city that met its approval and decided to host the 1925 National Air Race and Pulitzer at Mitchel Field, the site of the first Pulitzer Race. The army and the navy pooled their funds for racers and purchased three Curtiss R3C racers. Both services would fly an R3C-1 land plane in the Pulitzer. The navy would fly two R3C-2 seaplanes and the army would fly one in the Schneider Race to be held, two weeks after the Pulitzer, over the Chesapeake Bay near Baltimore.

The Pulitzer was scheduled for October 10 to take advantage of mild fall weather, and race organizers predicted total attendance of a half million people over the three days of the National Air Races, October 8, 9, and 10 (see Chapter Seven). Few things went right. The weather was frightful, with 70 mph winds and snow flurries and heavy rains causing postponement of the Pulitzer from Saturday, October 10, to Monday, October 12, which, fortunately for potential spectators, was Columbus Day, a holiday. But if that helped at all, it made little difference. Rather than the expected crowd of one or two hundred thousand, 25,000 spectators saw the Pulitzer.

Army lieutenant Cyrus "Cy" Bettis flew to a new world record for a closed-course flight, 249 mph, 5 mph faster than the 1923 record. A record, yes, but it was reported in a *New York Times* article with this deflating headline: "Pulitzer Race Won at 249-Mile Speed; Disappoints Fliers."[4] Navy lieutenant "Al" Williams finished second, two mph behind the 1923 winner, and 0.1 mph slower than the second finisher in 1923.

No one knew it when Bettis and Williams landed after completing the 1925 Pulitzer Race, but the Pulitzers were over. Surely NAA officials and army and navy officers knew well in advance of the general public, but almost a year passed before the public learned from a *New York Times* single-paragraph news story entitled "No Pulitzer Race This Year," on September 1, 1926[5] (see Chapter Eight). The Pulitzers were over. What had they done for aviation — in particular, for United States aviation?

Probably everyone even remotely associated with the Pulitzer Races would have agreed with an *Aviation* magazine editorial published a month before the 1925 race that "the Pulitzer speed races have played a major part in bringing aviation to the public attention."[6] To support his statement, the writer could have pointed to more than a million spectators that saw one or more Pulitzers with or without paying admission, the millions who read about them in magazines and newspapers — sometimes on the front page of the *New York Times* — and, probably, more millions who saw them in movie newsreels. The editorial writer's praise is especially noteworthy because the editorial, entitled "The Dangers of Public Speed Races," criticized races of military aircraft, which the Pulitzers had become three years earlier in 1922.

The Pulitzers generated some of the first foreign sales of American airplanes, engines, propellers and other aviation products. The sale of racing-developed Curtiss D-12 engines in Great Britain was such a breakthrough for the American aviation industry that the *New York Times* published news of the sale.[7] The D-12s were not mere curiosities in Britain. The Fairey Company combined Curtiss D-12s, Curtiss-Reed metal propellers, which had been perfected for the Curtiss racers, and a streamlined design into a light bomber, the Fairey Fox. An extremely good-looking biplane, the Fox flew faster than any contemporary RAF fighter and caused serious rethinking about airplane design and development in Britain.

The success of American marketing of engines, propellers, and other aviation products outraged some Britons. British aviation industry officials objected to their Air Ministry's purchase of American engines (Curtiss), American aluminum propellers (Curtiss-Reed), and American parachutes, and condemned the Air Ministry's decision to send a good-will flight from Cairo to Nigeria in airplanes powered with American engines.[8]

In words that would have been unimaginable in 1920, the *London Morning Post*, in 1925, printed, "America is aeronautically our strongest rival." The *Morning Post* went on to quote a "well known aeronautical authority":

> Our industry is in very grave danger of losing its foreign markets altogether. While the Ministry continues, in effect, to broadcast its opinion that British goods are inferior to American ones [by purchasing American products], you cannot expect foreigners to buy our products....
> Foreigners, relying more upon deeds than words, buy aircraft from those constructors who prove in open competition either in *air racing or in record breaking*, that their machines are equal to the world's best.[9] [Emphasis added.]

Racers and racing had expenses, but they attracted attention; they engendered sales. Mr. C.R. Fairey, who made the first foreign purchase of Curtiss engines,[10] said that his British countrymen incorrectly credited large-scale government financing as being responsible for United States progress in engines, propellers, and speed. He argued that racing, so important to improvements, had cost little in comparison to overall expenditures on army and navy aviation and that "the progress made was due largely to one private company, the Curtiss Company."[11]

Mr. Fairey was right to focus on the Curtiss Company. His conclusion that government funding had played a minor role is questionable. It's difficult to imagine that Curtiss would have designed, manufactured, and tested world-beaters without government contracts for Pulitzer racers. Equally difficult to imagine is what would have driven the government decision to purchase racing airframes, engines and propellers in the absence of the Pulitzer competition.

Through 1923, the Pulitzers, always purely United States affairs despite determined attempts to attract foreign racers and pilots, drew little attention in foreign countries except for generally short articles about Pulitzer racers in aviation journals. Another great race, the Schneider Trophy Race for seaplanes, also deserves credit of propelling United States aviation into international sales.

The Schneider Races, always—excepting the last race in 1931—were international competitions. The first two, flown in 1913 and 1914, before the beginning of World War I, attracted fliers from many countries including the United States. Suspended during the war, they resumed in 1919, and British, French, and Italian racers competed on a yearly basis through 1922. Curtiss racers made 1923 a year of change.

Curtiss CR-3s (seaplanes made by mounting 1922 CR-2s—which were, in turn, upgraded 1921 CRs—on floats or pontoons) flew away from the other racers in the 1923 Schneider (see Chapter Six). The victory, made possible by the availability of Pulitzer racers, brought international attention to United States aviation. The CR-3 seaplanes won first and second, leaving a British flying boat sputtering behind, a distant third. Their victory changed the type of airplanes flying in the Schneiders. From 1920 through 1922, flying boats—airplanes with boat-like fuselages that alighted on the water and floated on their fuselages—had done well in the Schneiders. The CR-3s were seaplanes, land planes attached by long struts to floats or pontoons that took the place of the land planes' wheels, and from 1923, seaplanes won all the Schneiders. (Seaplanes had also won the first Schneiders.)

The Curtiss D-12, twelve cylinder V-engine, which made its international debut in the Schneider, soon influenced aero engine development in Europe. The Italian Macchi Company purchased Curtiss D-12s to power its 1925 racers, the last flying boats in the Schneiders, which flew at Baltimore, and Italian engineers tore down and studied the D-12s to aid in their design and development of more powerful Italian V-12s.

No Curtiss engine powered a British racer, but a Curtiss engine played a major role in the development of the engines used in the last British Schneider racers (and, subsequently in World War II aircraft). In 1923, the British Air Ministry, which had bought two D-12s, loaned one to Rolls-Royce and ordered the company to design and build a better engine. According to Foxworth, the engine that emerged, resembling the D-12 rather than any earlier Rolls engine,[12] led to the Rolls-Royce "R" V-12s that powered Britain's 1929 and '31 Schneider-winning racers. An English engine expert, Edward Eves, wrote, "Curtiss-watchers could claim with some justification that there was a lot of D-12 in [Rolls-Royce engines]. On the other hand there was little that was revolutionary in the construction of the D-12, it simply worked better than well and made a lot of good ideas respectable."[13] The extent of the D-12's contribution to later Rolls-Royce engines may be debated (and in fact is). What is more significant, modified and improved Rolls V-12s—"Merlins" and "Griffons"—would power many of the most famous and important World War II warplanes.

From absolutely nowhere in international air racing, the United States flew to first place by winning the 1923 Schneider. As the winning country, the United States became responsible for hosting the next Schneider Race in 1924. Out of a sense of sportsmanship, the United States postponed the race when no foreign country had a racer ready. Had it not postponed the race, the United States would have had a certain victory in the 1924 Schneider. In the absence of a foreign competitor, a United States airplane could have completed the Schneider course at any speed, no matter how slow, and been declared the winner. The

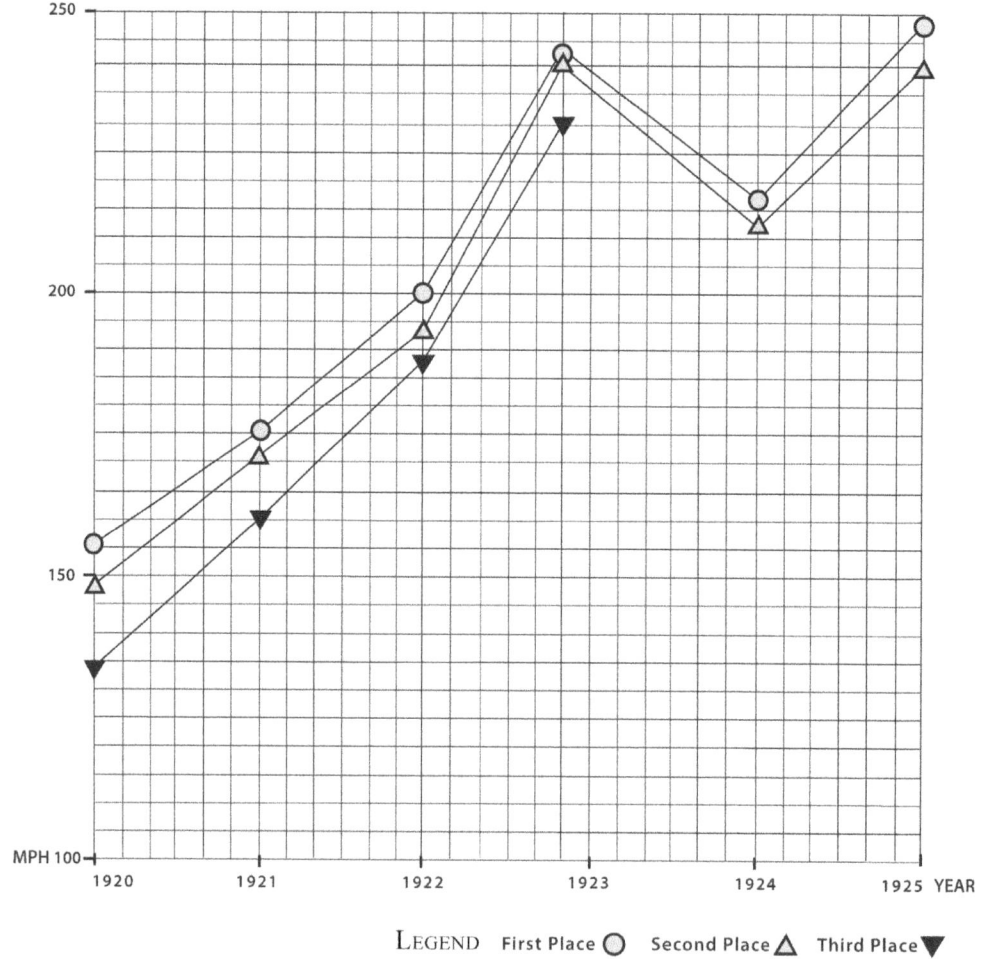

Figure 6. The speeds of the first three finishers in the 1920 through 1923 Pulitzers and of the first two finishers in 1924 and 1925, when only two racers finished. The speeds increased each year until 1924, when no new racers were made. Speeds bounced back in 1925, but only a little better than the speeds in 1923, indicating that the Curtiss biplanes, which had won the 1921, '22, '23, and '25 races, were approaching their inherent speed limits. No Curtiss biplane flew in another major landplane race after 1925, but the R3C-2 that won the 1925 Schneider was upgraded and flown in the 1926 Schneider. An Italian monoplane defeated it, and monoplanes won all the remaining Schneiders (figure by Lynn Springer).

"sporting gesture" of postponing the race would have great consequences. Had the United States held a 1924 Schneider, its second straight victory would have meant that one more victory would have given the United States permanent possession of the Schneider Trophy and ended the Schneider Races.

Curtiss biplane racers flew to their last hurrah against Great Britain and Italy in the 1925 Schneider, 12 days after the 1925 Pulitzer. During the 12 days between the Pulitzer and the Schneider, the army and navy racers were moved from Mitchel Field on Long Island to Bay Shore Park on the Chesapeake Bay near Baltimore, the site of the Schneider, where they were mounted on twin floats. The army assigned Lt. James "Jimmy" Doolittle to fly its R3C-2. Doolittle had been Cy Bettis' backup pilot in the Pulitzer, and Bettis was Doolittle's at Baltimore. Navy lieutenants George Cuddihy and Ralph Ofstie were assigned to fly the navy's racers.

The sleek British Supermarine S-4 monoplane seaplane—holder of the world's speed record for seaplanes and resembling the Spitfire that was ultimately derived from it—was the greatest challenge to the United States' racers. Unfortunately for the potential head-to-head Supermarine-Curtiss contest, the S-4 crashed into the Chesapeake before the Schneider.[14] All three Curtiss racers—two navy R3C-2s and the army R3C-2—were clearly faster than the British biplane seaplane and the Italian monoplane flying boat that flew against them in the Schneider, but engine problems forced both navy R3C-2s to alight before finishing the race. Doolittle won, followed by the British biplane and the Italian monoplane. On the next day, Doolittle flew the Curtiss racer to a new absolute speed record, 245.7 mph, for seaplanes, bettering the speed posted by the Supermarine S-4 only weeks earlier.

Had the United States held a Schneider Race in 1924 when there had been no foreign entry, that assured victory combined with United States wins in 1923 and '25 would have given the United States permanent possession of the Schneider Trophy. That opportunity was gone, but a United States victory in the 1926 race, again flown over the Chesapeake but this time from Hampton Roads, Virginia, would give the country three straight victories and permanent possession of the Schneider Trophy.

The 1926 Schneider was the Curtiss racers' swan song. United States Marine lieutenant Christian Schilt flew the same R3C racer powered by the same Curtiss V-1400 engine that Cy Bettis and Jimmy Doolittle had flown to first places in the 1925 Pulitzer and 1925 Schneider. Schilt finished second to an Italian Macchi M-39 monoplane seaplane powered by an 800-hp V-12 engine. Schilt's speed, 231.4 mph, was just over one mph slower than Doolittle's 232.6 mph in 1925 and 15.1 mph slower than the winning Macchi M-39's 246.5 mph.

The United States Army and the United States Navy entered no more Schneider racers. After retiring from the navy, Alford "Al" Williams, winner of the 1923 Pulitzer and second place finisher in the 1925 Pulitzer, raised sufficient funds to have a biplane racer—which looked a lot like Curtiss racers—built for the 1927 Schneider and a monoplane racer resembling the Italian Macchi monoplanes for the 1929 contest. Neither airplane proved promising as a racer, and neither raced.[15] After 1926, despite Williams' efforts and planned entries from France and Germany, only Italian and British racers flew in the Schneiders.

In 1923, some Europeans had raised strenuous objections to U.S. Navy sponsorship of Schneider racers. Until then, the Schneiders had been contests between privately held or company-owned racers. The prediction that the United States' victory would turn the Schneiders, like the Pulitzers, into contests among government-sponsored racers proved prescient. The Schneiders, as "sporting contests," sponsored and financed by aero clubs, individuals, and aviation companies, passed into history in 1923.

Two years later, in 1925, the British government agreed to pay for the design and manufacture of racers for the Baltimore Schneider and to "loan" the racers back to the manufacturers—Supermarine and Gloster—for racing when they met government specifications.[16] The Italian Macchi Company's 1925 racers, built by the company, were the last privately financed competitors. After 1925, the governments of both Britain and Italy funded racers until 1931 when the world-wide depression forced the British government to forego funding racers.

Following the Italian victory over the United States in 1926, a Britain Supermarine S-5 racer flew 276 mph to win the 1927 contest at Venice. Because of the cost and time necessary to develop new racers, the decision was made to hold the Schneiders at two-year intervals instead of annually after 1927. A new Supermarine, the S-6, powered by a Rolls-Royce "R" V-12, won the 1929 race at Calshot, England, and raised the winning speed to 323 mph.

Italy made plans for, and began to build, a racer for 1931, but the new British Socialist government decided it could not afford to pay for building and racing an airplane. Amid widespread dismay at the lack of a British racer, a wealthy British woman, Lucy, Lady Houston, a chauvinistic and Socialist-government–hating widow of a shipowner, stepped to the fore and donated 100,000 pounds (roughly $8,000,000 in 2012 dollars) to keep Britain in the race. Supermarine prepared a modification of the 1929 Supermarine S-6 racer, the S-6B, for 1931.

No foreign competitor was ready in time to face Britain in 1931. The British had complimented the United States for postponing the 1924 Schneider when no foreign contestant was ready, and there may have been some hesitation about holding the 1931 contest with no foreign entries. If there was any hesitation, it was put aside, and the "race" went off as scheduled. Pilot John Boothman completed the 350-km course to win the Schneider Trophy and take permanent possession of the trophy in fine style. Boothman's speed, 334 mph, in what would be better called a "flyover" than a "race," was a 101-mph increase over Doolittle's 233 mph in 1925. The 101-mph increase in speed in the last six years of the Schneiders was comparable to the 93-mph increase over the six years of the Pulitzer. Later in 1931, the Supermarine S-6B became the first airplane to fly more than 400 mph.

The 1925, '26, and later Schneider Races confirmed that monoplanes were shoving biplane racers, after glorious years in the spotlight, offstage. The 1925 Curtiss R3C-2 was the last biplane victor in the Schneider, and no biplanes competed after 1926.[17] Had the Pulitzer Races continued after 1925, or had the United States entered more Schneiders, Curtiss and other manufacturers would have built monoplanes or ceased competing in the races.

Government financing of racers, reluctantly adopted in Europe after 1923, had begun in the United States in 1920 with preparation for that year's Gordon Bennett Race. The army modified a prototype VCP pursuit into the VCP-R racer with a more powerful engine and other changes, and it further modified the VCP-R for the 1920 Pulitzer. Also in 1920, the navy modified a fighter, a Loening M-8, into the Loening M-81 S racer, for the Pulitzer. Neither the army nor the navy paid to build a racer from scratch in 1920. Government contracts for the building of racers from scratch began the next year, in 1921. The government ordered Curtiss racers every year between 1921 and 1925 except for 1924. It twice ordered Loening, Thomas Morse and Wright racers and ordered racers from the Aerial Engineering Corporation and Verville-Sperry one time. Only three companies—Curtiss, Packard, and Wright—produced all the racers' engines, with Curtiss engines powering all the winners from 1921 on, including 1924 when a Curtiss engine powered the Sperry-Verville R-3.

There's no denying that Curtiss built good racers and excellent engines. Nevertheless, questions can be raised about what kinds of engines and airplanes other companies would have developed if the government had contracted with them year after year. In October 1925, the month of the 1925 Pulitzer, Curtiss Company president C.M. Keys, apparently responding to criticisms that the company had benefitted from a too-cozy relationship with the government, wrote in *Aviation* magazine that "lack of a federal aviation policy" was responsible for Curtiss' favored position as a supplier of racers; "it [Curtiss] has been led by circumstances to work with the Government services in the development of racers to a greater extent than the rest of the industry combined."[18] Whatever Keys intended those words to mean, few readers would have interpreted them as saying a too-tight Curtiss-government relationship did not exist. Putting words into the mouth of an unspecified, anonymous, and amorphous manufacturer, Mr. Keys spelled out what was needed: "He would like to see the Government adopt a procurement policy which would permit every responsible builder in the country to confine his research and development work to certain types [pursuits, observation planes, light bombers, etc.]. We hope that such a system will be incorporated in a policy if it develops...."[19]

How could such a policy of government-established monopolies stimulate manufacturers to compete for racer contracts? More important, how would limiting pursuit contracts to a single manufacturer or a selected few promote better pursuit designs? More fundamentally confusing is why a capitalist—Mr. Key's "manufacturer"—working in a free-enterprise system, would "hope" for government control of research and development. The cynical and, likely correct, answer is a capitalist who benefits from a close relationship with the government.

However contracts were awarded, the Curtiss-government interactions provide positive specific answers to questions about what the United States government obtained from sponsoring races. Curtiss engines and Curtiss-Reed metal propellers, both developed and perfected for racing, were incorporated into military and civilian designs in the United States and penetrated foreign markets for aero engines and propellers.

General William Mitchell called the Pulitzers "a laboratory for the development of high-speed pursuits."[20] His comment leads to a question. Did the United States need high-speed pursuits in the 1920s? Some aerial defenses were needed. The sea-lanes around the United States were vulnerable to submarines and surface raiders—as they had been in World War I—but patrol planes and bombers were needed to counter those, not pursuits. A possible case can be made that high-speed pursuits were needed to defend against airplanes launched from aircraft carriers. Britain, France, Japan, and the United States were perfecting carrier operations in the 1920s, but carriers were small and few in number, and carrier aircraft had limited range, meaning carriers would have to venture close to the coast to launch their airplanes. Protecting against the carriers themselves, relying on shore-based patrol planes to find them and surface ships to sink them, did not require pursuits. Our immediate neighbors posed no threat. Canada, an ally during World War I, had no territorial ambitions. Mexico, poor, damaged in a 1916–17 war with the United States, and emerging from a revolution, could pose no threat even had it so desired.

Nevertheless, Keys conjured up a hypothetical war with Mexico to provide a national defense justification for, of all things, metal propellers, which had found immediate and wide acceptance in civilian markets. Metal propellers were lighter than wood, could be run at higher RPMs without worry about breakage or shattering, and were resistant to moisture—rain, snow, what have you, temperature changes, and insects. Manufactured in sizes from very small to very large, their commercial success fully justified their development.

Ignoring their commercial success, the Curtiss Company president praised the potential military usefulness of metal propellers in a preposterous scenario: "Had we gone to war with Mexico before this metal air screw was perfected, we should have been compelled to build propellers on the Rio Grande to insure efficiency in that climate."[21] Why stop there? Surely a potential opponent in the Antarctic can be imagined, and metal propellers would have proved handy there.

For airplane manufacturers in the tight economic climate of the early 1920s, a racer contract could make the difference between survival and collapse. After Thomas Morse's loss of the major production contract for its MB-3s in 1921, government contracts for four racers, two R-5s or MB-6s for the Air Service, and two MB-7s for the navy were important sources of funds and kept the company alive (but not for long).

By careful attention to costs, a manufacturer of one or two racers might make a small profit. Even without a profit, the potential publicity payoff from building a winner could make a racer contract an attractive proposition. In the case of Curtiss, racer successes led to United States government contracts for pursuits, engines, and propellers as well as foreign sales. No other racer manufacturer obtained a pursuit contract as a result of the Pulitzers, and the Boeing Company's success showed that participation in racing was not essential for contracting success.

Upstart Boeing, which never entered a Pulitzer Race and is now better known as a builder of bombers and commercial aircraft, broke into the pursuit business when it earned the production contract for Thomas Morse MB-3s in July 1921. It pulled even with Curtiss in 1924, when the Air Service awarded contracts to Boeing for the PW-9. Boeing would surge ahead in 1932, when it began producing P-26 all-metal, monoplane pursuits, effectively ending the era of the biplane fighter in the United States.

Three years earlier, Curtiss had ceased to exist as an independent company in 1929 when economic conditions had forced a merger between Curtiss and its former bitter rival Wright. The merged Curtiss-Wright Company moved back into prominence in the pursuit business when it secured a production contract for its P-36A in 1937, by which time Boeing was focused on multiengine cargo and bomber aircraft.

In all, Curtiss-Wright produced a total of about 1,100 P36s and exported variants of it. More important for the company, the P-36 led to the P-40, which was produced in larger numbers than any other United States pursuit in history. The P-40, however, was Curtiss-Wright's last design to result in a production contract, and the company went out of the airplane manufacturing business after World War II. Boeing, of course, continues as a dominant airplane producer to this day and is one of the two most important builders of passenger planes.

Government funding was essential to Curtiss success in building racers and racing engines, but it is difficult to imagine that Curtiss would have done so well if S.E.J. Cox, a private citizen, had not purchased the *Texas Wildcat* and *Cactus Kitten* racers from Curtiss for the 1920 Gordon Bennett. "Alphabet" Cox is better known for his shady and criminal dealings in promoting sales of nonproducing "oil fields" and "mineral lands" in this country and Mexico, for which he served about 10 years in the federal Leavenworth Penitentiary.[22] Cox is also, arguably, the single most important person to the Pulitzers. The engineers and designers that Curtiss assembled to build Cox's 1920 Gordon Bennett racers were the core of the design teams that developed the later Curtiss racers. His venture into racing in the 1920 Gordon Bennett ended dismally with the crash of his *Texas Wildcat* racer on a French airfield the day before the race, but he enjoyed a major success in 1921. His *Cactus*

Kitten triplane, the only individually owned racer to compete in any Pulitzer, took second place at Omaha. Exuberant in that success, Cox announced plans to return to the contest in following years, but a series of setbacks, including imprisonment for mail fraud, ended his racing career.

Testing and proving airplanes were changing in the mid–1920s. The era when races, including the Pulitzers, were the ultimate test of high-speed designs and when head-to-head competition could be the final arbiter in picking superior airplanes was ending. C.M. Keys, in another overstatement in his "explanation" for why Curtiss had received the bulk of racer contracts, wrote, "Our engineers have been able to predict, with absolute accuracy, the speed, stability and flying qualities of each new airplane designed, before building it."[23] His quote suggests that testing actual airplanes was an unnecessary frill, but surely not even Keys would have gone that far. But he would have been on solid ground if he had said racing was unnecessary for testing. The end of the races did not end the desirability of world records. As Keys wrote, the "average person ... glows with pride if another world's record is broken and lets it go at that. To those charged with our national defense in the air, however, this breaking of records is of vital importance. It indicates our potential strength in the air if we should be drawn into another war."[24]

But testing an airplane and setting a world record need not involve racing. Notification of the NAA, which certifies records to the FAI, that a record attempt was to be made and arranging for official measurements and timing were simple matters compared to organizing and hosting an air meet. If the weather turned bad, a postponement did not result in disappointed thousands who had planned to see the race, and, almost certainly, less-than-expected attendance at the rescheduled events. Record attempts, without the public fanfare of races, could be almost private affairs. Disappointing speeds were not widely broadcast, crashes were not so likely to be front-page news, and a world record, if certified by the timers and judges, was a world record regardless of the number of spectators.

Quite apart from changes in experts' evaluation of the value of racing, the failure of the 1924 and '25 Pulitzers to draw big crowds highlighted the fading allure of air meets and races. Hosting air meets was expensive and difficult and almost always a money-losing proposition. Requests from cities eager to host a Pulitzer had never flooded NAA offices, and in 1925, the organization complained about the limited response to its call for host cities. Regardless of public attendance, had military leaders considered racing essential, they would have found a way to continue building racers. They did not. The government's decision to buy no racers in 1926, made with little fanfare, terminated the Pulitzers.

Just as all racers are not created equal, neither are all races, and the Schneider Trophy Races for seaplanes are much more the subject of popular histories than are the Pulitzers. A Google search for "Schneider Trophy" in March 2011 yielded 2,080,000 hits, "Schneider Trophy Race" 937,000. Comparable searches for "Pulitzer Trophy" and "Pulitzer Trophy Race" produced fewer than one-tenth those hits, 168,000 and 98,800 respectively. A search for books with "Schneider Race" in the title at Amazon.com produced 15 hits, for "Pulitzer Trophy" none.

British authors have written the majority of the "Schneider books," turning the history of the Schneider Races, as well as the Schneider Trophy, into British property. Because the United States had been a good sport in 1924 and not hosted a race that it would have inevitably won, the Schneiders ended on a high note for the British. British seaplanes won three straight victories, 1927, '29, and '31, set new world speed records with each victory, and took permanent possession of the Schneider Trophy in 1931. Even better, Lady Houston's gift, which rescued the British Schneider effort from death expected from cessation of government funding, falls into the British tradition of "making do." And it all paid off for Britain. The development of the Supermarine racers led to the Spitfire, which debuted in 1936 and had lines immediately recognizable as stemming from the racers. Out of sight, underneath its smooth engine cowling, the Spitfire was powered by a Merlin engine, an offspring of the Supermarine racers' Rolls-Royce R engines.

Making connections between the Supermarine racers and the Spitfire can lead to speculation about what would have happened if the racers had not been built. Would the Spitfire have come along anyway, would there have been a better fighter, or would Britain have suffered grievous harm with no Spitfire and no comparable fighter? Such hypothetical questions have no testable or provable answer. What is known is the relationships between the airframes and engines of the racers and of the Spitfire and the importance of the Spitfire to Britain in World War II.

Contrasts between the last Schneider and the last Pulitzer make the Schneiders' greater visibility appear inevitable. Although the 1931 Schneider lacked any competition between airplanes of different nations, the new Supermarine S-6B, as expected, raced around the course to set a new world speed record. The Schneider Trophy was a much sought prize; individuals, companies, and countries had tried to capture it for almost two decades. The Supermarine S-6B flight finally brought it home to Britain. A happy crowd of 500,000, some fortified with a few pints of this or that, spread along the north shore of the Isle of Wight and the south shore of England at Portsmouth to watch racers flash by over the sea. Some paid admission to grandstands, others spread blankets and picnic baskets at vantage points from which they could see part of the action. They were not disappointed. The British flew off with a new speed record and permanent possession of the Schneider Trophy.

The last Pulitzer Race was no contest for permanent possession of a trophy or any other prize. It was simply the latest in an increasingly poorly attended series of races. No huge festive crowd watched. Attendance of 500,000, the number that would see the 1931 Schneider, had been predicted for the National Air Race, featuring the Pulitzer. About one-tenth that number showed up. Cy Bettis in a Curtiss R3C-1 flew to a new world record, but what kind of celebration could be expected for an event reported in a *New York Times* article headlined "Pulitzer Race Won at 249-Mile Speed; Disappoints Fliers"?[25] Another race, perhaps as ho-hum as 1925's, was expected to take place in 1926. Who in the small crowd at Mitchel Field could have known that the landing of the army's Cy Bettis and the navy's Al Williams in their Curtiss R3C-1s marked the end of the Pulitzers?

T.S. Eliot wrote the poem "Hollow Men" about significantly more profound subjects than air racing in 1925, the year of the last Pulitzer. The poem's last line, "This is the way the world ends, not with a bang but a whimper," captures the end of the Pulitzer Races. The Schneiders ended with a bang, a great triumph. History treats bangs better.

The announcement that no airplane was available to fly in the 1926 Pulitzer[26] sealed the end of the races. The Pulitzers were over. Had they been successful? Answering the following five questions provides some measure of their success:

1. Did the Pulitzer Races result in better-performing, faster airplanes? (Races are supposed to do that.)
2. Would Pulitzer racers have been competitive with other countries' racers if foreign racers had flown?
3. Were design features and innovations transferred from racers to other aircraft?

4. Did any of the racing-associated innovations achieve commercial success?
5. Did the races attract and hold the public's attention for the possible betterment of aviation?

The answers are yes:

1. Combinations of more powerful engines and better designs produced a 93-mph increase in winning race speeds (table 8). Better speeds were not limited to winners. Except in 1924, when no new racers were built, speeds for other racers as well as the winners increased each year (figure 6).
2. Beginning in 1921, Pulitzer racers set world records for closed-circuit flights of 100-, 200-, and 300-km and 500-km flights. In 1922 and 23, Pulitzer racers set world records for absolute speed. Pulitzer racers, mounted on pontoons, won the Schneider Races in 1923 and 1925 against international competition, setting world records for seaplanes.
3. U.S. Army pursuit aircraft provide the clearest evidence of "technology transfer" from racers to other aircraft. The U.S. Army Air Service purchased Curtiss PW-8 pursuits, closely resembling Curtiss racers, in 1922, and updated Curtiss biplanes and pursuits from other manufacturers that resembled the Curtiss racers that remained in army service until 1936.
4. An unanticipated byproduct of the Pulitzers, the sale of Curtiss engines and Curtiss-Reed propellers in foreign countries, was a significant event for United States aviation. The sales opened up foreign aviation markets and established United States companies as competitors in international markets.
5. The million or so people who saw at least part of a Pulitzer Race — whether after paying admission to an airfield, from a car parked outside, or sitting on the ground or a front porch watching racers fly over — surely thrilled to the sight and sound of small, fast airplanes near the ground. Far greater numbers read about the races, the racers, and the racer pilots whose pictures and achievements appeared in newspapers coast to coast. The grainy black and white films of fast-moving airplanes that appeared in movie theaters conveyed little detail about the racers and pilots but captured the races' speeds, turns, and thrills. Excitement about the races generated interest in faster airplanes, and that interest, even if only for a few years, spread to encompass all of aviation.

History, in general, ignores air races and, perhaps, ignores the Pulitzers in particular.[27] Nevertheless, for the few years that they were flown, the Pulitzer competitions marshaled government support for companies that built racers; engineers and designers that developed better airplanes and engines; and the pilots who flew the racers and the many people who serviced them. Civic organizations, motivated by the prestige and public acclaim that accompanied the Pulitzers, financed, sponsored, publicized and hosted air meets. Hundreds of thousands of spectators flocked to races to see head-to-head competition, the ultimate test of high-speed designs and the final arbiter in picking faster airplanes.

The Pulitzer Races, in concert with racing's importance to the development of aviation, came and went quickly. Five years after the first race, racing's allure had faded and its usefulness as a "laboratory" for high-speed flight was passing. But in those years, small, noisy, airplanes had flown flat out on straightaways and turned corners at a few hundred feet over spectators' heads. For those years, racing ruled United States skies and aviation.

Appendix: Pulitzer Racer Specifications
First Three Finishers in First Pulitzer, 1920; All Starters in Pulitzer Races Two Through Six, 1921–1925[a]

First Pulitzer, 1920
First three finishers. A total of about 36 airplanes started and 24 airplanes finished the race.

Place in race Manufacturer	Racer (No. in race)[b]	Engine (nominal/ maximum horsepower)[c]	Loaded Weight (lbs)	Wingspan[d]	Wing Area (sq. ft.)	Power Loading (lbs/hp)[e]	Wing Loading (lbs/sq.ft.)[f]	Speed in Pulitzer[g]	Top Speed[h]
1. U.S. Army	VCP-R (1)	Packard A-2025, V-12 (578/638)	3200	u. 27 ft 6 in l. 28 ft 2 in	228.5	5.5	14.0	156.5	186
2. Thomas Morse	MB-3 (1)	Wright H-2, V-8 (340/400)	2095	u. 26 ft 0 in l. 24 ft 6 in	250.5	6.54	8.36	148.2	163.7
3. Ansaldo	Balilla (1)	S.P.A. 6A, in-line 6 (205/225)	1950[i]	u. 25 ft 4 in[i] l. 25 ft 4 in	228[i]	9.28[i]	8.55[i]	134.0	139.9

Second Pulitzer, 1921
All starters.

Place in race Manufacturer	Racer (No. in race)	Engine (nominal/ maximum horsepower)	Loaded Weight (lbs)	Wingspan	Wing Area (sq. ft.)	Power Loading (lbs/hp)	Wing Loading (lbs/sq.ft.)	Speed in Pulitzer	Top Speed
1. Curtiss	CR (1)	Curtiss CD-12 V-12 (372/420)	2158	u. 22 ft 8 in l. 22 ft 8 in	168	5.26	12.85	176.8	185.3
2. Curtiss	Cactus Kitten (1)	Curtiss C-12 V-12 (NR[j]/427)	2406	u. 20 ft 0 in m. 20 ft 0 in l. 20 ft 0 in	179.5	5.53	13.40	170.3	196
3. Thomas Morse	MB-6 (R-2) (1)	Wright H-2 V-8 (340/400)	2023	u. 19 ft 0 in l. 19 ft 0 in	157.1	5.95	12.88	160.7	NR[j]
4. Ansaldo	Balilla (1)	Curtiss K-12 V-12 (415/428)	2367	u. 26 ft 0 in l. 26 ft 0 in	230	5.92	10.29	149.7	170
DNF[k] Ansaldo	SVA-9 (1)	S.P.A. 6A, in-line 6 (205/225)	2145	u. 29 ft 10 in l. 29 ft 10 in	289.5	10.21	7.41	105 (2 laps)	147.9
DNF ThomasMorse	MB-7	Wright H-2 V-8 (340/400)	1875	24 ft 0 in	112	4.69	16.74	(1 lap)	180.8

Third Pulitzer, 1922
All starters.

Place in race Manufacturer	Racer (No. in race)	Engine (nominal/ maximum horsepower)	Loaded Weight (lbs)	Wingspan	Wing Area (sq. ft.)	Power Loading (lbs/hp)	Wing Loading (lbs/sq.ft.)	Speed in Pulitzer	Top Speed
1. & 2. Curtiss	R-6 (2)	Curtiss D-12 V-12 (450/468)	2121	u. 19 ft 0 in l. 19 ft 0 in	135.9	4.37	15.61	205.9	240
3. & 4. Curtiss	CR-2 (2)	Curtiss D-12 V-12 (450/468)	2212	u. 22 ft 8 in l. 22 ft 8 in	168	5.40	13.17	193.7	198.8
5., 7. & DNF Verville-Sperry	R-3 (3)	Wright H-3 V-8 (NR/407)	2380	29 ft 6 in	144.3	5.95	16.50	180.7	191.1
6. Verville	R-1 (1)	Packard 1A-2025 V-12 (578/638)	3394	u. 23 ft 0 in l. 23 ft 0 in	222.7	5.32	15.24	178.8	NR
8. & 9. Loening	R-4 (2)	Packard 1A-2025 V-12 (578/638)	3050	26 ft 11 in	174.0	5.08	17.53	170.0	190
10. Thomas Morse	R-5[l]	Packard 1A-2025 V-12 (578/638)	2850[m]	25 ft 0 in	150	4.75[n]	19.0	155.4	187.7
11. Thomas Morse	R-5[o]	Packard 1A-2025 V-12 (578/638)	2850	29 ft 0 in	174	4.75[n]	16.38	149.2	NR
DNF.[q] Aerial Engineering Co.	BR-1	Wright H-3 V-8 (NR/407)	2020	30 ft 6 in	NA	5.06	NA	163.3 (1 lap)	188.5
DNF. Thomas Morse	MB-7	Wright H-2 V-8 (326/358)	1875	24 ft 0 in	112	4.69	16.74	139.8 (2 laps)	180.8
DNF. Wright	NW-1	Wright Tornado T-2 V-12 (NR/650)	3000	u. 30 ft 6 in l. 14 ft 6 in	180	4.62	16.67	186.0 (3 laps)	209

Fourth Pulitzer, 1923
All starters.

Place in race Manufacturer	Racer (No. in race)	Engine (nominal/ maximum horsepower)	Loaded Weight (lbs)	Wingspan	Wing Area (sq. ft.)	Power Loading (lbs/hp)	Wing Loading (lbs/sq.ft.)	Speed in Pulitzer	Top Speed
1. & 2. Curtiss	R2C-1	Curtiss D-12A V-12 (507/520)	2112.3	u. 22 ft 0 in l. 19 ft 3 in	144.2	4.22	14.64	243.7	266.6
3. Wright	F2W-1	Wright Tornado T-3 V-12 (680/780)	2858[p]	u. 22 ft 6 in l. 22 ft 6 in	174	3.81	16.42	230.1	247.7
4. Wright	F2W-1	Wright Tornado V-12 (680/780)	3086[q]	u. 22 ft 6 in l. 22 ft 6 in	174	4.11	17.74	230.0	NR
5. Curtiss	R-6	Curtiss D-12A V-12 (507/520)	2121	u. 19 ft 0 in l. 19 ft 0 in	135.9	4.18	15.61	218.9	NR
6. Curtiss	R-6	Curtiss D-12 V-12 (450/468)	2121	u. 19 ft 0 in l. 19 ft 0 in	135.9	4.42	15.61	216.4	240

Fifth Pulitzer, 1924
All starters.

Place in race Manufacturer	Racer (No. in race)	Engine (nominal/ maximum horsepower)	Loaded Weight (lbs)	Wingspan	Wing Area (sq. ft.)	Power Loading (lbs/hp)	Wing Loading (lbs/sq.ft.)	Speed in Pulitzer	Top Speed
1. Verville-Sperry	R-3	Curtiss D-12A V-12 (507/520)	2578	30 ft 7 in	146.5	5.39	16.91	216.6	235
2. Curtiss	R-6[r]	Curtiss D-12A V-12 (507/520)	2121	u. 19 ft 0 in l. 19 ft 0 in	135.9	4.18	15.61	214.4	240
3. Curtiss	XPW -8B[s, t]	Curtiss D-12 (470/478)	2280	u. 30 ft 0 in l. 30 ft 0 in	255.0	NR	NR	168.0	172

Sixth Pulitzer, 1925
All starters.

Heat 1: Racers

Place in race Manufacturer	Racer (No. in race)	Engine (nominal/ maximum horsepower)	Loaded Weight (lbs)	Wingspan	Wing Area (sq. ft.)	Power Loading (lbs/hp)	Wing Loading (lbs/sq.ft.)	Speed in Pulitzer	Top Speed
1. & 2. Curtiss	R3C-1 (2)	Curtiss V-1400 V-12 (565)	2181	u. 22 ft 0 in l. 22 ft 0 in	144	3.64	15.15	249.0	285

Heat 2: Pursuits[v]

Place in race Manufacturer	Racer (No. in race)	Engine (nominal/ maximum horsepower)	Loaded Weight (lbs)	Wingspan	Wing Area (sq. ft.)	Power Loading (lbs/hp)	Wing Loading (lbs/sq.ft.)	Speed in Pulitzer	Top Speed
3. Curtiss	P-1 (1)[u]	Curtiss D-12 (470)	2846	u. 31 ft 6 in l. NA	250	6.06	11.38	169.9[v]	163[v]
4., 5. & DNF Curtiss	PW-8[u] (3)	Curtiss D-12 (470)	3155	u. 32 ft 0 in l. 32 ft 0 in	279	6.72	11.31	168.8	168[v]

Notes

[a] Data are from T.G. Foxworth, *The Speed Seekers* (New York: Doubleday, 1974), Appendices 3 and 4, pp. 480–529 unless otherwise specified.

[b] Number of aircraft of that type that flew in that year's Pulitzer.

[c] Foxworth "usually" calculated "power loading," the weight of the airplane divided by horsepower using maximum horsepower ratings (T.G. Foxworth, *The Speed Seekers* [New York: Doubleday, 1974], 534 @ 205). In addition to "nominal horsepower," which is used throughout this book to describe engines, "maximal horsepower," is also used, both written in the form nominal/maximal

[d] u denotes upper wing; l lower wing; m middle wing. Monoplanes have no letter.

[e] Power loading is calculated as weight of airplane divided by horsepower of engine. Foxworth "usually" calculated power-loadings based on maximum horsepower of the engines (T.G. Foxworth, *The Speed Seekers* [New York: Doubleday, 1974], 534 @ 205). In general, aircraft with lower power loadings are faster. [Dividing the loaded weight of an airplane by the maximum horsepower, as presented by Foxworth, does not, in many cases, produce the number for power loading reported by him (T.G. Foxworth, *The Speed Seekers* [New York: Doubleday, 1974], 480–529).

[f] Wing loading is calculated as weight of airplane divided by wing area. In general, aircraft with lower wing loading are more maneuverable, can glide farther, and fly at lower speeds.

[g] Speed of the faster or fastest racers of that model that completed the race.

[h] "Top speeds" were obtained under varying conditions at different times, making comparisons among top speeds uncertain.

[i] Calculated from metric values reported by Foxworth (T.G. Foxworth, *The Speed Seekers* [New York: Doubleday, 1974], 508–509).

[j] NR, Not Reported by Foxworth (T.G. Foxworth, *The Speed Seekers* [New York: Doubleday, 1974], 480–529).

[k] Did Not Finish. Pulitzer Speeds calculated for the number of laps completed.

[l] Short-wing version.

[m] T.G. Foxworth, *The Speed Seekers* (New York: Doubleday, 1974), 496–497, provides only one weight for both the short- and long-wing versions, which is surely incorrect.

[n] Both the short- and long-wing versions had the same engine, but their weights were surely different. It is impossible that the power loading value of 4.75 is correct for both airplanes.

[o] Long-wing version.

[p] 30 gallon fuel tank.

[q] 60 gallon fuel tank.

[r] A second R-6 crashed at the start, killing Capt. Burt Skeel, its pilot. Skeel's was the only fatality in the Pulitzers.

[s] T.G. Foxworth, *The Speed Seekers* (New York: Doubleday, 1974) @ 458 identifies this airplane as a PW-8A. It was not. The PW-8B was a PW-8 from which the wing radiators were removed and a "tunnel" radiator was installed under the nose (see photo in Chapter Six). The airplane was variously called, perhaps in slightly different versions, the "XPW-8A, second modification," the "XPW-8B" and the "PW-8B." The airplane was the prototype for the Curtiss P-1 pursuit. The PW-8A was an earlier PW-8 modification that had a conventional radiator mounted in the upper wing (see photo in Chapter Six).

[t] Specifications for the PW-8B, PW-8, and P-1, except for identification of the engine, horsepower, and speed in the Pulitzer, are from E. Angelucci and P. Bowers, *The American Fighter* (New York: Orion, 1985), 125–131.

[u] Standard army pursuits. They were not competitive with the racers.

[v] The discrepancies between the measured top speeds in the Pulitzer and the published straightaway top speeds, which should be measurably faster, reflect differences in reporting rather than reality.

Chapter Notes

Chapter One

1. F. Kafka, "Die Aeroplane in Brescia," *Prague Bohemia*, September 29, 1909 (translation in P. Demetz, *The Air Show at Brescia, 1909* (New York: Farrar, Straus, and Giroux, 2002, 79–80).
2. *New York Times*, "Pulitzer Air Trophy to Be Perpetual Prize," August 29, 1920, p. 21.
3. J.M. Morris, *Pulitzer: A Life in Politics, Print, and Power* (New York: Harper, 2010), 476.
4. S. Shulman, *Unlocking the Sky: Glenn Hammond Curtiss and the Race to Invent the Airplane* (New York: HarperCollins, 2002).
5. *New York Times*, "Conditions Fixed for St. Louis Flight," July 17, 1910, p. 1.
6. "St. Louis to New York, Cross-Country Flight," *Flight*, September 23, 1911, p. 832.
7. M. Maurer, *Aviation in the U.S. Army, 1919–1939* (Washington, DC: Office of Air Force History, 1987), 30.
8. *New York Times*, October 8, 1919, quoted in Leary, "W.M. "Billy Mitchell and the Great Transcontinental Air Race of 1919," *Air University Review*, May-June 1984, http://www.airpower.maxwell.af.mil/airchronicles/aureview/1984/may-jun/leary.html.
9. Leary, 3.
10. Manufacturers Aircraft Association, *Aircraft Year Book* (New York: Doubleday, Page, 1920), 273.
11. Office of Director of Air Service, "Report on First Transcontinental Reliability and Endurance Test," *Air Service Information Circular* I, 5 February 1920, p. 6 (cited in R.L. Bowers, "The Transcontinental Reliability Test," *Airpower Historian* 8: 45–54).
12. R.L. Bowers, "Aviation on Trial: Success and Catastrophe; Transcontinental Reliability Test—Part II," *Airpower Historian* 8: 88–100.
13. Leary, 6.
14. Bowers, "Aviation on Trial," 99.
15. Ibid.
16. M.G. Rawls, "Belvin W. Maynard and Maynard Field," *The Past Is Present* 1, June 2008, Winston-Salem, NC: Preserve Historic Forsyth.
17. See Fédération Aéronautique Internationale (see FAI, http://www.fai.org/, for information about the charter, history, and activities of the FAI).
18. See H.S. Villard, *Blue Ribbon of the Air: The Gordon Bennett Races* (Washington, DC: Smithsonian, 1987), and M. Gough, "Blue Ribbon of the Air," *Aviation History*, January 2011, pp. 54–59.
19. M.J. Gross, Jr., "Afterword: The Future of the NAA; A Forecast from the President," in B. Robie's *For the Greatest Achievement* (Washington: Smithsonian Institution Press, 1993), 219–225.
20. Unless otherwise referenced, the information about the Aero Club of America, the American Flying Club, and the National Aeronautics Association in this chapter is from B. Robie's *For the Greatest Achievement* (Washington: Smithsonian Institution Press, 1993), especially chapter 11, "Times of Change," and chapter 12, "A New Beginning," 93–117.
21. Belvin Maynard won the New York to Toronto Race. He and the first four military fliers who followed him across the finish line flew DH-4s.
22. "Plan First Flight Around the World," *Aerial Age Weekly*, August 9, 1920, p. 736.
23. Robie, 99.
24. C.C. Grey ("Who's Who in American Aviation," *Aeroplane*, February 4, 1920, pp. 240–41) called for the FAI to recognize the Flying Club or some completely new organization as its representative in the U.S.
25. *New York Times*, "Seeks to Oust Aero Club," January 6, 1920, p. 14.
26. *New York Times*, "Our Aero Beginning Outstrips Europe's," October 29, 1916, p. 1.
27. *New York Times*, "Officers Uphold Aero Club Fund," June 9, 1920, p. 32.
28. W. Mitchell, quoted in Robie, 100.
29. "Minutes of a Meeting of the Contest Committee of the ACA, Held at 11 East 38th Street, Tuesday, 24th August, 1920, at Five o'clock P.M.," Archives, National Air and Space Museum, Smithsonian Institution, Washington, DC, acquisition number XXXX-0627.
30. *New York Times*, "Pulitzer Air Trophy to Be Perpetual Prize," August 29, 1920, p. 21.
31. In this book, "first Pulitzer race" refers to the 1920 race. The 1919 cross-country contest to and from Atlantic City is ignored in the numbering of the races.
32. *New York Times*, "World's Swiftest Planes to Fly Here," November 21, 1920, pp. 1 and 3.
33. *New York Tribune*, "Aero Club Closed-Circuit Race Set for Thanksgiving," August 29, 1920.
34. *First Airplane Race for Pulitzer Trophy and*

Valentine Liberty Bond Prizes, from the printed program for the 1920 Pulitzer Race (NASM Archives).

35. *New York Times*, "Army Pilot Wins Pulitzer Air Race," November 26, 1920, pp. 1, 3. Henry Woodhouse's *Aerial Age Weekly* reported the race results in "The Pulitzer Race," December 6, 1920, pp. 343–344. The magazine article is a near word-for-word copy of the *New York Times* article.

Chapter Two

1. M. Rubenstein and R. Goldman, *To Join with the Eagles: Curtiss-Wright Aircraft, 1903–1965* (New York: Doubleday, 1974), 68.
2. The number of airborne racers is variously reported, with both 36 and 38 commonly occurring. Either would be the largest number ever in a closed-course race.
3. J.E. Kordes, *Visions of Garden City* (Garden City, NY: John Ellis Kordes, 2007), 5.
4. Interview with John Ellis Kordes, Garden City Village, Long Island, New York, village historian, January 3, 2010.
5. G.C. Dade and F. Strand, *Picture History of Aviation on Long Island, 1908–1938* (New York: Dover, 1989), 5.
6. M.H. Smith, *Garden City, Long Island, in Early Photographs, 1789–1919* (New York: Dover, 1998), 70.
7. *Brooklyn Daily Eagle*, "France May Enter Pulitzer Air Race," October 25, 1920, p. 1.
8. *New York Times*, "World's Swiftest Planes to Fly Here," November 21, 1920, pp. 1 and 3.
9. Subheading in "World's Swiftest Planes to Fly Here," *New York Times*, November 21, 1920, pp. 1, 3.
10. See chapter 9 for more information on nominal and maximum. Also see T.G. Foxworth, *The Speed Seekers* (New York: Doubleday, 1974), Appendix 4, note 6.
11. Some authors write that "VCP" stands for "Verville, Clark, Packard," representing, respectively, Verville, a famous designer of wing airfoils, Virginius Evans Clark, who designed the VCP's wing, and the Packard engine. Certainly, "P" cannot have stood for "Packard," The first version of the airplane was powered by a Wright-built Hispano-Suiza engine. "VCP-1" was painted on its rudder.
12. Foxworth (*Speed Seekers*, 172) reports that the car's occupants were attempting to time the flight. Another source ("Verville VCP/R-1," *Skyways* 25, January 1993, pp. 1–15) states that the occupants were making a motion picture of the flight. It is possible that the car's occupants were attempting to do both.
13. "The Packard 500–600 Hp. Aircraft Engine," *Aviation*, July 15, 1920, pp. 470–472.
14. R.W. Schroder, "Why We Lost," letter to Alfred Verville, reprinted in *Slipstream*, November 1, 1920, pp. 8–9.
15. "Thanksgiving Day Race for Pulitzer Trophy Arousing Great Interest," photocopied typed page, no date; there appears a notation, "V-2778, A.S. [Air Service]," at the bottom right of the page (National Air and Space Museum Archives).
16. "The Pulitzer Trophy Race," *Aviation and Aircraft Journal*, December 6, 1920, pp. 378–381.

17. "Verville VCP-R/R-1," *Skyways* 25, January 1993, pp. 1–15.
18. "The Curtiss Model 18-T Triplane," *Flight*, May 29, 1919, pp. 698–700; "The Curtiss Model 18-B Biplane," *Flight*, July 10, 1919, pp. 902–904.
19. "The Pulitzer Trophy Race," *Aviation and Aircraft Journal*, December 6, 1920, pp. 378–381.
20. "Thanksgiving Day Race for Pulitzer Trophy Arousing Great Interest," photocopied typed page, no date; the notation "V-2778, A.S," appears at the bottom right of the page (National Air and Space Museum Archives).
21. Lord Semphill of the Royal Air Force made this comment in a report to the British government (quoted in Foxworth, *Speed Seekers*, 289).
22. The bitter, bloody conflict between Austria and Italy during World War I, sometimes overlooked, rivaled the war in France for brutality and may have exceeded it in futility. Germany and Great Britain contributed forces to the "Italian Front" in the incredibly rough and difficult Alps, but the war was largely fought between the armies of Austria and Italy. Ernest Hemingway's novel *A Farewell to Arms* is set at the Italian Front.
23. *New York Times*, "Army Pilot Wins Pulitzer Air Race," November 26, 1920, pp. 1, 3.
24. Glenn Curtiss had hired Thomas, an Englishman from Sopwith, and Thomas designed the famous Curtiss JN-4 "Jenny," He subsequently moved to Thomas Morse (he was not a relative of the "Thomas" for whom the company was named).
25. "Conditions of Contest" in "First Airplane Race for Pulitzer Trophy and Valentine Liberty Bond Prizes," in the program for the 1920 Pulitzer Race (National Air and Space Museum Archives).
26. "Inside Comment by Captain Hartney, Who Finished Second: Tribute to Moseley," *U.S. Air Service*, December 1920, pp. 15–16, 34–35.
27. "The Pulitzer Trophy Race," *Aviation and Aircraft Journal*, December 6, 1920, p. 378–381.
28. Foxworth, *Speed Seekers*, 290.
29. "Army Flyer Winner of America's Greatest Race," *U.S. Air Services*, December 1920, pp. 12–16, 34.
30. Unless noted otherwise, all speeds are corrected speeds, not the speeds that were announced and published immediately after the conclusion of the race.
31. See chapter 3.
32. Foxworth, *Speed Seekers*, 290.
33. Foxworth (*Speed Seekers*, 173) and *Aviation and Aircraft Journal* (December 6, 1920, pp. 378–381) credit Moseley with shooting down one German airplane. Moseley is mentioned several times in the history "1st Pursuit Group History" for August-September, October, and November, 1918 (http://www.acepilots.com/wwi/us_1st_3.html), but the unit history does not confirm the aerial victory.
34. "Verville VCP-R/R-1," *Skyways* 25:1–15, p. 7, January 1993, p. 7.
35. Ibid., 9.
36. Ibid., 7.
37. Foxworth, *Speed Seekers*, 174.
38. R.L. Trimble, "Fred Verville and His Flying Eggshell," *Air Classics Quarterly Review* (vol 4, no. 4, Winter 1977), pp. 54–61.
39. "Army Pilot Wins Pulitzer Air Race," 6.
40. "New Speed Record," *Aviation*, November 29, 1920, p. 327.

41. A. Klemin, "Notes on the Pulitzer Air Race," *Aviation and Aircraft Journal*, December 6, 1920, pp. 382–384.
42. "Inside Comment by Captain Hartney, Who Finished Second: Tribute to Moseley," *U.S. Air Service* 4:15 (1920), 16, 34, and 35.
43. Klemin, "Notes on the Pulitzer Air Race," 382–384.
44. *New York Times*, "Maynard Questions Lt. Mosley's Speed," November 27, 1920, p. 8.
45. T.G. Foxworth, "Thomas Morse Military Racing Airplanes," *AAHS Journal* 12 (1967), 235–253.
46. "The Pulitzer Trophy," *Aviation and Aircraft Journal*, December 6, 1920, p. 377. The "record book" remark was published when it appeared the VCP-R had flown at 178 mph.
47. "American Pilots," *Aviation and Aircraft Journal*, December 6, 1920, p. 377.
48. "The Pulitzer Trophy Race," *Aviation and Aircraft Journal*, December 6, 1920, pp. 378–381.
49. Roughly $3,000,000 in 2012 dollars.
50. "The Pulitzer Trophy Race," *Aviation and Aircraft Journal*, December 6, 1920, pp. 378–381.
51. "The Pulitzer Trophy," *Aviation and Aircraft Journal*, December 6, 1920, p. 377.
52. G. Curtiss, quoted in *Aircraft Year Book* (Boston: Small, Maynard, 1921), 135.

Chapter Three

1. "The Second Pulitzer Trophy Race," *Aviation and Aircraft Journal*, November 14, 1921, p. 559.
2. *New York Times*, "Maynard Questions Lt. Mosley's Speed," November 27, 1920, p. 8.
3. *New York World*, "Pulitzer Trophy Air Race Goes to Omaha; Fly Nov. 3," August 14, 1921, pp. 1,
4. *New York Times*, "Lecointe Cables Entry," July 21, 1921.
5. *New York Times*, "Cancel Pulitzer Air Race," July 19, 1921, p. 11.
6. *New York World*, "Pulitzer Trophy Air Race Goes to Omaha; Fly Nov. 3," August 14, 1921, p. 1.
7. Astonishing amounts of money were awarded to flyers in the 1909–1910 period. Curtiss won $10,000 for a flight of any kind at the 1910 air meet in Omaha. First prize in the 1920 Pulitzer, after a decade of inflationary pressures and the First World War, was worth only $3,000.
8. "Army Balloonists Wanted," *U.S. Air Service*, November 1920, p. 18.
9. *Omaha Daily News* "He's Acclaimed as Omaha's Only Flying Ace," Sunday Magazine and Story Section, March 16, 1919, pp. 1, 2, 10.
10. R.E. Adwers, *Rudder, Stick and Throttle: Research and Reminiscences on Flying in Nebraska* (Omaha: Making History, 1994), 230.
11. *New York World*, "Pulitzer Trophy Air Race Goes to Omaha; Fly Nov. 3," August 14, 1921, p. 3.
12. *New York World*, "Omaha Glad It Got Pulitzer Air Race," August 15, 1921. This article exists as a clipping in the Archives of the National Air and Space Museum. The masthead has been clipped off, and the identification of the newspaper is based on "Special to the World" as the byline.

13. "Pulitzer Air Race for Detroit in 1922," newspaper clipping dated "Aug 1921" with no reference to its source from NASM archives.
14. *Omaha Evening Sun*, "She Wants to Sell Aero Stamps, Too," October 4, 1921, p. 2. I've found no information about the selling price for the stamps, but it may have been a penny. The lowest-priced "saving stamps" sold in schools to raise money for the war effort in World War II cost one penny.
15. *New York Times*, "Offer 500 Mails Fliers," October 19, 1921, p. 2.
16. *Omaha World-Herald*, "Enter Seventy-Three Planes in Air Races" and "Fly for These Trophies at Omaha Air Meet," October 25, 1921. The "Trophies" were the silver Pulitzer Trophy and the silver and gold Larsen Trophy, to be awarded to the "commercial, cargo-carrying" plane that made the most economical round-trip flight between Omaha and Des Moines, Iowa.
17. *Omaha World-Herald*, "Will Determine Fate of Aerial Congress Today," October 26, 1921, p. 1.
18. *Omaha World-Herald*, "Big Aerial Congress Saved for Omaha," October 28, 1921, p. 1.
19. The long-feared railroad strike was called off on October 28.
20. "Air Mail Performance Improved," *Aviation*, November 14, 1921, p. 570.
21. *Omaha World-Herald*, "The Omaha Spirit," October 29, 1921, editorial.
22. R.E. Adwers, *Rudder, Stick and Throttle: Research and Reminiscences on Flying in Nebraska* (Omaha: Making History, 1994).
23. T.G. Foxworth, *The Speed Seekers* (New York: Doubleday, 1974), 33.
24. *New York World*, "Omaha Glad It Got Pulitzer Air Race."
25. There is disagreement about whether the windmill was the marker at the first or second turn. Adwers (*Rudder, Stick and Throttle*, 230) reports that the windmill was at the first turning point at Loveland, Iowa. The author has relied upon Adwers' account. Thomas Foxworth (*Speed Seekers*, 33) writes that the windmill was at the second turn.
26. Foxworth, *Speed Seekers*, 33.
27. Originally the course was to have been flown in the opposite direction, to Calhoun then Loveland and return.
28. Other airplanes—a Thomas Morse MB-3, two Ansaldos, a SPAD, and a Miller monoplane—were formally entered in the race, but all were withdrawn.
29. "The Aerial Derby," *Flight*, July 14, 1921, pp. 468–471.
30. "The Winner of the Derby," *Flight*, July 21, 1921, p. 490.
31. "The 'Alula' Wing Demonstrated," *Flight*, October 22, 1921, p. 687;
32. R.K. Bagnell-Wild, "The Progress of Research," *Flight*, February 23, 1922, pp. 122–124.
33. T.G. Foxworth, "Thomas Morse Military Racing Airplanes," *AAHS Journal* (Winter 1967), 235–253.
34. Ibid.
35. The 18-T could be fitted with either of two wings. For speed work, a short-span, single-bay wing was mounted, and the airplane was designated 18-T-1. A longer, two-bay wing was employed for high-altitude work on the 18-T-2. Rudolf Rohlfs flew the 18-

T-2 to a world altitude record of 32,450 feet on September 18, 1920. In 1919, another "Curtiss Wasp," as the 18-Ts were called, made the first flight from LaPaz, Bolivia, which is 13,000 feet above sea level. Purchased by Bolivia, the Wasp was the single airplane in the Bolivian Air Force until it crashed a year later.

36. Attributed to the Federal Trade Commission (Olien and Olien, *Easy Money*, p. 104).
37. Olien and Olien, *Easy Money*, 106.
38. "World Records (in Competition)," advertisement placed by the Curtiss Aeroplane and Motor Company in *Organization of Aeronautic Contests and Contest Rules* (Contest Committee, Aero Club of America: New York City, 1922), no page number but located toward the back of the book.
39. *New York Times*, "Claims World's Air Record," August 30, 1920, p. 6.
40. Foxworth, *Speed Seekers*, 183.
41. Ibid., 184.
42. Ibid., 185.
43. *Omaha World-Herald*, "World's Fastest Planes Will Fly at Big Meet," October 28, 1921, p. 1.
44. *Omaha World-Herald*, "'Texas Wildcat' Will Not Race; Said Too Speedy," October 31, 1921, p. 1.
45. *Omaha Bee*, "Rohlfs Going to See Air Meet," November 1, 1921, p. 2.
46. I refer to both 1921 Curtiss racers as "CR," Curtiss designated the first of the 1921 racers as "CR-1" and the second "CR-2." Both the "CR-2" and the "CR-1," which flew in the 1920 Pulitzer, flew in the 1921 Pulitzer.
47. T.G. Foxworth, "The Curtiss R-6 Racer: Army's Bid for Speed," *Journal American Aviation History Society* (Summer 1970), 73–91.
48. "The Aviation Meeting at Omaha," *Aviation*, November 14, 1921, pp. 565–570.
49. *Omaha Daily Bee*, "Nine Fastest Speed Planes Ever Built Scheduled to Start in Pulitzer Trophy Race Today," November 3, 1921, p. 1.
50. *Omaha Daily Bee*, "Acosta Is Air Classic Favorite," November 2, 1921, p. 1.
51. *Omaha World-Herald*, "Will Need Many Stop Watches at Aerial Congress," October 31, 1921, p. 2.
52. Adwers, *Rudder, Stick and Throttle*, 241.
53. "The Aviation Meeting at Omaha," *Aviation*, November 14, 1921, pp. 565–570.
54. Ibid.
55. A flying wire is a bracing wire between the lower part of the fuselage (or in some cases the landing gear) and the bottom of a wing, designed to counter the upward forces on the wing. A landing wire runs from near the top of the fuselage (or a kingpost) to the top of a wing to counter downward forces on the wing.
56. Adwers, *Rudder, Stick and Throttle*, 241.
57. "The Aviation Meeting at Omaha," *Aviation*, November 14, 1921, pp. 565–570.
58. *Omaha Daily Bee*, "Coombs' 'Taming' of *Cactus Kitten* Thrills Mail Pilots, Veterans of Air," November 4, 1921, p. 2.
59. Adwers, *Rudder, Stick and Throttle*, 243.
60. Foxworth, "Thomas Morse Military Racing Airplanes," 242.
61. Adwers, *Rudder, Stick and Throttle*, 243. How to turn the pylons was a hot topic in the 1920s. Bertaud, for instance, started his turn low and climbed as he turned, continued to climb between the pylons, lost altitude as he neared the next one, and zoomed up and around it. James Doolittle, and others, preferred to dive into the turns, climb between the pylons, and then dive into the next one. At some time, everyone apparently became convinced that the best way was to maintain level flight and bank as sharply as possible.
62. Adwers, *Rudder, Stick and Throttle*, 240–246.
63. This recounting of the crash differs from Foxworth's. Foxworth writes that the troublesome fuel pump caused the engine to quit as Hartney was flying about 180 mph at low altitude about two miles short of the first turn. Trading speed for altitude, he was able to climb to about 1,000 feet, working desperately to switch on the MB-7's gravity-fed reserve fuel tank. Unsuccessful at that effort, he looked for a landing place and had to settle for a small plowed field. He made a good upwind approach but hit the ground at high speed. The landing gear was ripped off, and the racer slid on its belly, hit a gully, and threw Hartney from the cockpit. His helmet was ripped from his head as he slid along the ground, his thigh bone was broken, and his hip was severely fractured (Foxworth, "Thomas Morse Military Racing Airplanes," 244).
64. *New York Times*, "Bert Acosta Wins Air Race Trophy," November 4, 1921, p. 17.
65. Foxworth, "Thomas Morse Military Racing Airplanes," 235–253. (Beauty is in the eye of the beholder. The Alula wing is far from pretty in mine.)
66. Foxworth, *Speed Seekers*, 539.
67. *Omaha Daily Bee*, "First Thought of Injured Pilot Is of Wife," November 4, 1921, pp. 1–2.
68. *Omaha Daily Bee*, One Dead, One Hurt at Air Meet," November 4, 1921, p. 1.
69. *New York Times*, "Bert Acosta Wins Air Race Trophy," November 4, 1921, p. 17.
70. Foxworth, *Speed Seekers*, 186. Why right-hand turns would have made a difference isn't clear.
71. Foxworth, "Thomas Morse Military Racing Airplanes," 235–253.
72. "Transport: Pilot's Pilot," *Time*, June 10, 1935, http://www.time.com/time/magazine/article/0,9171,883487-1,00.htm.
73. Quoted in "Bertrand 'Bert' Blanchard Acosta," http://www.earlyaviators.com/eacosta1.htm.
74. Check-Six.com, "The Trans-Atlantic Flight of the 'America,'" www.check-six.com/Crash_Sites/America-NX206.htm.
75. *New London (CT) Day*, "Byrd's Nerves Overstrained; Acosta's Collarbone Broken; Paris Gives Wild Welcome," July 2, 1927, p.1.
76. Foxworth, *Speed Seekers*, 199–200.
77. "The Second Pulitzer Trophy Race, *Aviation*, November 14, 1921, p. 559.
78. B. Robie, *For the Greatest Achievement* (Washington: Smithsonian, 1993), 101.
79. *Omaha Bee*, "Promoters Not to Get Control of Air Association," November 4, 1921, p. 2.
80. P.M.H. Lewis, "Air Racing in the U.S.A. Part 1," *Flight*, July 20, 1956, pp. 125–128.
81. "The Second Pulitzer Trophy Race," *Aviation*, November 1, 1921, p. 550.

Chapter Four

1. "Detroit Race Impressions," *Aviation*, October 23, 1922, pp. 550–557.

2. "The Curtiss Model CD-12 400 H.P. Aero Engine," *Flight*, January 5, 1922, pp. 7–9.

3. Detroit's population peaked at 1,850,500 about 1960 and declined to 870,000 by 2000. As measured by school enrollment, Detroit's loss of population between 2000 and 2011 was especially dramatic. School enrollment dropped 50 percent "as residents move[d] elsewhere in search of jobs" ("School's Out," *The Economist*, March 12 -18, 2011, pp. 38–40).

4. "The National Airplane Races," *Aircraft Year Book, 1923* (New York: Aeronautical Chamber of Commerce of America, 1923), 56–81.

5. *New York Times*, "206 Miles an Hour for 156 Miles Sets New Flying Record," October 15, 1922, pp. 1, 20.

6. "The National Airplane Races," 56–81.

7. T.G. Foxworth, "The Racing Curtiss Triplanes," *Journal American Aviation Historical Society* (Spring 1976), 32–37, p.36.

8. (Capt.) W.H. Sitz, USMC, "Navy Department Bureau of Aeronautics Technical Note No. 229. Navy Entries in National Airplane Races, Detroit, Mich," photocopied typescript, dated Oct. 5, 1922 (Library, Navy Yard, Washington, DC).

9. William A. Moffett (1869–1933) graduated from the United States Naval Academy in 1890. He was awarded the Medal of Honor for heroism at Vera Cruz in the war with Mexico in 1913, and after service on various surface vessels and commands in the navy was appointed director of Naval Aviation in 1921. He vigorously opposed General Mitchell's goal of having a single air force to include both army and navy aviation. Moffett was a champion of dirigibles. He died in the crash of the airship *Akron* off the New Jersey coast in 1933.

10. S.L. Brown, *A Genesis Workshop: Five Generations of Engineering Enterprise from the Birthplace of Aviation* (Wright Patterson AFB, OH: History Office, Air Force Material Command, www.ascho.wpafb.af.mil/Genesis/GENESIS.HT (p. 12).

11. Charles L. Lawrance (1892–1950) had built automobile engines before World War I, became interested in air-cooled engines in 1914, and managed to sell his J-1 radial engine to the navy just as the war ended. The navy was unconvinced that the small Lawrance Company could be a major supplier of engines and encouraged ("forced" might be a more appropriate word) a merger between that enterprise and the Wright Company. The Wright Company went on to produce the famous "Whirlwind" radials, and Lawrance, who became the company's president in 1925, was awarded the Collier Trophy in 1927 for his development of radial, air-cooled engines.

12. "The Curtiss Marine Flying Trophy Race," *Flight*, November 2, 1922, p. 637.

13. Don Green, talk delivered at a meeting of the Rochester-Avon (MI) Historical Society, Rochester Hills (MI) Public Library, May 7, 2009.

14. M.L. McDougal, *Mount Clemens: Bath City, U.S.A., in Vintage Postcards* (Chicago: Arcadia, 2000), i.

15. N.D. Longstaff, "The Mineral Bath Era," in *Centennial History of Mount Clemens, MI, 1879–1979*, ed. D.M. Magee (Mount Clemens: MI: Mt. Clemens Public Library, 1980), 81–96, and McDougal, *Mount Clemens: Bath City*, ii.

16. Joseph Pulitzer, his wife, and other members of his family traveled often to Europe to take the waters at various expensive and trendy spas and clinics.

17. Green talk delivered at the Rochester-Avon Historical Society.

18. Interview with Deborah J. Larsen, Mt. Clemens Public Library, May 9, 2009.

19. Longstaff, "The Mineral Bath Era," 81–96.

20. Interview with Larsen.

21. D.J. Larsen and L.J. Nigro, *Selfridge Field* (Charleston, SC: Arcadia, 2006), 14.

22. Ibid., 16.

23. J.S. Ketchum, "Selfridge Field" in *Centennial History of Mount Clemens, MI, 1879–1979*, ed. D.M. Magee (Mount Clemens, MI: Mount Clemens Public Library, 1980), 165–175.

24. "Army to Buy Selfridge Aviation Field," *Aerial Age Weekly*, April 26, 1922, p. 227.

25. Carl A. Spaatz (born "Spatz"; he added the extra "a" in 1937) was born in 1891. He graduated from West Point in 1914, became an army pilot in 1916, and commanded a training depot in France in 1918 before transferring to a combat command where he shot down three German aircraft in three weeks of combat. He commanded the 1st Pursuit Group, 1922–24. In 1929, he and Ira Eacker set a flight endurance record of 150 hours in the "Question Mark" airplane using aerial refueling. He was commander of the U.S. Army Air Forces in Europe during World War II and commanded the Army Air Forces in the Pacific after the surrender of Germany. He became commander in chief of the Army Air Force in 1946 and the first chief of staff of the United States Air Force in 1947. He retired from military service in 1948 and remained active in various aviation arenas before his death in 1974.

26. D.R. Mets, *Master of Airpower: General Carl A. Spaatz* (Novato, CA: Presidio, 1988), 51.

27. Ibid., 52.

28. "The National Airplane Races," *Aircraft Year Book, 1923* (New York: Aeronautical Chamber of Commerce of America, 1923), 56–81.

29. "New World Speed Records at Detroit," *Aeronautical Digest*, November 1922, pp. 178–184.

30. "Detroit Race Impressions," *Aviation*, October 23, 1922, pp. 550–557.

31. T.G. Foxworth, "Thomas Morse Military Racing Airplanes," *AAHS Journal*, Winter 1967, pp. 235–253.

32. *New York Times*, "206 Miles an Hour for 156 Miles Sets New Flying Record," October 15, 1922, pp. 1, 20.

33. *Detroit News*, "Michigan Flier Is Race Winner," October 14, 1922, p. 1.

34. H.V. Wilcox's comments about the upcoming Pulitzer Race appeared after the bulletin titled "Michigan Flier Is Race Winner" in the *Detroit News*, October 14, 1922, pp. 1–2.

35. Detroit Aviation Society, Inc., *Rules and Entry Blank: Detroit's Aerial Contests, 1922* (Detroit: Detroit Aviation Society, 1922), 5 (Archives, National Air and Space Museum).

36. Foxworth, "Thomas Morse Military Racing Airplanes," 244–5.

37. T.G. Foxworth, *The Speed Seekers* (New York: Doubleday, 1974), 294.

38. Wing radiators had been "patented in France several years" before 1922, but the Curtiss Company

"put it into actual use" (A. Black, "Record-Breaking Pulitzer Racers Show Advance in Plane Design," *Automotive Industries*, October 19, 1922, pp. 755–759).

39. F.H. Johnson, "The Navy Bee-Line Racer (BR1), *Aviation*, October 30, 1922, pp. 604–605.

40. Ibid.

41. "World Records Established in Pulitzer Race," *Aerial Age*, November 1922, pp. 533–536.

42. Foxworth, *Speed Seekers*, 453.

43. R.L. Bliss, "The American SPAD," *Skyways*, July 1989, pp. 20–38.

44. "The National Airplane Races," *Aircraft Year Book, 1923* (New York: Aeronautical Chamber of Commerce of America, 1923), 56–81. The decisions to invite particular companies to compete were not exactly "industrial policy" of the sort in which the government picks winners and losers, but it seems close.

45. "Famous Record Breaking and Racing Aircraft: Verville-Sperry R-3," http://www.airracinghistory.freeola.com/aircraft/Verville-Sperry%20R-3.htm.

46. A May 11, 1922, memo from Major T.H. Bane, McCook Field, to General Patrick states that the army would contract with Loening, Curtiss, and Thomas Morse for racers. A month later, a memo from W.E. Donnelly, assistant to the chief engineer, informed the divisional plans officer at McCook that no project engineers would be assigned to the racers from the three companies. The shortness of time between completion of the racers and the date of the Pulitzer made it necessary to deliver the racers directly to Selfridge without any wringing out at McCook (McCook Field papers, National Archives and Record Administration II, College Park, MD).

47. *New York Times*, William Wait, Jr., an Airplane Designer," October 19, 1993 (obituary).

48. G. Rossano, "When Curtiss Racers Ruled the Roost," *Journal American Aviation Historical Society* (Summer 1985), 109–113, and "A Conversation with Bill Wait When Curtiss Racers Ruled the Roost," *Curtiss Aeroplane and Motor Company Magazine*, June 1982, pp. 3–4.

49. William Wait, quoted in Rossano, "When Curtiss Racers Ruled the Roost," 110.

50. Throughout the 1920s, the Curtiss Company sold OX-5 engines, which it had bought as surplus from the government after World War I. The government had paid $1,800 for the engines; Curtiss repurchased them for about $230, or thirteen cents on the dollar, and initially sold them at $1,000 in 1919. The price dropped to $550 in 1921 and $250 in the late 20s (P.M. Bowers, *Curtiss Aircraft, 1907–1947* (London: Putnam, 1979), 168–169).

51. T.G. Foxworth, "The Curtiss R-6 Racer: The Army's Bid For Speed," *AAHS Journal* (Summer 1970), 73–91.

52. "Curtiss Model D-12 Aeronautical Engine," *Aerial Age*, November 1922, pp. 542–545.

53. Foxworth, "The Curtiss R-6 Racer," 74.

54. "Detroit Race Impressions," *Aviation*, October 23, 1922, pp. 550–557.

55. "The World's Speed Record," *Flight*, September 27, 1922, p. 244.

56. Foxworth, "The Curtiss R-6 Racer," 79.

57. See J. Libby's "Lester Maitland and Albert Hegenberger and the First Nonstop Flight to Hawaii" (*AAHS Journal* (Summer 2010), 101–116). Lester Maitland (1899–1990) was a first lieutenant in 1922 when he flew in the Pulitzer. In 1927, he and Lt. Alfred Hegenberger were the first men to fly from California to Hawaii. Maitland was the commander of Clark Field in the Philippines when the Japanese attacked and seriously damaged it at the beginning of the war. He rose to brigadier general, retired from the air force in the late 1940s, and remained active in aviation until 1957 when he was ordained an Episcopal priest ("Lester J. Maitland Is Dead at 91; First to Fly from U.S. to Hawaii," *New York Times*, March 30, 1990).

58. Russell L. Maughan (1893–1958) flew in France during World War I and downed four German aircraft. After the war, he was detailed to McCook Field as a test pilot. He was a first lieutenant when he won the 1922 Pulitzer, and was still a first lieutenant when he completed the first coast-to-coast flight between dawn and dusk on June 23, 1924. He served in the Philippines in the 1930s, selected sites for air bases in Greenland before World War II, and commanded various transport command units during the war. He retired as a colonel in 1946 and died in 1958 ("Russell Lowell Maughan," Utah History to Go, http://historytogo.utah.gov/people/utahns_of_achievement/russelllowellmaughan.html).

59. W. Wait quoted in G. Rossano, "When Curtiss Racers Ruled the Roost" (*Journal American Aviation Historical Society* (Summer 1985), 109–113).

60. Foxworth (*Speed Seekers*, 210) writes that the snap roll at about 100-foot altitude took place when Maughan (not Maitland) was making a practice flight over Selfridge Field (not over Curtiss Field) a few days later.

61. W. Wait quoted in Rossano, "When Curtiss Racers Ruled the Roost," 110.

62. "Army Flier Speeds 220 Miles an Hour," *Aviation*, October 16, 1922, p. 594; "Army Flier Speeds 220 Miles an Hour, New York *Times*, October 9, 1922; "Summarized History of Curtiss Army R-6 Racing Airplane P-278, A.S. No. 68564 and P-279, A.S. No. 68563," photocopied typescript attached to photocopy of "Report on Curtiss R-6 Racing Airplane" from "John F. Curry, Chief of [Engine] Division, McCook Field, to Chief of Air Service, February 24, 1925" (National Archives II, Suitland, MD).

63. Rossano, "When Curtiss Racers Ruled the Roost," 109.

64. T. Huntington, "Racetracks in the Sky," *American Heritage* (Spring 2007), http://www.americanheritage.com/articles/magazine/it/2007/4/2007_4_42.shtml.

65. Foxworth, *Speed Seekers*, 200.

66. *New York Times*, "Oil Promoter Sued for Two Airplanes," July 19, 1922, p. E2.

67. Foxworth, *Speed Seekers*, 187.

68. Ibid., 200.

69. See J.S. Berdahl, "Jerome Hunsaker's Dark Shadow," in *Scope: The Student Publication of the Graduate Program in Science Writing at MIT*, posted November/December 2009, (http://scopeweb.mit.edu/?p=510), for a report on Hunsaker's regrets about persisting in the development of dirigibles after their vulnerability to weather had been recognized.

70. J.L. Kerrebrock, "Jerome Clark Hunsaker, 1886–1984," in *Biographical Memoirs* (vol. 78) (Washington: National Academy Press, 2000), 93–107, http://www.nap.edu/html/biomems/jhunsaker.

71. Foxworth, *Speed Seekers*, 245.
72. Brown, *A Genesis Workshop*, 2.
73. "Detroit Race Impressions," *Aviation*, October 23, 1922, pp. 550–557.
74. W. Wait quoted in Rossano, "When Curtiss Racers Ruled the Roost," 110.
75. A. Black, "Record-Breaking Pulitzer Racers Show Advance in Plane Design," *Automotive Industries*, October 19, 1922, pp. 755–759.
76. Hunter left the army in 1918 to complete college, rejoined in 1920, and remained in the army until 1946. During World War II, he commanded the Eighth Air Force Fighter Command and then the First Air Force.
77. Bissell's contribution to the cover-up of Soviet responsibility for the murder of some 10,000 Polish officers, soldiers, and civilians after the Soviet invasion of Poland in 1939 mars his record, but he did not act alone or without the knowledge of his military and civilian superiors. The cover up extended upwards to President Roosevelt, who may have been reluctant to accuse our wartime ally, the Soviet Union, of war crimes (B.J. Fischer, "The Katyn Controversy: Stalin's Killing Field," Central Intelligence Agency, Center of the Study of Intelligence, winter 1999-2000, https://www.cia.gov/library/center-for-the-study-of-intelligence/csi-publications/csi-studies/studies/winter99-00/art6.html). Bissell was also accused of black market activities (flying coffee from postwar Britain to Germany while serving as air attaché in the London embassy) and he was ordered back to the United States in 1948 ("Army Probe," *Chester (PA) Times*, August 14, 1948).
78. W. Wait, quoted in Rossano, "When Curtiss Racers Ruled the Roost," 110.
79. *New York Times*, "Army Airmen Race Despite High Winds," October 14, 1922.
80. "The Third Pulitzer Race," *Aviation*, October 23, 1922, pp. 544–549. The Dayton-Wright RB-1 flown at the 1920 Gordon Bennett had retractable landing gear, but it had not been demonstrated in public in the United States.
81. *Aircraft Year Book, 1923* (New York: Aeronautical Chamber of Commerce, 1924), 77.
82. Foxworth, *Speed Seekers*, 283.
83. *New York Times*, "206 Miles an Hour for 156 Miles Sets New Flying Record," October 15, pp. 1, 20.
84. W. Wait quoted in Rossano, "When Curtiss Racers Ruled the Roost," 111.
85. "World Records Established in Pulitzer Race," *Aerial Age*, November 1922, 533–536.
86. "206 Miles an Hour for 156 Miles Sets New Flying Record."
87. Associated Press, "Air Race Winner Made Unconscious by Great Speed," *Atlanta Constitution*, October 15, 1922, p. 1.
88. Foxworth, *Speed Seekers*, 210.
89. Detroit Aviation Society, "Official Timers Report of Records Made in the Pulitzer Race," signed by Harry H. Knepper, official timer for the American Automobile Assn., and Odis A. Porter, official timer for the Indianapolis Motor Speedway. There is no date on this two-page document, but it was notarized on November 29, 1922, for Knepper, and November 24, 1922, for Porter (Archives, National Air and Space Museum).

90. "206 Miles an Hour for 156 Miles Sets New Flying Record."
91. "Air Race Winner Made Unconscious By Great Speed."
92. "Detroit Race Impressions," *Aviation*, October 23, 1922, pp. 550–557.
93. Foxworth, *Speed Seekers*, 200.
94. W. Wait quoted in Rossano, "When Curtiss Racers Ruled the Roost," 110. Foxworth (*Speed Seekers*, 245) writes that the lower sesqui-wing was entirely plywood-covered. Whether fabric or plywood-covered, it was near the ground and susceptible to damage.
95. Foxworth, *Speed Seekers*, 247.
96. "World Records Established in Pulitzer Race," *Aerial Age*, November 1922, 533–536.
97. Foxworth, *Speed Seekers*, 210.
98. Ibid., 288.
99. "New World Speed Records at Detroit," *Aeronautical Digest*, November 1922, pp. 178–184.
100. *New York Times*, "Lieut. Maughan Beats All Speed Records By Flying at Rate of Four Miles a Minute," October 17, 1922, p. 1.
101. Ibid.
102. W. Wait quoted in Rossano, "When Curtiss Racers Ruled the Roost," 111.
103. "Gen. Mitchell Sets New World's Speed Record," *Aviation*, October 23, 1922, p. 558.
104. "Record Flight of General Mitchell," *Aviation*, October 30, 1922, p. 585.
105. Wait quoted in Rossano, "When Curtiss Racers Ruled the Roost," 111.
106. Letter from J.G. Vincent, president of NAA, to secretary-general of FAI, October 19, 1922 (Archives of the National Air and Space Museum), rf. According to Foxworth (*Speed Seekers*, 211), Mitchell's actual speed was 1.5 mph higher, 224.5 mph. The difference resulted from there being two different ways to average the speed of the four passes. The FAI added the times for each pass together and divided by four to establish an average time and calculated the average speed from it. In the other method, the speed for each pass is calculated; those four speeds are added and divided by four to arrive at an average. The second method is more accurate and results in the higher speed.
107. "Record Flight of General Mitchell," *Aviation*, October 30, 1922, p. 585.
108. Letter from secretary general of FAI to J.G. Vincent, president of NAA, January 5, 1923 (Archives, National Air and Space Museum).
109. "Army Pilot to Try for New Record," *Aerial Age*, December, 1922, p. 605.
110. "The National Airplane Races," *Aircraft Year Book, 1923* (New York: Aeronautical Chamber of Commerce of America, 1923), 56–81.
111. "New World Speed Records at Detroit," *Aeronautical Digest*, November 1922, pp. 178–184.
112. "Editorial Comment," *Flight*, October 19, 1922, pp. 601–602.
113. "Detroit Race Impressions," *Aviation*, October 23, 1922, pp. 550–557.
114. "*Curtiss FIRST again!*" (advertisement), *Aerial Age*, November 1922, p.532.
115. See, for example, "Faster than Human ever flew — and Valspar helped!" (advertisement), *Aerial Age*, November 1922, p. 554.

116. B. Robie, *For the Greatest Achievement* (Washington: Smithsonian Institution Press, 1993), 103.
117. Ibid., 107, 113–117.
118. "The National Airplane Races," 73.

Chapter Five

1. Secretary of the navy Edwin Denby and General Mason Patrick quoted in "206 Miles an Hour for 156 Miles Sets New Flying Record," *New York Times*, October 15, 1922, pp. 1, 20.
2. M.M Patrick, "The American Army in the Air," *International Air Races, October 1, 2, 3: Official Program, Saint Louis, 1923* (St. Louis: St. Louis Aeronautic Corporation, St. Louis Air Board, and Flying Club of St. Louis, 1923), 55, 109, 112.
3. "What the French Think of American Air Records," *Slipstream*, May 1923. pp. 23, 31.
4. Ibid.
5. "New Pursuit Plane for the Army Air Service," *Aerial Age*, July 1923, p. 338.
6. The Missouri Historical Society's "Fly with Us through Decades of St. Louis History" (http://www.mohistory.org/Flight_City/HTML/htmlTimeline.html) provides a useful time line for aviation in St. Louis.
7. James J. Horgan's *City of Flight: The History of Aviation in St. Louis* (St. Louis: Patrice Press, 1990) has been invaluable in writing this chapter. The information here is found on page ix.
8. T. Crouch, *The Eagle Aloft: 200 Years of Ballooning in the United States* (Washington: Smithsonian Institution Press, 1983), 250–254 and elsewhere. N. Moehlmann's "John Wise, A Pioneer Aeronaut: 1808–1879" was reprinted from the first issue of the John Wise Society's *Journal* and is available at http://www.johnwise.net.
9. Horgan (*City of Flight*, 50) quotes an article in the *Daily Missouri Republican*, July 7, 1859, p. 2.
10. John Wise and another man ascended from St. Louis on a flight made "in the interests of meteorological science" (Horgan, *City of Flight*, 5). They disappeared during the flight. The body of the other man was recovered from Lake Michigan. No trace was found of John Wise, but it is almost certain that the balloon crashed in the lake and that both men died there. John Wise was 71 when he made his 463rd, and last, ascension.
11. Horgan, *City of Flight*.
12. "Historical National Population Estimates" for July 1, 1900, to July 1, 1999, can be found at http://www.census.gov/popest/archives/1990s/popclockest.txt.
13. Photocopy of untitled and unpaginated article from the *Saint Louis Globe-Democrat*, hand-dated November 16, 1929 (Mercantile Library, University of Missouri at St. Louis).
14. "Thomas Scott Baldwin (1854–1923)," U.S. Air Force Fact Sheet, http://www.hill.af.mil/library/factsheets/factsheet_print.asp?fsID=5837&page=1. Baldwin's purchase of an engine from Glenn Curtiss led to Curtiss' involvement in aviation and the inventions and innovations he brought to the new industry.
15. Knabenshue built the first passenger-carrying dirigible in the United States. At the time of his death, however, "Roy and Mrs. Knabenshue ha[d] been living on old age pensions; they were in a destitute condition. Roy's two sons and two daughters were in the same condition" ("A. Roy Knabenshue, 1876–1960," http://www.earlyaviators.com/eknabens.htm; see also "A. Roy Knabenshue Collection," National Air and Space Museum, Archives Division, http://www.nasm.si.edu/research/arch/findaids/knabensh/ark_print.html).
16. Early pilots, whether of balloons, dirigibles, or airplanes, had to be mechanics.
17. "Coupe Aéronautique Gordon Bennett," http://www.coupegordonbennett.org/.
18. J.J. Horgan, "The International Aero Tournament of 1907," *Bulletin*, Missouri Historical Society, April 1965, pp. 216–235.
19. C. Munsey, "Often a Bridesmaid but Never a Bride," *Bottles and Extras* (May-June 2007), 58–59, http://www.fohbc.com/PDF_Files/Listerine_CMunsey.pdf.
20. S.D. Levitt and S.J. Dubner, *Freakonomics: A Rogue Economist Explores the Hidden Side of Everything* (New York: William Morrow, 2005), 91–92.
21. L. McMurtry, *Rhino Ranch* (New York: Simon & Schuster, 2009), 214.
22. Erin A. Metro, e-mail to Michael Gough, August 24, 1910.
23. D. Terry, "Town's Most Air-Minded Citizen," *St. Louis Post-Dispatch*, June 16, 1946.
24. *St. Louis Globe-Democrat*, "Air Story of St. Louis—First International Race Was One of Most Successful Ever Held," January 26, 1930.
25. Horgan, "The International Aero Tournament of 1907," 220.
26. Three of the airplanes were ornithopters, flapping-wing aircraft. The fact that one of the ornithopters was powered by a seven-horsepower Curtiss engine underlines the ubiquity of Curtiss engines to early aviation.
27. Horgan, "The International Aero Tournament of 1907," 8.
28. See P. Demetz's *The Air Show at Brescia, 1909* (New York: Farrar, Straus and Giroux, 2003) for a detailed and entertaining history of the Brescia air meet with colorful descriptions of aviators—Blériot, Curtiss and others—as well as spectators, including Franz Kafka.
29. Horgan, *City of Flight*, 129, quoting article in *St. Louis Republic*, October 8, 1909, p.1.
30. *New York Times*, "Break Stops Curtiss in St. Louis Flight," September 23, 1909, p.5.
31. *New York Times*, "Curtiss Flies Over Trees," October 10, 1909, p. 20.
32. Horgan, *City of Flight*, 115.
33. H.S. Villard, *Blue Ribbon of the Air* (Washington: Smithsonian Institution Press, 1987), 60.
34. Horgan, *City of Flight*, 141, quoting from the *St. Louis Post-Dispatch*, September 11, 1910, p. 2.
35. Ibid., October 12, 1910, p. 4.
36. D. Terry, "Town's Most Air-Minded Citizen," *St. Louis Post-Dispatch*, June 16, 1946.
37. *New York Times*, "Le Blanc Flies Fast Mile," October 15, 1910, p. 2.
38. Horgan, *City of Flight*, 149, quoting from *St. Louis Globe-Democrat*, October 15, 1910, p. 2.

39. The balloon was named for its sponsor, a club that was determined to raise the population of St. Louis to one million, an idea that may seem a bit odd 100 years later. The city's population never reached that number.

40. *New York Times*, "Abercron to Quit Race," October 16, 1910, p. 15.

41. *New York Times*, "Ten Balloons Off for World's Title; Three American Entrants Start from St. Louis in Fifth International Race," October 18, 1910, p. 1.

42. *New York Times*, "Airmen Lost in Canadian Wilds," October 22, 1910, p. 1.

43. *New York Times*, "Canadian Wilds Give Up Airmen," October 27, 1910, p. 1

44. Alex Hawley, as the pilot of *America II*, was awarded the Lahm Cup. No one outdid his 1,173 mile flight in the next three years, and he was awarded permanent possession of the cup. In 1976, a United States balloon did fly farther.

45. *New York Times*, "Protests Balloon Award," December 30, 1910, p. 7.

46. Legislation prevents St. Louis from annexing surrounding land areas, and the city's population cannot be increased in that way.

47. City-Data.com, "St. Louis: Population Profile," http://www.city-data.com/us-cities/The-Midwest/St-Louis-Population-Profile.html.

48. "The National Air Races at St. Louis...," *Aircraft Year Book, 1924* (New York: Aeronautical Chamber of Commerce of America, 1924), 150.

49. Ibid., 149.

50. Terry, "Town's Most Air-Minded Citizen."

51. A.B. Birge, "The St. Louis Aeronautic Corporation," *International Air Races, October 1, 2, 3: Official Program, St. Louis, 1923* (St. Louis Aeronautic Corporation, St. Louis Air Board, Flying Club of St. Louis, 1923].

52. Horgan (*City of Flight*, 194) quoting W.F. Carter, "What St. Louis Field Means to St. Louis" *Cherry Diamond*, September 1923, p. 13.

53. A Boeing 747-400ER (extended range) has a fuel capacity of 63,700 gallons.

54. *St. Louis Post-Dispatch*, "500 Acres of Woods and Fields Turned into Well-Equipped Air Terminal," September 30, 1923, p. 5f.

55. Ibid.

56. Ibid.

57. *St. Louis Post-Dispatch*, "Air Meet Field Thrown Open to Public Today," September 30, 1923, p. 1.

58. "Flying Club Names Field for Albert Bond Lambert," *Aerial Age*, April 10, 1922, p. 100.

59. "St. Louis, 1923," *Aviation*, October 15, 1923 pp. 486–489.

60. Foxworth, *Speed Seekers*, 275.

62. "New Pursuit Airplane for the Army Air Service," *Aerial Age*, July 1923, p. 338.

62. *Aircraft Year Book, 1924*, 154.

63. "The Giuseppe M. Bellanca Collection," National Air and Space Museum, Archives Division, http://www.nasm.si.edu/research/arch/findaids/bellanca/gmb_print.html.

64. Foxworth (*Speed Seekers*, 283) quoting from an unspecified source. Deliveries of Boeing P-26s began to the U.S. Army in December 1933. It was the first all-metal monoplane pursuit, but it retained heavy spatted fixed landing gear and external bracing wires.

65. The James Doolittle Collection at the North Texas State University library, Denton, Texas, contains some letters from Verville to Doolittle. Verville's tone is often overly thankful, fawning and apologetic about bothering Doolittle, despite the fact that they had known each other for decades. His nature in those letters means, I think, that he had difficulty in standing up for himself, indicating to me that he was convinced that there was a conspiracy against the R-3. I expect that he had evidence for it.

66. Foxworth, *Speed Seekers*, 212, no source given.

67. J.F. Curry, "Summarized History of Curtiss Army R-6 Racing Airplane P-278, A.S. No. 68564 and P-279, A.S. No. 68563," photocopied typescript attached to photocopy of a memo from "John F. Curry, Chief of [Engine] Division, McCook Field to Chief of Air Service," "Report on Curtiss R-6 Racing Airplane," February 24, 1925 (National Archives II, Suitland, MD).

68. "Test Over 3-Kilometer Course," *Aerial Age*, July, 1923, p. 338.

69. "Service Ships Versus Racers," *Aviation*, October 1, 1923, p. 301.

70. "Wright Aeronautical to Build Planes, *Aerial Age*, April, 1923, p. 192.

71. "Army and Navy Entries in the Pulitzer Trophy Race," *Aviation*, October 1, 1923, pp. 400–403.

72. Ibid., 402.

73. T.G. Foxworth, "The Curtiss R2C Racers," *AAAHJ [American Association of Aviation History Journal]* (Spring 1966), 7–12.

74. A quote from a 2009 article about the debutante ball captures the flavor of the occasion. The author wrote, "Doing research for this article was truly an enjoyable experience. I would invite *ladies of quality* to lunch or small groups to my home for tea to discuss their time in the sun" (emphasis added) (J. Sullivan, "Rites of Passage," *St. Louis Seasons*, August-September, 2009, pp. 45–49).

75. *St. Louis Post-Dispatch*, "100 Maids in Purple Knickers to Greet Prophet Tomorrow," September 30, 1923, p. 1.

76. *St. Louis Post-Dispatch*, "Heavy Rains Force Postponement of Air Race Program," October 1, 1923, p. 1

77. *St. Louis Globe-Democrat*, "Aero Meet Postponed Three Days Due to Flooded Flying Field," October 1, 1923, pp. 1 and 3.

78. *St. Louis Post-Dispatch*, "Veiled Prophet Arrives for the Coronation Party," October 1, 1923, p. 2.

79. *St. Louis Globe-Democra*t, "ZR-1, Giant of Air, Speeding Westward, Arrives Here at 7:00 A.M.," October 2, 1923, p. 1, 2. As explained in the text of the article, at its best speed the airship was flying at 53 mph.

80. "The Barling Bomber: An American Six Engined Giant," *Flight*, December 13, 1923, pp. 749–751. A giant in its day, the Barling's wingspan of 120 feet was roughly the same as a Boeing 737 when equipped with winglets, about 118 feet, and 90 feet shorter than a 747's 211-foot span.

81. "A Monumental Ship," *Aviation*, October 15, 1923, p. 489.

82. "The Barling Bomber," 749–751.

83. "Witteman-Lewis XNBL-1 Barling Bomber Fact Sheet," National Museum of the United States Air Force, http://www.nationalmuseum.af.mil/factsheets/factsheet.asp?id=2434.

84. After a few flights in 1923, the Barling was largely ignored at Wilbur Wright Field until H.H. Arnold ordered it destroyed in 1928 (H.H. Arnold, *Global Mission* (New York: Harper Brothers, 1949), 128–129).

85. *St. Louis Post-Dispatch*, "85,000 Paid to See Pulitzer Air-Race," October 7, 1923, p. 1.

86. R. Kinert, *American Racing Planes and Historic Air Races* (Chicago: Wilcox and Follett, 1952), 51

87. *St. Louis Post Dispatch*, "Lieut. Williams of Navy Wins Pulitzer Race, Shatters All Speed Records," October 7, 1923, pp. 1–2.

88. In contradiction to the newspaper report of a metal propeller, Foxworth (*Speed Seekers*, 284–285) wrote that the "hasty substitution of a suitable wood prop postponed fashioning a new spinner until just one hour remained before race time.... [T]he replacement wood prop and new spinner proved to be badly out of balance," However, in accord with the newspaper account, in his tabulation of the Pulitzer Race results (p. 456), Foxworth ascribed the R-3's problems to an "unbalanced Reed dural (aluminum) propeller which 'sprung crankshaft' causing 'extreme vibration.'" Whatever the cause — an unbalanced metal propeller seems likely — Pearson was forced down and out of the race.

89. *St. Louis Post-Dispatch*, "How It Feels to Be Speed King of World in Words of Pulitzer Trophy Winner," October 7, 1923, pp. 1–2.

90. A. Williams quoted in "Pulitzer Trophy Is Presented to Lieut. Williams," *St. Louis Post Dispatch*, October 7, 1923, p. 1.

91. "Prizes Presented to Contestants at Smoker," *Aviation*, October 22, 1923, pp. 519–520.

92. Foxworth, *Speed Seekers*, 245.

93. Ibid., 255.

94. "St. Louis—1923," *Aviation*, October 15, 1923 pp. 486–489.

95. *St. Louis Post-Dispatch*, "Sanderson Crash Due to Lack of Gasoline," October 7, 1923.

96. *St. Louis Globe-Democrat*, L. Sanderson, quoted in "'Whoopee! Navy!'" Gasps Lieut. Williams After His Victory in Spectacular Race," October 7, 1923, p. 1.

97. "Sanderson Crash Due to Lack of Gasoline."

98. "Whoopee! Navy!," p. 1.

99. "Navy Wins Two Most Coveted Trophies," *U.S. Air Service*, November 1923, pp. 11–14.

100. R.J. Brown, "Four Miles a Minute!" *Slipstream*, November 1923, pp. 5–9.

101. *St. Louis Post-Dispatch*, "How It Feels to Be Speed King of World in Words of Pulitzer Trophy Winner," October 7, 1923, p. 1.

102. Foxworth, *Speed Seekers*, 220.

103. A. Williams quoted in "How It Feels to Be Speed King of World in Words of Pulitzer Trophy Winner."

104. "How It Feels to Be Speed King of World in Words of Pulitzer Trophy Winner," p. 1.

105. "Whoopee! Navy!," p. 1.

106. S. Calloway quoted in "Whoopee! Navy!," p. 1.

107. W. Miller quoted in "Whoopee! Navy!," p. 1.

108. Horgan, *City of Flight*, 215.

109. Women now seldom faint at such times, or, perhaps, crowding is not so bad now as it was in 1923.

110. "St. Louis—1923," *Aviation*, October 15, 1923, pp. 486–489.

111. "Prizes Presented to Contestants at Smoker," *Aviation*, October 22, 1923, pp. 519–520.

112. M. Patrick quoted in "Whoopee! Navy!," p. 1.

113. "Prizes Presented to Contestants at Smoker," *Aviation*, October 22, 1923, pp. 519–520.

Chapter Six

1. "The Dayton Air Races," *Air Craft Year Book, 1925* (New York: Aeronautical Chamber of Commerce, 1926), 189–192.

2. *New York Times*, "Plane Speed Tests Delayed Two Weeks," October 8, 1923, p. 2.

3. The Giants baseball team began as the "Gothams" in 1883 and played in New York until they moved to San Francisco in 1957.

4. Newspaper reports consistently published speeds with great precision to one one-hundredth of a mile per hour, i.e., 257.42. In this book, speeds are generally rounded to whole mphs, i.e., "257" or nearest tenth mph, i.e., "257.4,"

5. *New York Times*, "Naval Airman Flies 259 Miles an Hour for World's Record," November 3, 1923, p. 1.

6. The *New York Times* article ("Wins, Loses, Rewins Air Record in a Day; 266 Miles an Hour," November 5, 1923, pp. 1–2) provides no information about who was among the 20,000 that greeted him when he landed in front of the park grandstand other than that there were hundreds of graduates of Fordham University where Williams had gone to college.

7. "Wins, Loses, Rewins Air Record," p. 1.

8. Ibid.

9. Ibid.

10. *New York Times*, "Airmen to Match Speed Again Today," November 6, 1923, p. 12.

11. *New York Times*, "Navy Flier Climbs 5,000 Feet in a Minute," November 7, 1923, p. 19.

12. *New York Times*, "Plane Speed Tests Delayed Two Weeks," October 8, 1923, p. 2.

13. *New York Times*, "High-Speed Hazards in Aviation," October 7, 1923, p. 16; "400-Mile Speed in Air Is Foreseen," November 11, 1923, p. xx2.

14. (Maj.) L.H. Bauer, "1,000-Mile Speed Likely to Kill," *The Slipstream*, April 1924, pp. 32, 36.

15. The distances were 500, 1,000; 1,500; 2,000; 2,500; 3,000; 3,500; 4,000; 5,000; and 5,300 km.

16. *New York Times*, "Our Aviators Set 33 Records in 1923," December 31, 1923, p. 6.

17. Macready had flown a Thomas-Morse MB-6 to third place in the 1921 Pulitzer.

18. "Our Aviators Set 33 Records in 1923," p. 6.

19. The three very large "Patterson scrapbooks" at "Dayton History!," the official history organization of Montgomery County, Ohio, where Dayton is located, contain many newspaper and magazine clippings about aviation. The scrapbooks, probably put together by someone in Frederick Patterson's office, are a great resource for anyone interested in Dayton aviation in 1923 and '24. They are also frustrating, as they are stuffed with redundancies, identical United Press and Associated Press news releases published in many different newspapers having been clipped out and pasted in. Worse, the person(s) who kept the

scrapbooks often cut off any information that identified the newspapers from which the clippings came (Patterson Scrapbooks, Dayton History!).

20. From the Patterson Scrapbooks, Dayton History!.

21. *Minneapolis Tribune*, "Twin Cities Working for Pulitzer Events in 1925," editorial reprinted in *National Aeronautic Association Review*, January 1925, p. 1.

22. H.H. Robertson (assistant engineer of the air races), "Preliminary Plans for the 1924 Races," *The Slipstream*, May 1924, pp. 5–6, 24.

23. "The population of Dayton, Ohio, on October 16, 2010, was approximately 166,179, http://www.trueknowledge.com/q/population_of_dayton_ohio_2010.

24. The population was 153,000 in 1920 and 201,000 in 1930; assuming the population increased at a constant rate during the decade, it would have been about 173,000 in 1924.

25. C.H. Paul, "Industrial History of Dayton," in R.C. Conover, *Dayton and Montgomery County: Resources and People* (New York: Lewis Historical, 1932), 463–557.

26. In 1924, the Dayton Chamber of Commerce, Dayton, Ohio, printed "Dayton: The Birthplace of the Airplane and the Nation's Center of Aviation; Pictures, Map, Statistics" as a map that could be folded and carried in a jacket or shirt pocket (Archives National Air and Space Museum).

27. "Deeds and the Cash," *Time*, December 4, 1932, http://www.time.com/time/magazine/article/0,9171,882471,00.html#ixzz1Y21BAg5z.

28. Ibid.

29. F. Deford, *The Old Ball Game* (New York: Atlantic Monthly Press, 2005), 91.

30. T. Zumwalt, *For the Love of Dayton* (Dayton Daily News, 1966), 132.

31. B. Robie, *For the Greatest Achievement* (Washington: Smithsonian Institution Press, 1993), 106.

32. Readers of a certain age will recall that Jack Benny, a popular radio entertainer in the 1940s, supposedly owned a Maxwell. For whatever reason, the mere mention of "Maxwell" generated a laugh from the studio audience or from the laugh track.

33. Paul, "Industrial History of Dayton," 494.

34. The Dayton-Wright Company's revolutionary RB-1 cantilever monoplane racer with retractable landing gear and variable camber wing did not complete a single lap in the 1920 Gordon Bennett Race, but it was a harbinger of the future (see chapter one). The company closed in 1923, when General Motors, which had purchased it in 1919, left the airplane business.

35. The U.S. Centennial of Flight Commission, "The U.S. Aircraft Industry During World War I," http://www.centennialofflight.gov/essay/Aerospace/WWi/Aero5.htm.

36. Diana G. Cornelisse (historian, Wright-Patterson Air Force Base), "The World's First Flying Field," 1990, http://www.456fis.org/THE_WRIGHT_BROTHERS_AND_THE_FIRST_AIR-FIELD.htm.

37. McCook Field was named for a local family that sent 17 men to fight for the Union during the Civil War.

38. "The Army Experimental Station," *Aviation*, November 15, 1920.

39. *The Slipstream*, June 1924, p. 6.

40. "Government Accepts New McCook Field Site at Dayton, Ohio," *The Slipstream*, September 1924, pp. 8–9.

41. "The International Air Races," *The Slipstream*, September 1924.

42. *The Slipstream*, June 1924, p. 23.

43. Foxworth, *Speed Seekers*, 35.

44. M. Rubenstein and D.M. Goldman, *To Join with Eagles* (Garden City, NY: Doubleday, 1974), 70.

45. Foxworth, *Speed Seekers*, 201.

46. *The Slipstream*, June 1924, p. 23.

47. *New York Times*, "Schneider Air Race Has Been Canceled," September 26, 1924, p. 24.

48. Navy Historical Center, "Naval Aviation Chronology, 1920–1929," http://www.history.navy.mil/branches/avchr3.htm.

49. Memo No. 16403 to A.C. of S., G-2, Washington, D.C., for "The Chief of the Air Service," photocopy of carbon copy lacking identification of the originating office, dated March 13, 1924 (Archives, National Museum of the United States Air Force, Dayton, Ohio).

50. *New York Times*, "Lecointe Won't Race Here," September 13, 1924, p. 17.

51. Foxworth, *Speed Seekers*, 458.

52. "The Development of High-Speed Aircraft," *Flight*, December 6, 1923, pp. 743–745.

53. *New York Times*, "Fears U.S. Air Force Will Be Nil by 1927," April 24, 1924, p. 7.

54. Ibid.

55. Foxworth, *Speed Seekers*, 223.

56. Ibid., 285.

57. Curtiss Aeroplane and Motor Company, advertisement, "Speed with Safety," "International Air Race," *The Slipstream*, October 1924, p. 20.

58. "The Curtiss-Reed Metal Airscrew," *Flight*, January 24, 1924, p. 54.

59. W.J. Boyne, "'P' for Pursuit: The Army's Biplane Hawks," *Airpower*, March 1976, pp. 26–35, and *The Best of Wings Magazine* (Dulles, VA: Brassy's, 2001).

60. Ibid.

61. "The Curtiss 'PW-8' Biplane," *Flight*, November 6, 1924, pp. 705–706.

62. Dayton's estimated 1924 population of 173,000 was about 20,000 fewer than Omaha's 191,000 and much less than Detroit's and St. Louis' populations. Garden City, New York, the site of the 1920 and 1925 races was far smaller, but it drew on the population of New York City.

63. H.H. Robertson (assistant engineer of the air races), "Preliminary Plans for the 1924 Races," *The Slipstream*, May 1924, pp. 5–6 and 24.

64. "The International Air Races," *The Slipstream*, September 1924, pp. 3–12.

65. Robertson, "Preliminary Plans for the 1924 Races," 5–6, 24.

66. "Reduced Fare Is for All Members of Association," *National Aeronautic Association Review*, September 2, 1924, p. 1.

67. "Notes Concerning the Air Races of 1924," *The Slipstream*, March 1924, pp. 21–22.

68. "The International Air Races," p. 7.

69. "Largest Gathering in History of Sport Expected to Witness Three Days' Program," *National Aeronautic Association Review*, September 2, 1924, p. 1.

70. "The International Air Races," *The Slipstream*, September 1924, pp. 3–12.

71. Foxworth, *Speed Seekers*, 285. The tandem-

mounted, two-engine design was used in the Italian Macchi-Castoldi MC-72 seaplane that set the world speed record, 440 mph, for piston-engine seaplanes in 1933. That record stands to this day and will almost certainly never be bested.

72. *New York Times*, "Air Racers Open Meet at Dayton," October 3, 1924, p. 10.

73. Clipping in Patterson scrapbooks, Dayton History!

74. Clipping from page 1, *Dayton Journal*, December 28, 1923, in Patterson scrapbooks. The author estimated overall attendance at the 1924 races would be 225,000, with 180,000 attending the last day. Dividing the remaining 70,000 by two produces an estimated attendance of 35,000 on each of the first and second days of the meet.

75. "Largest Gathering in History of Sport Expected to Witness Three Days' Program," *National Aeronautic Association Review*, September 2, 1924, p. 1.

76. *New York Times*, "Air Racers Open Meet at Dayton," October 3, 1924, p. 10.

77. *New York Times*, "Airship Launches Plane While Aloft," October 4, 1924, p. 1.

78. "The International Air Races, Dayton," *Flight*, October 30, 1924, pp. 693–695.

79. D. Shearin, "Air Thrills Galore in '24," *Dayton News Camerica* (Sunday magazine), September 5, 1954, pp. 10–12.

80. "The International Air Races, Dayton," *Flight*, October 30, 1924, pp. 693–695.

81. "Air Racers Open Meet at Dayton."

82. *New York Times*, "Airship Launches Plane While Aloft," October 4, 1924, p. 1.

83. Shearin, "Air Thrills Galore in '24," p. 11.

84. *New York Times*, "Baby Airplane Here; Mass Production Next," October 26, 1924, p. XX6.

85. "The Mitchell and Pulitzer Trophy Races," *Aviation*, October 13, 1924, pp. 1118–1119.

86. Foxworth, *Speed Seekers*, 39.

87. *Mt. Clemens Daily Leader*, "Captain Skeel, Selfridge Flier, Meets Death When Racing Plane Collapses," October 6, 1924, pp. 1 and 6.

88. "The Mitchell and Pulitzer Trophy Races," *Aviation*, October 13, 1924, pp. 1118–1119.

89. *New York Times*, "Capt. Skeel Killed in Dayton Air Race," October 5, 1924, pp. 1, 25.

90. *Aviation* (October 30, 1924, p. 695), *Flight* magazine (October 13, 1924, p. 1118) and the *Mt. Clemens Daily Leader* (October 6, 1924, p. 1) report the takeoff sequence as Skeel, Brookley, Mills, and Stoner. The *New York Times* (October 5, 1924, p. 1) reports the sequence was Mills, Stoner, Brookley, and Skeel, which is likely an error because race regulations called for takeoffs in the order of race numbers.

91. This description of the start of the Pulitzer and Skeel's dive is taken from Foxworth's *Speed Seekers*, 214–215.

92. *New York Times*, "Capt. Skeel Killed in Dayton Air Race," October 5, 1924, pp, 1, 25.

93. Ibid.

94. Foxworth, *Speed Seekers*, 215.

95. Quotes are from Foxworth's *Speed Seekers*, 216, no citation for their origin.

96. *Mt. Clemens Daily Leader*, "Captain Skeel, Selfridge Flier, Meets Death When Racing Plane Collapses," October 6, 1924, pp. 1, 6.

97. D. Waller, *A Question of Loyalty: Gen. Billy Mitchell and the Court-Martial That Gripped the Nation* (New York: HarperCollins, 2004), 302–303.

98. M. Rubenstein and D.M. Goldman, *To Join with Eagles* (Garden City, NY: Doubleday, 1974), 80.

99. Foxworth, *Speed Seekers*, 215.

100. Clipping from first page, *Dayton Journal*, December 28, 1923, in Patterson scrapbooks.

101. Foxworth, *Speed Seekers*, 214.

102. Ibid., 285.

103. *Dayton News*, September 24, 1924, from the Patterson clippings at History Dayton!

104. "Dayton's Big Race Meet Scores Unusual Success," *National Aeronautical Association Review*, October 8, 1924, p. 1.

105. "Four World Records Were Set in Races," *National Aeronautical Association Review*, October 8, 1924, p. 2.

106. "The One Sad Note in Successful Air Meet," *The National Aeronautical Association Review*, October 8, 1924, p. 6.

107. F.F. Marshall, "The Dayton Air Race Meet: Was It a Failure and Why?" *The Slipstream*, November 1924, pp. 19–21.

108. Photocopy of poster from the National Museum of the United States Air Force, Dayton, Ohio.

109. Quoted by Foxworth (*Speed Seekers*, 39) with no reference.

110. *Aircraft Year Book, 1925* (New York: Aeronautical Chamber of Commerce of America, 189).

111. Marshall, "The Dayton Air Race Meet, 19–21," *National Aeronautic Association Review*, October 8, 1924, pp. 1, 6.

112. H.W. Robertson (assistant manager, International Races), "What Will the Air Races Mean to Dayton?" in *Dayton: The Nation's Air Center; Official Publication of the Dayton Chamber of Commerce*, May 1924, pp. 1, 14–15.

Chapter Seven

1. F.F. Marshall, "The Dayton Air Race Meet: Was It a Failure and Why?" *The Slipstream*, November 1924, pp. 19–21.

2. E.P. Warner, "Lessons from the Air Races," an editorial or op-ed piece from a newspaper dated October 26, 1925, in "1925 Event File," NASM. Name and date of the newspaper have been clipped off.

3. "The Danger of Public Speed Races," *Aviation*, September 21, 1925, p. 345.

4. *New York Times*, "Asks Bids for Pulitzer Air Race," March 15, 1925, p. 21. Foxworth (*Speed Seekers*, 19), on the other hand, writes that the Twin Cities and Galveston, Texas, also wanted to host the 1925 National Air Race.

5. "Twin Cities Working for Pulitzer Events in 1924," *National Aeronautic Association Review*, April 1, 1924, p. 1.

6. Short article with no title, *New York Times*, April 24, 1925, p. 17.

7. *New York Times*, "Test Pulitzer Course Over Mitchel Field," April 25, 1925, p. 8

8. *New York Times*, "Pulitzer Air Race Plans," April 26, 1925, p. 14.

9. *New York Times*, "Will Hold Air Races Here," May 1, 1925, p. 2.
10. *New York Times*, "Races for Records to Mark Air Meet," July 13, 1925, p. 8.
11. *New York Times*, "Our Airplane Motors Invade Foreign Field," October 25, 1925, p. 5.
12. "The Danger of Public Speed Races," *Aviation*, September 21, 1925, p. 345.
13. Foxworth, *Speed Seekers*, 227.
14. The U.S. Navy, or at least its press department, was either ignorant of the V-1400s' rated horsepower or wanted to keep it from the public. A navy press release (no date and headed only "Navy Department") that describes the navy's plans for the 1925 Pulitzer and Schneider said that the V-1400 was capable of approximately 470 hp (photocopy of press release in air racing files of Navy Department Library, Navy Yard, Washington, DC).
15. *Aircraft Year Book, 1926*, 130.
16. Foxworth, *Speed Seekers*, 494–495.
17. "The Curtiss R3C — Contemporary Description," *Aviation*, September 28, 1925, available at Holcomb's Aerodrome, http://www.airminded.net/.
18. *Aircraft Year Book, 1926*, 128.
19. Foxworth, *Speed Seekers*, 228.
20. E. Vance, "The Life of Cyrus Bettis," http://www.earlyaviators.com/ebettis4.htm.
21. See M. Gough, "Doolittle Wins in Baltimore," *Airpower*, November 2005, pp. 44–56.
22. *New York Times*, "Army Flier Lost; Planes Hunt in Vain," August 25, 1926, p. 12.
23. *New York Times*, "A Career Full of Thrills," September 2, 1926, p. 21.
24. *New York Times*, "Pulitzer Race Won at 249-Mile Speed; Disappoints Fliers," October 13, 1925, p. 1.
25. William Wait, quoted in Rossano, "When Curtiss Racers Ruled the Roost," *Journal American Aviation Historical Society* (Summer 1985), 109–113.
26. Ibid., 112.
27. Paul Garber of the Smithsonian National Air Museum wrote to Williams in 1949 to inquire about the claimed 302 mph flight. Williams replied that he wanted to test the possibility of 300 mph over the electrically timed speed course and that he achieved it. He believed it was important to do so because "the medical profession and all the intelligentsia along with the common herd were solidly possessed of the opinion there were limits to the speeds at which men could travel," and he wanted to show them wrong. Williams also wrote that he achieved the 300 mph with "a slight glide angle," making no mention of a diving start (which is sometimes mentioned in connection with the 300 mph flight) (Letter from Alford J. Williams to Paul E. Garber, curator of the National Air Museum, April 27th, 1949, photocopy of the letter in National Air and Space Museum Archives).
28. *New York Times*, "Army Flyer Beats Pulitzer Records," September 21, 1925, p. 3.
29. Ibid.
30. The famous and many-times-reproduced photograph of Doolittle standing on the pontoon of a R3C-2 is a photo of the first R3C-2 after its repairs and not of the army's R3C-2 that Doolittle flew in the Schneider Race.
31. *New York Times*, "Pulitzer Air Racers Out for a New Speed Record," September 29, 1925, p. XXII.
32. Foxworth (*Speed Seekers*, 40) identifies Doolittle as the "swooping" pilot.
33. *New York Times*, "Night Fliers Show Fireworks to City," October 7, 1925, p. 12.
34. *New York Times*, "Airmen Will Cavort Over Rialto Tonight," October 6, 1925, p. 27.
35. *New York Times*, "Flier Dies in Crash as Thousands See Opening Day Races," October 9, 1925, pp. 1, 4.
36. "The New York Air Meet," *Flight*, October 29, 1925, p. 704.
37. C. Caldwell, "Seen Through the Bayonets," *Aviation*, October 19, 1925, p. 532.
38. "Flier Dies in Crash," 1.
39. The Air Service and navy, in 1925, required aircrew to wear parachutes, but concern for fliers' safety had not extended to the point that adequate seat belts were installed in airplanes such as Chamerblain's.
40. "The New York Air Meet," *Flight*, October 29, 1925, p. 704.
41. "Flier Dies in Crash," 1.
42. "High and Low Lights: New York Air Races," *Aviation*, October 29, 1925, p. 544.
43. Ibid.
44. "The Events of the New York Air Races," *Aviation*, October 19, 1925, pp. 533–543.
45. "High and Low Lights," 544.
46. Ibid.
47. Caldwell, "Seen Through the Bayonets," 532.
48. "Flier Dies in Crash," 1, 4.
49. Marshall, "The Dayton Air Race Meet," 19–22.
50. "Flier Dies in Crash," 4.
51. Ibid.
52. "City's Coldest Oct. 10 Brings 72-Mile Gale; Plane Races Put Off," headline in newspaper clipping from "1925 Events File," NASM. The name of the newspaper has been cut off.
53. *New York Times*, "65-Mile Gale Halts Pulitzer Air Event," October 11, p. 7.
54. "Gale Damages 10 Planes," headline in newspaper clipping from "1925 Events File," NASM. The name of the newspaper and date line have been cut off.
55. "City's Coldest Oct. 10 Brings 72-Mile Gale."
56. A.J. Williams, "'Bumps' Imperil Racing Planes; Danger in Proportion to Speed," clipping from "1925 Events File," NASM. The name of the newspaper has been cut off.
57. "65-Mile Gale Halts Pulitzer Air Event," 7
58. *New York Times*, "Speed Fliers Ready for Pulitzer Race," October 12, 1925, p.8.
59. *New York Times*, "Pulitzer Race Won at 249-Mile Speed; Disappoints Fliers," October 13, 1925, p. 1.
60. Foulois enlisted in the army, earned a commission and was one of the first three army officers to operate the first military airplane purchased from the Wright brothers. He commanded various air units in the South and the Southwest of the United States through the 1910s until he moved to Washington and was put in charge of all "American aeronautical material in the United States" (Official Website of the United States Air Force, http://www.af.mil/information/bios/bio.asp?bioID=5445). In November 1917, he was named chief of air service, American Expeditionary Force, and served in other high positions in France. After World War I, he continued in high-level staff

appointments and retired as chief of the Air Service in 1935.

61. "Pulitzer Race Won at 249-Mile Speed; Disappoints Fliers," 1.

62. "The Powell Light Plane," *Flight*, December 31, 1925, pp. 852–855.

63. William Faulkner, *Pylon*, quoted by William Caverlee in "Flyboy Faulkner," *Aviation History*, January 2011, pp. 19–20.

64. *New York Times*, "'Casey' Jones Wins Detroit Air Trophy," October 10, 1925, p. 4.

65. William Wait, quoted in Rossano, "When Curtiss Racers Ruled the Roost," 109–113.

66. Rossano, "When Curtiss Racers Ruled the Roost," 111–112.

67. Unlike the sometimes "acerbic" Williams, Doolittle "was always noticeably modest, unassuming, and patient.... [H]e possessed these qualities to such a degree that he endeared himself to almost everyone ... and generated deep loyalty" (Foxworth, 230). Doolittle had worked as a miner and had been an acrobat and boxer. He earned a doctorate in aeronautical engineering at MIT in two years. During World War II, he led the "Doolittle raid" on Tokyo in 1942 and later commanded the 8th Air Force in Europe. He has been described as the prototype of the modern test pilot: athletic, talented engineer, and especially able pilot. The title of his biography, *I Could Never Be So Lucky Again*, reflects his unassuming nature; a reviewer of the book said that Doolittle emerges as "a thoroughly decent human being whose relative unconcern for his military reputation is especially refreshing" (quoted at Amazon.com, http://www.amazon.com/Could-Never-Be-Lucky-Again/dp/0553584642).

68. M. Gough, "Doolittle Wins in Baltimore," *Airpower*, November 2005, pp. 44–56.

69. C. Bettis, quoted in *New York Times*, "Army Flyer Beats Pulitzer Records," September 21, 1925, p. 3.

70. "Pulitzer Race Won at 249-Mile Speed; Disappoints Fliers," 1.

71. Foxworth (*Speed Seekers*) identifies the airplanes flown by the 1st Pursuit Group and in the Mitchell and Pulitzer races as "Curtiss PW-8Bs." There was only one PW-8B, more accurately the XPW-8B, which was a modification of the PW-8A that had flown in the 1924 Pulitzer, and it was the prototype for the Curtiss P-1, which was entering service in 1925. The 1st Pursuit Group flew PW-8s in 1925.

72. "Mitchell Trophy Air Races at Selfridge Field," Mount Clemens Public Library Local History Sketches, http://www.libcoop.net/mountclemens/local%2history/mitchell%20trophy.pdf.

73. "Pulitzer Race Won at 249-Mile Speed; Disappoints Fliers," 1.

74. Ibid.

75. Foxworth, *Speed Seekers*, 231.

76. Rossano, "When Curtiss Racers Ruled the Roost," 109.

77. Foxworth, *Speed Seekers*, 232.

78. Williams, while on active duty as a navy pilot, completed law school at Georgetown University in 1925 and was admitted to the New York bar a year later. During the late 1920s, he was the navy's chief test pilot. He left the navy in 1930, and in 1933 became manager of the Gulf Oil Company's aviation division. He also accepted a commission in the Marine Corps Reserve. He wrote widely and made radio broadcasts about aviation during the 1930s. His criticisms of U.S. aviation policy before World War II annoyed navy and Marine officials, and he resigned his commission in 1940. Many people saw him fly aerobatic performances in the bright orange Gulfhawk (Grumman F3F) biplane (now in the Udvar-Hazy Museum of the National Air and Space Museum) until 1948, when he switched to a Grumman F8F. During World War II, he took leave from Gulf and demonstrated fighter techniques to Army Air Force cadets and pilots. He retired from Gulf in 1951 and died in 1957 at age 61.

79. William Wait, quoted in Rossano, "When Curtiss Racers Ruled the Roost," 112.

80. Promotions were slow in the 1920s Air Service. Dawson had won two Distinguished Service Crosses during World War I, and he was still a second lieutenant in 1925.

81. Foxworth, *Speed Seekers*, 41.

82. "Pulitzer Race Won at 249-Mile Speed; Disappoints Fliers," 1, 4.

83. One British visitor, designer H.P. Follard, was delighted with the slow speed. Before the race, C.R. Fairey had bet him "a fair amount of American currency" that the winning speed would be at least 255 mph ("The New York Air Meet," *Flight*, October 29, 1925, p. 704).

84. *Aircraft Year Book, 1926* (New York: Aeronautical Chamber of Commerce of America, 1926), 127–128.

85. "The New York Air Races," *Aviation*, October 29, 1925, p. 531. The editors of *Aviation* wrote that 5,000 people had seen the first day of the races, and that the "light plane races [held on the last day] were witnessed by less than half that number."

86. "The British Seaplane World's Record," *Flight*, October 15, 1925, p. 668.

87. So far as I am aware, there is no evidence that the freak first V-1400 was installed in Doolittle's plane as suggested in "The New York Air Meet" (*Flight*, October 29, 1925, p. 704). Smarting from losing the Pulitzer, the navy would have had no incentive to make such a switch.

88. "American Aviation," *Flight*, December 3, 1925, pp. 794–796.

89. "Pulitzer Race Won at 249-Mile Speed; Disappoints Fliers," 1.

90. *New York Times*, "Pulitzer Air Trophy to Be Perpetual Prize," August 29, 1920, p. 21.

Chapter Eight

1. *New York Times*, "No Pulitzer Race This Year," September 1, 1926, p. 3.

Chapter Nine

1. T. Huntington, "Racetracks in the Sky," AmericanHeritage, http://www.americanheritage.com/articles/magazine/it/2007/4/2007_4_42.shtml, reproduced from *American Heritage Invention and Technology Magazine*, Spring 2007.

2. W.J. Boyne, "P" for Pursuit: The Army's Biplane Hawks," *Airpower*, March 1976, pp. 26–35.

3. The Fédération Aéronautique Internationale [International Aeronautics Federation], which now calls itself the "World Air Sports Federation," was founded in 1905. "It is a non-governmental and non-profit making international organization with the basic aim of furthering aeronautical and astronautical activities worldwide, ratifying world and continental records and coordinating the organization of international competitions" (FAI, http://www.fai.org/about-fai).

4. *New York Times*, "Pulitzer Race Won at 249-Mile Speed; Disappoints Fliers," October 13, 1925, p. 1.

5. *New York Times*, "No Pulitzer Race This Year," September 1, 1926, p. 3.

6. "The Dangers of Public Speed Races," *Aviation*, September 21, 1925, p. 345.

7. *New York Times*, "Our Airplane Motors Invade Foreign Field," October 25, 1925, p. 5.

8. "The Indignant British," *Aviation*, December 14, 1925, p. 850.

9. Ibid.

10. "Our Airplane Motors Invade Foreign Field," 5.

11. "American Aviation," *Flight*, December 3, 1925, pp. 794–796.

12. T.G. Foxworth, *The Speed Seekers* (New York: Doubleday, 1974), 71.

13. Edward Eves, *The Schneider Trophy Story* (St. Paul MN: MBI, 2001), 224.

14. See M. Gough, "Doolittle Wins in Baltimore," *Airpower*, November 2005, pp. 44–56.

15. See Eves, *Schneider Trophy Story*, especially pp. 146–7, 152–3, and 178.

16. This arrangement is a bit reminiscent of the United States Army and United States Navy "loaning back" the racers they had purchased to the airplanes' manufacturers to race in the 1921 Pulitzer.

17. R. Barker, *The Schneider Trophy Races* (London: Chatto and Mundus, 1971); D. Mondey, *Schneider Trophy* (London: Robert Hale, 1974). On the World Wide Web, "The Schneider Trophy and Vintage Seaplanes," http://www.hydroretro.net/indexen.html, is especially useful and concise.

18. C.M. Keys, "New Planes for Pulitzer Trophy Race Indicate America's Strength in the Air," *Aviation*, October 5, 1925, pp. 433–436.

19. Ibid., 433.

20. Foxworth, *Speed Seekers*, 205.

21. Keys, "New Planes for Pulitzer Trophy Race Indicate America's Strength in the Air," 436.

22. R.M. Olien and D.D. Olien, *Easy Money: Oil Promoters and Investors in the Jazz Age* (Chapel Hill: University of North Carolina Press, 1990). S.E.J. Cox is mentioned throughout *Easy Money* and is the major subject of pp. 104–179.

23. Keys, "New Planes for Pulitzer Trophy Race Indicate America's Strength in the Air," 433.

24. Ibid.

25. "Pulitzer Race Won at 249-Mile Speed; Disappoints Fliers," October 13, 1925, p. 1.

26. *New York Times*, "No Pulitzer Race This Year," 3.

27. For instance, Frederick Lewis Allen's popular histories of the 1920s and '30s, *Only Yesterday* and *Since Yesterday*, have no mention of air racing.

Bibliography

Books

Adwers, R.E. *Rudder, Stick and Throttle: Research and Reminiscences on Flying in Nebraska.* Omaha: Making History, 1994.

Allen, F.W. *Only Yesterday.* 1931. New York: Perennial Library, 1964.

_____. *Since Yesterday.* 1940. New York: Perennial Library, 1968.

Angelucci, E. *The American Fighter.* With P. Bowers. New York: Orion, 1985.

Arnold, H.H. *Global Mission.* New York: Harper, 1949.

Barker, R. *The Schneider Trophy Races.* London: Chatto and Mundus, 1971.

Berliner, D. *Airplane Racing: A History, 1909–2008.* Jefferson, NC: McFarland, 2010.

Biographical Memoirs, Vol. 78. Washington: National Academy Press, 2000.

Bowers, P.M. *Curtiss Aircraft: 1907–1947.* London: Putnam, 1979.

Boyne, W.J. *The Best of Wings Magazine.* Dulles, VA: Brassy's, 2001.

Brown, S.L. *A Genesis Workshop: Five Generations of Engineering Enterprise from the Birthplace of Aviation.* Wright-Psatterson AFB, OH: History Office, Air Force Material Command, 2001.

Conover, R.C. *Dayton and Montgomery County: Resources and People.* New York: Lewis Historical, 1932.

Crouch, T. *The Eagle Aloft: 200 Years of Ballooning in the United States.* Washington: Smithsonian Institution Press, 1983.

Dade, G.C., and F. Strand. *Picture History of Aviation on Long Island, 1908–1938.* New York: Dover, 1989.

Deford, F. *The Old Ball Game.* New York: Atlantic Monthly Press, 2005.

Demetz, P. *The Air Show at Brescia, 1909.* New York: Farrar, Straus, and Giroux, 2002.

Doolittle, J.H. *I Could Never Be So Lucky Again.* With C.V. Glines. New York: Bantam, 1992.

Dumenil, L. *Modern Temper: American Culture and Society in the 1920s.* New York: Hill and Wang, 1995.

Eves, Edward. *The Schneider Trophy Story.* St. Paul, MN: MBI, 2001.

Foxworth, T.G. *The Speed Seekers.* New York: Doubleday, 1974.

Gwynn-Jones, T. *The Air Racers: Aviation's Golden Era, 1990–1936.* London: Pelham, 1983.

_____. *Farther and Faster: Aviation's Adventuring Years, 1909–1939.* Washington: Smithsonian, 1991.

Horgan, J.J. *City of Flight: The History of Aviation in St. Louis.* St. Louis: Patrice Press, 1990.

Kinert, R. *American Racing Planes and Historic Air Races.* Chicago: Wilcox and Follett, 1952.

Kordes, J.E. *Visions of Garden City.* Garden City, NY: John Ellis Kordes, 2007.

Kyvig, D.E. *Daily Life in the United States, 1920–1940.* Chicago: Ivan R. Dee, 2001.

Larsen, D.J., and L.J. Nigro. *Selfridge Field.* Charleston, SC: Arcadia, 2006.

Levitt, S.D., and S.J. Dubner. *Freakonomics: A Rogue Economist Explores the Hidden Side of Everything.* New York: William Morrow, 2005.

Magee, D.M., ed. *Centennial History of Mount Clemens, MI, 1879–1979.* Mount Clemens, MI: Mount Clemens Public Library, 1980.

Manufacturers Aircraft Association. *Aircraft Year Book* [1921 through 1926]. Various publishers.

Matthews, B. *Race with the Wind: How Air Racing Advanced Aviation.* Osceola, WI: MBI, 2001.

Maurer, M. *Aviation in the U.S. Army, 1919–1939.* Washington: Office of Air Force History, 1987.

McDougal, M.L. *Mount Clemens: Bath City, U.S.A., in Vintage Postcards.* Chicago: Arcadia, 2000.

McMurtry, L. *Rhino Ranch.* New York: Simon & Schuster, 2009.

Mets, D.R. *Master of Airpower: General Carl A. Spaatz.* Novato, CA: Presidio, 1988.

Miller, N. *New World Coming: The 1920s and the Making of Modern America.* New York: Scribner, 2003.

Mondey, D. *Schneider Trophy.* London: Robert Hale, 1974.

Morris, J.M. *Pulitzer: A Life in Politics, Print, and Power.* New York: Harper, 2010.

Olien, R.M., and D.D. Olien. *Easy Money: Oil Promoters and Investors in the Jazz Age.* Chapel Hill: University of North Carolina Press, 1990.

Robie, B. *For the Greatest Achievement.* Washington: Smithsonian Institution Press, 1993.

Rubenstein, M., and R. Goldman. *To Join with the*

Eagles: Curtiss-Wright Aircraft, 1903–1965. New York: Doubleday, 1974.
Shulman, S. *Unlocking the Sky: Glenn Hammond Curtiss and the Race to Invent the Airplane.* New York: HarperCollins, 2002.
Smith, M.H. *Garden City, Long Island, in Early Photographs, 1789–1919.* New York: Dover, 1998.
Villard, H.S. *Blue Ribbon of the Air: The Gordon Bennett Races.* Washington: Smithsonian, 1987.
Waller, D. *A Question of Loyalty: Gen. Billy Mitchell and the Court-Martial That Gripped the Nation.* New York: HarperCollins, 2004.
Wohl, R. *The Spectacle of Flight.* New Haven, CT: Yale University Press, 2005.
Zumwalt, T. *For the Love of Dayton.* Dayton: Dayton Daily News, 1966.

Journals and Magazines

American Aviation History Society Journal [*AAHS Journal*]
Aerial Age
Aerial Age Weekly
Aeronautical Digest
Aeroplane
Air Trails
Air University Review
Airpower
Airpower Historian
American Heritage
American Magazine of Aeronautics
Automotive Industries
Aviation
Aviation and Aircraft Journal
Aviation History
Bulletin, Missouri Historical Society
Curtiss Aeroplane and Motor Company
Economist
Flight
Flight International
National Aeronautic Association Review
Newsletter, Preserve Historic Forsyth, Winston-Salem, NC
Skyways
The Slipstream
St. Louis Seasons Magazine
Time
U.S. Air Service

Newspapers

Atlanta Constitution
Brooklyn Daily Eagle
Chester (PA) Times
Dayton Journal
Dayton News Camerica
Detroit News
Mt. Clemens (MI) Daily Leader
New London (CT) Day
New York Times
New York Tribune
New York World
Omaha Daily Bee
Omaha Daily News
Omaha Evening Sun
Omaha World-Herald
St. Louis Globe-Democrat
St. Louis Post-Dispatch

Index

Numbers in ***bold italics*** indicate illustrations.

Acosta, Bertrand Blanchard "Bert" 33, 51–52; duration of flight record 58, 17; "ladies' man" 574; 1920 Pulitzer ***29***, 30, 34–35, 45, 57, 80, 93, 195; 1921 Pulitzer 39, 54–56, 58, ***60***–61, 94–97, 99, 195; praise of 58–59; problem-plagued life 58, 59
Adwers, Robert E. 43, 57
Aerial Age magazine 13, 82, 98
Aerial Derby Around the World 14
Aerial Engineering Company 72, 77–81, ***79***, 93, 102, 121, 202; *see also* Bee-Line airplanes
Aéro Club de France 16, 20
Aero Club of America (ACA) 6, 8, 12–16, 18–20, 39–41, 45, 55, 57, 59–61, 74, 101, 107, 109, 112, 193; amalgamation with American Flying Club 12–15; National Air Association as ACA's successor 59–60
Aero Club of California 39
Aero Club of Omaha 40–41, 61
Aero Club of St. Louis 105, 107, 109–111
Aero Club of Texas 48
Aero Corporation 28, 45
Aero Digest Trophy Race, 1925 176, 180
Air Power magazine 13
Air Service *see* United States Air Service (USAS)
Aircraft Year Book (various years) 9, 64, 74, 117, 118, 154, 171, 197, 189
Albany (NY) to New York City flight, first 7
Allison engines 24
Alula wing 46–***47***, 54
American Flying Club 10, 12–16, 59
"The Ancient and Dishonorable Members of the Montmatre and Rue Brey Club" 14
Ansaldo airplanes: Balilla 28, ***29***; Fighters 20; 1920 Pulitzer 28, 32–35, 54, 195, 209; 1921 Pulitzer 45, 48, 55, 56, 195, 209; Not otherwise specified 21, 28, 32, 45; SVA-5 28; SVA-9 28, 45
Ansaldo Company 28; *see also* Ansaldo airplanes
Anti-Saloon League 11
Army *see* United States Army
Army Air Service *see* United States Air Service (USAS)
Army aviation appropriation 72, 146
Army Aviation Engineering Division 141–142; *see also* McCook Field

Atkinson, Jack 116
Atlantic S-3 airplane 153
Automobile industry 62, 72
Aviation Country Club of Detroit Trophy Race, 1922 63, 74
Aviation Country Club of Detroit Trophy Race, 1923 116, 118
Aviation Town and Country Club of Detroit Trophy Race, 1924 153, 176
Aviation Town and Country Club of Detroit Trophy Race, 1925 175, 189

"Baby" airplanes 156, 165
Baltimore, Maryland 4, 58, 144, 172–173, 181, 189, 197, 199, 201, 202
Barksdale, Lt. Eugene, USAS 173; 1922 Pulitzer 77, ***82***, 93, 99, 196; R-3 time trial 118
Beau, Lt. Lucas, USAS 121
Becker, Fred 175, 180
Beech, W.H. 153
Bee-Line airplanes (Aerial Engineering Company): Booth Racer, BR-1 77–78, ***79***, 80, 93, 210; Booth Racer BR-2 78, ***79***, 80, 93
Bellanca airplanes 58, 114; Bellanca Sedan (CF) 116, 118, 180; Wright-Bellanca WB-1 175, 180
Bertaud, Lloyd 55, 56, 195
Bettis, Lt. Cyrus, "Cy," USAS 82, 172; backup pilot 1925 Schneider 201; death 172; end of Pulitzers 206; 1924 Mitchell 154, 158, 162, 176; 1925 Pulitzer ***170***, 171–173, 176, 181–183, 185–187, ***187–188***, 188–191, 197–198, 201, 206
Bishop, Courtlandt Field 14
Bissell, Lt. Clayton, USAS 92; 1922 Pulitzer 77, 99, 196
Blackburn, Lt. T.W., USAS 117, 158
Blum, Julia ix
Boeing Company 26, 28, 45, 74, 178, 204; P-26 airplane 119, 204; PW-9 airplane 149, 204
Booth, Harry 78–80, 121
Booth Racer 121; *see also* Bee-Line airplanes
Boothman, Flight Lt. John, Royal Air Force 202
Borja, Elizabeth ix
Boston Globe 8
Boyne, Walter 193

231

Bradley, Lt. Benjamin, USMC: 1920 Pulitzer 25, *27*, 31, 32
Bragg, Caleb 39
Breguet, Jacques 178
Breguet Company 178
Breguet XIX airplanes 175, *178*
Brescia, Italy 5, 109
Bristol Cherub engine 176, 182
British Schneider (1925) Racing Team 173, 174
Brookley, Lt. Wendell H., USAS 147; 1924 Pulitzer 154, 158, 159, 161, 162, 196
Brooklyn Daily Eagle 20
Brow, Lt. Harold J., USN 145; 1922 Pulitzer 63, 77, *85*, 94, 96, 97, 145, 195; 1923 Pulitzer 117, 129–133, 196; world speed record 136, 137, 144, 145, 147
Brow-Williams competition for world speed record (R2C airplanes) 136–138
Brown, Charles E. ix
Brown, Squire ix

Cactus Kitten 49, *50*, 51, 61, 78, 204; 1921 Pulitzer 52, *53*, 54, 55, 143, 205, 209; 1922 Pulitzer 86, *87*, 195; unconfirmed absolute speed record, 1921 195
Caldwell, C.C., "Cy" 153, 174, 179
California Arrow dirigible 106
Calloway, Lt. Steven, USN 114; 1922 Pulitzer 77, 114, 123, 129–131, 133, 196
Camp Skeel *184*
Campisi, Dr. Judith x
Case, Charles D. 48
Central Labor Union Trophy Race, 1924 153, 175
Chase, Dr. Anne ix
City of Flight: The History of Flight in St. Louis 105
Clayton, Richard 105
Col. Henry J. Damm Airfield 19, 36
Cook, Capt. Harvey, USAS: 1925 Pulitzer 176, 186, 188, 197
Coombs, Clarence *50*, 52, 59; 1921 Pulitzer *53*, 55, 56, 58, 86, 195
Corkille, Lt. John, USAS 121; 1923 Pulitzer 128–131, 133, 196
Cox, Nelda 48, 49, 86
Cox, S.E.J. 48; contract with Curtiss for Gordon Bennett racers 48, 49, *50*–51, 121; critical importance to Curtiss and Pulitzer Races 204, 205; legal and financial problems and imprisonment 59, 86; 1921 Pulitzer 52, 53, 56, 59
Cox-Curtiss airplanes *see Cactus Kitten*; *Texas Wildcat*
CR: 1921 Pulitzer 54–57, 59, *60*, 61, 62, 78, 83, 84, *85*, 93, 94, 97, 121, 195, 199, 209
CR-2 145, 163; conversion to CR-3s *143*, 199; 1922 Pulitzer 63, 77, *85*, 86, 93, 94, 96–98, 144, 195, 210
CR-3: 1923 Schneider Trophy Race *143*, 144, 145, 163, 199
Cradle of Aviation Museum ix, 19, 169
Cribbs, Deborah E. ix
Cuddihy, Lt. George, USN: 1925 Pulitzer 186, 188, 201, 171
Culbertson, Lt. W.D., USMC 33
Cummings, Lt. C.M., USAS *12*, 63
Curtiss, Glenn H. 5, 21, 33, 40, 109; Albany-to-New York flight 7; Curtiss Marine Trophy Race 64, 67; importance of second Pulitzer race 38; rivalry and legal battles with Wright Brothers 47, 81, 132, 156; S.E.J. Cox 48–49, 50–51, 121, 204, 205; *Scientific American* trophy 19; *see also* Curtiss airplanes; Curtiss Company; Curtiss engines
Curtiss airplanes: Curtiss-built airplanes from other companies' designs 27, 28, 106; foreign praise 102; *see also Cactus Kitten*; CR; CR-2; CR-3; 18-B; 18-T; JN-4; Mitchell, Gen. William; Oriole; P-1; PW-8; PW-8A; PW-8B; R2C; R2C-2 seaplane; R3C; R3C-2 seaplane; Skeel, Capt. Burt; *Texas Wildcat*; XPW-8B
Curtiss Company (Curtiss Aeroplane and Motor Company) 19, 24, 33, 48, 50–55, 59, 77, 78, 80, 81–83, 85, 86, 91, 93, 101, 102, 119–121, 123, 124, 134, 142, 144–146, 149, 152, 168–170, 172, 180, 183, 192, 193, 198, 202–205; cessation of airplane production 28; Curtiss-Wright rivalry 47, 93, 97; factory *35*, *60*, 61, 172; government-Curtiss relationship 202–205; merger with Wright Company 28, 36, 202; *see also* Curtiss airplanes
Curtiss engines: C-6 153; C-12 48, 49, 52, 55, 82, 195, 209; CD-12 54–55, 59, 62, 65, 82, 85, 195, 209; D-12 81–83, 85, 96, 119, 121, 129, 148, 154, 169, 188, 195–200, 210–211; D-12A 121, 124, 129, 170, 196, 210; K-12 24, 45, 48, 54, 55, 65, 82, 195, 209; OX-5 153, 171, 192; OXX-6 153; V-1400 169–171, 186, 189, 197, 201, 211; V-1400, "freak" 170, 172, 186, 189
Curtiss Field, Long Island, NY 19, 51, 84, 85, 123, 173
Curtiss Marine Trophy Race, 1922 63–*65*, *66*, *67*, 88, 97, 130, 144, 194
Curtiss Trophy Race *see* National Air Race, 1925
Curtiss-Cox airplanes *see Cactus Kitten*; *Texas Wildcat*
Curtiss-Kirkham airplanes *see* 18-B; 18-T
Curtiss-Reed propellers 119–120, 121, 125, 149, 169, 170, 198, 203, 207
Curtiss-Wright Company 28, 204
Curtiss-Wright P-40 204

Dack, Jerry 176
Daniels, Josephus 16
Dawson, Lt. Leo, USAS: 1925 Pulitzer 176, 186, 188, 197
Dayton, Ohio 4, 7, 16, 107, 127, 136–166; "keep McCook" 141–142; in the 1920s and 2010s 138–141; *see also* National Air Race, 1924
Dayton Bicycle Club and Engineers' Club Trophy Race, 1924 154, 176
Dayton Chamber of Commerce Trophy Race, 1924 153
Dayton Daily News Light Airplane Contest, 1924 154
Dayton Daily News Trophy Race, 1925 176, 182, 189
Dayton-Wright Company 63; government contracts during World War I 141; 1920 Gordon Bennett racer 143
Deeds, Edward A. 141; *see also* Dayton-Wright Company
De Havilland DH-4 (airplane) *12*, 41, 138, 141; *Honeymoon Express* wins Aviation County Club of Detroit Race, 1922 63, 74; 1919 Transcontinental Race *11*, 192–193; 1920 Pulitzer *12*, 20, 21,

24, 29, 32, 36–37; postcard for 1925 Pulitzer 168, **169;** *see also* National Air Race, 1922; National Air Race, 1923; National Air Race, 1924; National Air Race, 1925
Delange, Gustave 87–88
Detroit, Michigan 4, 39, 65; in the 1920s and 2010s 62; *see also* National Air Race, 1922
Detroit Aviation Society 40–41, 62–64
Detroit Chamber of Commerce 62, 155
Detroit News 63, 74–76, 116, 154, 177
Detroit News Aerial Mail Trophy Race, 1922 63, 74
Detroit News Air Mail Trophy Race, 1923 116
Detroit News Air Mail Trophy Race, 1924 153–155; "decidedly out-of-date" airplanes 155
Detroit News Trophy Race, 1925 177
Dewey, Dawne ix
Donahue, Cynthia S. ix
Doolittle, Lt. James H., "Jimmy," USAS 138, 171, 183; 1925 Schneider 4, 171, 172, 189, 201, 202; seaplane speed record 201; testing of R3C 172, 173, 189
Dormoy, E. 153, 154, 176
Dormoy "flying bathtub" 153, 154, 176
Douglas airplanes: C-1 177; DT-4 airplane 116, 177; O2 178; World Cruisers (DWC) 152; XO-2 178
Driggs-Johnson "baby airplane" 153, 154, 156, **157,** 176

Eads Bridge, St. Louis 105
Eagle Flying Corporation 28
Eibe, Harry 57–60
18-B 24, 48
18-T 24, 48, 151; Curtiss Marine Trophy Race **65,** 66, 68, 69, **70,** 71, 72, 88, 97, 130; 1920 Pulitzer 24, **25,** 33, 151
Elliott, Lt. H.A., USN 63
Emrick, Clyde 175, 176
Étampes, France 16
Eves, Edward 199

FAI *see* Fédération Aéronautique Internationale
Fairey, C.R. 198–199
Fairey Fox airplane 189
Farquhar, Raymond 41
Faulkner, William 182
Fédération Aéronautique Internationale (FAI) 13–14
1st Pursuit Group 30, 33, 63–64, **73**–75, 117–118, 154, 158, 177, 179, 183–185
Flying Club of America *see* American Flying Club
Flying Club of Baltimore 144
Flying Club of St. Louis 113, 114, 116
Flying Club of St. Louis Trophy Race, 1923 116, 118
Flying magazine 13
Fokker airplanes: CO-4 116, 118, 154; D-VII 20, 26, 30; T-2 63, 74, 138, 155
Ford, Edsel 73–74
Ford, Henry 73, 156
Foxworth, Thomas G. 1, 32, 34, 43, 64, 77–78, 84, 86, 93, 97–99, 117, 120, 144–145, 157, 167, 177, 188, 197, 199, 211–212
Frank Lahm Cup *see* Lahm Cup
Fred Marshall Collection of Dayton Aviation History x, 142
French praise of United States aviation 103–104

Gallaudet airplanes 65–**66,** 68, 71
Garden City, New York 4, 16, 60, 61, 80, 84, 142, 172; A.S. Stewart 18–19; airfields 19, **35,** 37; *see also* Pulitzer Trophy Race, 1920; National Air Race, 1925
Gaukler Point, Lake St. Clair, Michigan 75, 94, **95,** 97
Gilmore, Jim 57
Gilmore, William 48, 82
Glenn H. Curtiss Trophy Race, 1925 175, 182
Glenn Martin Commercial-70 airplane 153
Gloster Aircraft Company 145, 202
Gordon Bennett Air Races 13, 16, 20; 1912 race 16; 1913 race 16; 1920 race 16, 17, 193
Gordon Bennett Balloon Races 106–**108,** 110–112
Gorton, Lt. W.A., USN 63, 69
Gough, Dr. Laura ix
Gould, Jay 7
Grumman airplanes 86, 172
Gulf Oil Company 86, 171

Hampton Roads, Virginia 201
Harmon, Lt. Norton, USMC: 1925 Pulitzer 176, 186, 188, 197
Hartney, Capt. Harold, USAS 30; crash in 1921 Pulitzer 57–58, 60; 1920 Pulitzer 30–32, 34, 37, 195; 1921 Pulitzer 47, 55–58, 60, 78; speed record attempt 35–36
Hartney, Mrs. Harold 57
Hartzell airplane 116, 153, 175
Hawley, Alan 13, 112
Hazelhurst Aviation Field, Long Island, NY 19, 33
Hazelhurst Aviation Field #2, Long Island, NY 19
Hearst, William Randolph 7
Hello Frisco (De Havilland DH-4 airplane) **11**
Hempstead Plains, Long Island, New York 18
Hempstead Plains Aerodrome 19
Henderson, Lt. George, USN 175, 178
Henderson (motorcycle) engine 153, 154
Hensley, Maj. William N., Jr, USAS 146
Hispano-Suiza ("Hisso") engines 21, 24, 30, 35, 47, 48, 54, 80, 86, 192
Honeymoon Express 63, 74
Honeywell, H.P. 107
Horgan, C.C. 105, 110–111, 114
Horlacher, Nancy R. x
Huffman Prairie (airfield), Dayton, Ohio 7, 139, 141
Hunsaker, Jerome 86–88, 115
Hunter, Capt. Frank O'D., USAS 92; 1922 Pulitzer 76–77, **92,** 99, 196
Hutton, Perry 116, 153

International Aero (or Air) Congress, Omaha, 1921 41–42, 62
International Aviation Federation 13
Irvine, Lt. Rutledge, USN 68, 144
Irwin, Bill 48, 100

James, Mansell R. 8, 12
JN-4 "Jenny" 32, 88, 142
John L. Mitchell Trophy Race, 1922 63–64, 74–76
John L. Mitchell Trophy Race, 1923 **115,** 116–118, 158
John L. Mitchell Trophy Race, 1924 149, 152, 154, 156–158, 162

John L. Mitchell Trophy Race, 1925 171, 172, 175–177, 182–185, 188, 190
Johnson, Lt. Fonda, USAS 118; 1922 Pulitzer 77, 93, 99, 193, 196
Johnson, James M. 153, 154, 156
Jones, C.S., "Casey" 63, 74, 116, 153, 175, *180*, 183
Journal-American, New York 7
Joy, Henry Bourne 72–73
Joy Aviation Field 72

Kafka, Franz 5
Kelly, Lt. O.G., USAS 138
Kerger, Brent, Dr. x
Kettering, Charles F. 141
Kiel, Lt. Emil, USAS 10, 11
Kinney, Dr. Jeremy ix
Kirkham, Charles 24, 48, 82
Kirkham-Curtiss Airplanes see 18-B; 18-T
Klemin, Alexander 156
Kline, Sgt. William, USAS 9, *11*, 12
Knabenshue, Roy 106
Korbel, Mario 7, *8*
Kordes, John E. ix

Lahm, Lt. Frank, USA 106, 108, 112; *see also* Lahm Cup
Lahm Cup 109, 112
Laird Commercial airplane 153
Laird Swallow airplane 116
Lake Listerine 107
Lake St. Clair 64, *67*–68, 72–75, 77, 88, 94, *95*, 96–97, 130
Lambert, Albert Bond 107, 109, 111, 113, 114, 164, 194, 196; ballooning 109
Lambert Field, St. Louis 103, 113–*115*, 136
Lamblin radiator 54, *60*, 78, *79*, 80–81, 83, *85*, *88*
Larsen, Deborah ix, 71
Laverents, Lt. A., USN 32, 35
Le Blanc (Leblanc), Alfred 110–111
Lecointe, Joseph Sadi 40, 80, 83, 87, 94, 145, 152, 154
Lees, Walter E. 116–117, 153
Leisenring, Rick ix
Lemaitre, Capt. Henri, French Air Force 175, 178
Liberty Engine Builders' Trophy Race, 1922 63, 68, 74, 93
Liberty Engine Builders' Trophy Race, 1923 116
Liberty Engine Builders' Trophy Race, 1924 153
Liberty Engine Builders' Trophy Race, 1925 175, 178, 179; criticism of the race 178; entry and victory of French airplanes 178–179
Liberty engines 29, 65–66, 117–118, 127, 153–155, 175, 178
Lindberg, Charles 3, 58, 152, 174, 192
Listerine 105, 109
Loening, Grover 25, 90, 99
Loening airplanes: M-8 20, 28, 33, 202; M-81 S Special (Loening Special) 20, 25, *26*, *27*, 31, 32, 202; 1922 Pulitzer 77, 89, 90, *91*, 93, 99, 102, 196, 210; R-4 76, 90, 91
Loening Company 25, 76, 80–81, 90–91, 292; *see also* Loening airplanes
London Morning Post 198
Lufberry Airfield 19

Macchi aircraft 199–201
Macchi Company 199; *see also* Macchi aircraft
MacReady, Capt. John, USAS first non-stop transcontinental flight 128; 1921 Pulitzer 46, 55, 56, 195; 1923 Pulitzer 155
Maitland, Lt. L.J., USAS 58, 84, 85, 121, *132*; 1922 Pulitzer 63, 77, *83*, 94–96, 101, 195; world speed record 101, 120
Maitland-Maughan competition 120
Marshall, Fred 142, 143, 146, 164–167, 179
Martin Bomber 63, 74, 116, 121, 137, 153, 181
Martin Transport 63, 74
Matthews, Lt. T.K., "Thomas," "Tom," USAS 117, 154, 156, 175, 185, 190
Maughan, Lt. R.I., USAS 84, 85, 123; first "dawn-to-dusk" transcontinental flight 138, 151; 1922 Pulitzer 63, 76, 79, 94–97, *98*, 99, 133, 161, 195; 1925 National Air Race, 186; world speed record 100, 101, 120; *see also* Maitland-Maughan comptetition
Maynard, Lt. Belvin, USAS 12; death 11; officiates at first aerial wedding 11; resigns from Air Service, 11; role in 1920 Pulitzer 11, 36–37; wins Transcontinental Race 9, *10*, *11*, 12–13
Maynard, Mrs. Belvin 11
Maynard Field 11, 12
Mayo, Maj. R.H., Royal Air Force 146
Mayor of Detroit (1922) 73–74
MB-3 26, *27*, 32, 45–47; 1920 Pulitzer 20, 30, 32, 34, 37, 45, 46, 72, 78, 193, 195, 204, 209
MB-3A (Boeing-built) 45; 1922 Mitchell 63, 74, *75*, 76; 1923 Mitchell 117, 118
MB-6 45, *46*, 48, 54, 204; name change to R-2 76, 209; 1922 Pulitzer 55, 56, 61, 195
MB-7 46, *47*, 54, 78, 204; 1921 Pulitzer 47, 48, 54–57, 209; 1922 Pulitzer 77, 92, 93, 204, 210
McCook Field 21, 46, 56, 118, 139, 141, 142, 165; air-cooled engines 70; Barling bomber 121, *127*, 155, 194; central role in 1920s aviation 89; costs and criticisms 89, 119; propellers 81, 148–149, 159–161; R-3s 118, 119, 148; R-6s 120; R-8 147, 148; VCP/VCP-R/R-1 21, 36, 89; *see also* National Air Race, 1924
McMurty, Larry 107
Menoher, Gen. C.T., USAS 9
Merchants Association of New York Race, 1925 175
Merchants Exchange of St. Louis Trophy Race, 1923 116
Michigan Air National Guard 72
Miller, Lt. Walter, USAS: 1923 Pulitzer 121, 129, 131, 133, 196
Mills, Lt. Harry, USAS: 1923 Pulitzer 121; 1924 Pulitzer 148, 154, 158, 159, 161, 162, 176, 196
Mitchel, John Purroy 19
Mitchel Field 5, 6, 15, 18, 19, 30, *35*, 37, 78, 114, 136, 137, 145, 167, 188, *170*, 172–174, 175, 177, 179, 181, *183*, *184*, 186, *187*, *188*, 193–195, 197, 201, 206
Mitchell, Gen. William, "Billy," USAS 9, 15, 73, 92, 103–104, 117, 142, 152, 163, 203; court-martial 160–161; design of R-3 80–81; design of R-6 81–82, design of VCP-R 22, 80; fatal crash 147, 161; John L. Mitchell Race 63–64; 1920 Pulitzer 22, 34–38; 1922 Mitchell 74; 1922 Pulitzer 63, 76, 77, 81, 82, *83*, *84*, 93, 94, 96, *98*, 99, 100, 144, 195, 210; 1923 Mitchell 117; 1923 Pulitzer 118, 121,

124, 128–131, 133, 196, 210; 1924 Pulitzer 120, 147, 151, 154, 158, 159, 162, 196, 210; record flights 101, 101, 118, 120; world speed records 98, 100–101, 118, 141, **147**, 148, 149, 194, 195
Mitscher, Cmd. Marc, "Pete," USN 86, 134
Moffett, Adm. W.A., USN 70, 73, 134, 136–137, 147, 164, 174, 179, 186
Morane-Saulnier AR airplane 20, 21, 28
Morane-Saulnier Company airfield 51
Moseley, Lt. Corliss, USAS 24, 39; 1920 Pulitzer 31, 33, **34**, 35–37, 195; 1922 Pulitzer 36, 46, 77, 89, 99, 196; post-1920 Pulitzer Race attempt at world speed record 36
Mt. Clemens, Michigan 4, 16, 64, **95**; "baths, booze, and babes" 71; General Mitchell sets world speed record 100–102; Joy Aviation and Selfridge Field 71–73; National Air Race 73–100
Mt. Clemens *Daily-Leader* 157, 161
Mulcahy, Capt. Francis P., USMC 77, 78, 83
"Mystery plane," British *see* Supermarine S-4
"Mystery plane" or "mystery ship" *see* Wright NW-1

NAA *see* National Aeronautic Association
NAF *see* Naval Aircraft Factory
National Advisory Committee for Aeronautics (NACA) 119
National Aeronautic Association (NAA) 8, 13, 30, 60, 61, 101, 102, 113, 115, 119, 136, 142–146, 151, 152, 164–168, 171, 174, 189, 194, 197, 198, 205; establishment of 101; goals of 101–102
National Aeronautic Association Review 151, 167, 168, 191, 194
National Aeroplane Fund 14
National Air Association 59–60
National Air Race, 1922 62–102; events and results 63; *see also* Aviation Country Club of Detroit Trophy Race, 1922; Curtiss Marine Trophy Race, 1922; *Detroit News* Aerial Mail Trophy Race, 1922; John L. Mitchell Trophy Race, 1922; Liberty Engine Builders' Trophy Race, 1922; Pulitzer Trophy Race, 1922
National Air Race, 1923 103–136; Barling bomber 121, **127**; best of the National Air Races and of the Pulitzers 194; events and results 116–117; weather-forced delay of start 125–126, 134; *see also* Aviation Country Club of Detroit Trophy Race, 1923; *Detroit News* Air Mail Trophy Race, 1923; Flying Club of St. Louis Trophy Race, 1923; John L. Mitchell Trophy Race, 1923; Liberty Engine Builders' Trophy Race, 1923; Merchants Exchange of St. Louis Trophy Race, 1923; Pulitzer Trophy Race, 1923; St. Louis, Missouri
National Air Race, 1924 136–166; Barling bomber 138, 155, 194; disappointing crowds and performances 152, 155, 163–167, 194; events and results 153–154; *see also* Aviation Town and Country Club of Detroit Trophy Race, 1924; Central Labor Union Trophy Race, 1924; Dayton, Ohio; Dayton Bicycle Club and Engineers' Club Trophy Race, 1924; Dayton Chamber of Commerce Trophy Race, 1924; *Dayton Daily News* Light Airplane Contest, 1924; *Detroit News* Air Mail Trophy Race, 1924; John L. Mitchell Trophy Race, 1924; Liberty Engine Builders' Trophy Race, 1924; National Cash Register Company Trophy Race, 1924; Pulitzer Trophy Race, 1924
National Air Race, 1925 167–190; disappointing attendance 179, 182; events and results 175–177; over-all disappointment 197–198; selection of Mitchel Field 168; terrible weather forces rescheduling 180–182, 197; *see also Aero Digest* Trophy Race, 1925; Aviation Town and Country Club of Detroit Trophy Race, 1925; *Dayton Daily News* Trophy Race, 1925; *Detroit News* Trophy Race, 1925; Garden City, New York; Glenn H. Curtiss Trophy Race, 1925; John L. Mitchell Trophy Race, 1925; Liberty Engine Builders' Trophy Race, 1925; Merchants Association of New York Race, 1925; Pulitzer Trophy Race, 1925; *Scientific American* Trophy Race, 1925
National Cash Register Company (NRC) 140–141, 146
National Cash Register Company Trophy Race, 1924 153, 175
National Museum of the United States Air Force 127, 139, 159
Naval Aircraft Factory (NAF) seaplanes: TR-1 63, 69, 70; TR-2 69; TS-1 69, 70; TS-2 69
Navy *see* United States Navy
Nelson, Lt. Eric. H., USAS 63, 186
New Swallow airplane 153
New York Evening Post 14
New York Giants 86, 136–137
New York Herald 8
New York Journal 7
New York Times 7, 9, 16, 20, 35, 36, 58, 64, 85, 94, 96, 136–138, 145, 152, 154–156, 159, 168–174, 177, 182, 186, 191, 198, 206; evaluation of 1923 United States' aviation successes 138
New York-to-Toronto Air Race, 1919 13
New York World 7, 39, 40, 43
Nieuport-Delange sesquiplane 87
Nieuport 29V 53–54
Nigro, Louis ix
Niklas, Brian ix
"1910 International Tournament" *see* St. Louis
Nominal horsepower 21, 49, 54, 56, 129, 156, 197, 211
North Field, Omaha, Nebraska 39, 42–**44**, 45, 55
Nutt, Arthur 26, 57, 82, 169–170

Ogilive, Alexander 110
O'isy, Capt. Pelletier, French Air Force 178
Omaha, Nebraska 4, 16, 38, 40; airplanes in 1921 Pulitzer 45–51; death of parachutist 57–60; early aviation in 40, 41; formation of National Air Association 59–60; Hartney crash 56, 57; makes 1921 Pulitzer possible 39–46 Pulitzer Race 55–61
Omaha Bee 58
Omaha Daily Bee 55, 57
Omaha World Herald 42, 43
Opt, Jeff x
Ordnance D airplane 20 *see* Orenco D airplane
Orenco Company 27, 28
Orenco D airplane 26–28, 32, 33, 35, 99
Oriole 48, 63, 74, 116, 153, 175, **180**

P-1 **150**, **184**; 1925 Pulitzer 186, 188, 197, 211
Packard Airplane Engines 21–24, 38, 89, 90, 92; Packard engine 1A-2025 209, 210

Packard-Lepere airplane 63
Packard Motor Company 72
Patrick, Gen. Mason, USAS 103, 105, 155, 162, 164; 1922 Pulitzer 73, 98; 1923 speed record 101; 1924 Pulitzer 114, *132*, 134, 135, 163; 1925 Pulitzer 168, 187, ***188***
Patterson, Frederick B. 142–143, 145–146
Patterson, John H. 140, 142
Paustenbach, Dennis, Dr. x
Pavuk, Marian, D. x
Pearson, Lt. Alexander, Jr., USAS 119, 120, 148, 167; death in R-8 147, 159, 161, 162, 167; 1923 Pulitzer 129, 130; world speed record for 500 km 148
Pershing, Gen. John J., USA 16
Porter, Earl W. 40–43, 60, 62
Porter, Finley 48, 54, 82
Post-1920 Pulitzer Race attempts at world's speed record 35
Powell, C.H. 182
Powell Racer 176, 182, ***183***, 187
Presidio, San Francisco, California 9
Professional questions of value of air racing 167, 169, 205
Prohibition 71, 73, 164
Pulitzer, Herbert 7
Pulitzer, Joseph, Jr. 7, 112
Pulitzer, Joseph, Sr. 6, 104; and *New York World* 7; and *St. Louis Post* 7; and *St. Louis Post-Dispatch* 7; sponsors prize for first Albany (NY) to New York City flight 7
Pulitzer, Ralph 6, 7, 15
Pulitzer Family 8, 12
Pulitzer Race-Schneider Race comparison 199, 205–206
Pulitzer Sons 193
Pulitzer Transcontinental Race (planned, never flown) 6, 8
Pulitzer Trophy Race, 1920 18–38; entries 18–*22*, *23*, 24, *25*, *26*, *27*, 28, *29*, 30; judged a success 37–38; mis-measurement of course and revision of speeds 36–*37*, 193; organization 18–21; race 30–*34*, *35*, 36
Pulitzer Trophy Race, 1921 39–61; Acosta and Curtiss airplane victory 65–66, 58–59, ***60***, 61; death of parachutist 57–60; Detroit plans to host race but pulls out 39–40; Hartney crash 56, 57; Omaha steps up to host race 40–42; preparation for race 43, *44*, *46*, *47*, *49*, *50*, *52*, *53*, 54, 55; race 55–***60***, 61; unofficial world closed-course speed record 55, 195
Pulitzer Trophy Race, 1922 62–64, 68, 69, 74–78, ***79***, 80–81, ***82***, ***83***, 84, ***85***, 86, ***88***, 90, ***91***, ***92***, 93, ***95***, 96, 97, 98, 99, 100, 102; entries and performances 77; establishment of NAA 101; goals of NAA 191–202; Maughan attempt at world speed record 100; Mitchell's world speed record 100–102; Sanderson crash 97; world record closed-course speed 98; *see also* Detroit, Michigan; Mt. Clemens, Michigan
Pulitzer Trophy Race, 1923 103, 104, 113, 114, ***115***, 116–122, ***122***, ***123***, 124, 128–***132***; world record closed-course speed 131
Pulitzer Trophy Race, 1924 142–166; Capt. Skeel's death, 1924 158, 159, ***160***, 161–164, 167, 195
Pulitzer Trophy Race, 1925 167–169; advertising 168–169; 173; delivery and testing of R3Cs 172–173; "Heat B" for service pursuits 171; two-airplane race 185–186; world closed course record 185, 188–190; 197
Pulitzer Trophy Race, accomplishments 192, 194, 198, 206–207; disappointing attendance 1924, 1925 205; deaths in 1919 transcontinental race and changing Pulitzer to closed-course race 9, ***10***, ***11***, ***12***, 13–15; establishment of 5–17, ***8***, 64, 74; Joseph Pulitzer's three sons plan annual transcontinental race to honor their father 6–8; plans for first Pulitzer 15–17; summary of Pulitzer Race results 193–199; *see also* Aero Club of America; National Air Race 1922, 1923, 1924, 1925; Pulitzer Trophy Race 1920, 1921
PW-8 138, 149, ***150***, 155, 173, ***184***, 207; foreign praise of 150–1, 189; 1924 Mitchell 154, 157, 158, 162; 1925 Mitchell 176, 185; 1925 Pulitzer 176, 182, 186–189, 197, 211, 212; similar to R-6 racer 104, 118
PW-8A (also PW-8B, XP8-A and XPW-8B) 149; second modification ***150***, 197, 211
PW-8B (also XPW-8A, second modification, and XPW-8B) 149, ***150***, 154, 158, 159, 171, 197, 210, 211
Pylon 182

R2C (also R2C-1) ***124***, 125, 148, 152, 169–171; conversion to seaplane 146; 1923 Pulitzer 117, ***122***, 128–131, ***132***, 133, 159, 172, 196, 210; short racing career 147; transfer of one to USAS 146, ***147***, 162; world speed records 137, 137, 194
R2C-2 seaplane 146, 147, 170; fatal crash 147
R3C (also R3C-1) 169, ***170***, 171–173, 198, 202; 1925 Pulitzer 176, 182, 185, 186, ***187***, ***188***, 189, 206, 211; record closed-course 189, 197
R3C-2 seaplane 169, 170, 202; 1925 Schneider 186, 189, 197, 201; 1926 Schneider 201; world-record speed 189
R-5 92, 93, 204; criticisms of 92–93, 99; 1922 Pulitzer 76, 77, 89, ***92***, 93, 99, 102, 196, 210
R-6 83, ***84***, 85, 92, 104, 118 125, 138, 148, 158; attempts at speed record 120
Ray, J.C. 153
Reed, Dr. Sylvanus Albert 82, 83, 149
Richter, Lt. J.P., USAS 138
Rittenhouse, Lt. David, USN 80, 144
Rodgers, Cal 7
Rohlfs, Roland 49, ***50***–51, 53
Rolls-Royce engines 24, 199; Griffon 200; Merlin 200, 206; R 206
Roos, Frederick W., Dr. Boeing Company (St. Louis) ix
Roosevelt, Pres. Theodore 111
Roosevelt, Quentin 19
Roosevelt, Theodore Jr. 162
Roosevelt Field, Long Island, New York 9, ***10***, 11, 19, 49, 58
Rosenberg, Gary R. ix
Rowe, B.I. (Basil) 153, 175
Royal Aero Club 142, 189
Royal Air Force 9, 146
Royal Aircraft Factory S.E. 5 29, 30

St. Louis, Missouri 4, 7, 16, 40, 68; ballooning 105–108; first airplane flight west of Dayton

109–110; great success of 1923 National Air Race 134–135; 1907 airplane race 108; 1910 international tournament, air meet 110–112; 1923 Mitchell Race 117–118; 1923 National Air Race 113–135, *115*, *121*, *123*, *126–128*; 1923 Pulitzer Race 118–*132*, *133*, 134; 1924 speed trials of R6s 120–121; rain delay of National Air Race 125–126; Veiled Prophet 125–126
St. Louis Aero Club *see* Aero Club of St. Louis
St. Louis Post-Dispatch 7
St. Louis World's Fair, 1904 105–106
Sanderson, Lt. Lawson, USMC 16, 76, 123, 134; crashed or forced landing 130; ditching in Lake St. Clair 97, 130; 1922 Curtiss Marine Trophy Race *65*, 68, 69; 1922 Pulitzer 68–69, 76, 77, *86*, 97, 117, 124, 128–130, 133, 196; 1923 Pulitzer *123*, 128–130, 132–134, 196
Santos-Dumont, Alberto 106
Schilt, Lt. C.F., USMC 177, 201
Schneider Trophy Races 1, 3, 4, 86, 144, 146, 147, 149, 197, 199, 201–202, 205–207; 1922 Schneider 70; 1923 Schneider 143, 143, 144, 149, 163, 197, 162, 199, 200, 206; 1924 Schneider (cancelled) 144, 145, 200–202; 1925 Schneider 145, 169, 171, 173, 183, 186, 189, 199–202, 205; 1926 Schneider 201; 1931 Schneider 202, 206; *see also* Pulitzer Race-Schneider Race Comparison
Schriver, Mayor Harry M. 9, *10*
Schroeder, Maj. Rudolf, "Shorty," USAS 22–24, 31, 33; comments on all-military Pulitzer 171
Schulgen, Lt. George, USAS 176, 185
Schulze, Lt. Lester, USAS 90; 1922 Pulitzer 77, 91, *99*, 196
Scientific American Trophy 19
Scientific American Trophy Race, 1925 175, 182
Scott, Dr. Paul x
Second Pan-American Aeronautic Convention, 1919 8
Selfridge, Lt. Thomas, USA 72
Selfridge Field, Mt. Clemens, Michigan 16, 62–64, 71–*73*, 76, 77, 93, 94, *95*, 100, 157–159, 161, *184*, 194, 195
Shenandoah dirigible 126; *see also* ZR-1 dirgible
Skeel, Capt. Burt, USAS 117, 147, 154, 158; fatal crash 158–159, *160*, 161–164, 167, 195; 1924 Pulitzer 158, 211
Skeel, Burt Jr. 159
Skeel, Mrs. Burt 159
Skin radiators *see* wing radiators
The Slipstream 142, 144, 146, 151, 164, 166, 167
Smith, Lt. D.C., USAS 116
Smith, Lt. L.H., USAS 116, 138
Sopwith Camel airplane 8
Sopwith Dolphin airplane 20
SPA 6A airplane engine 195, 209
Spaatz, Maj. Carl A., "Tooey," USAS 73
SPAD S. 13 airplane 20
Sperry Company 76, 77, 80, 81, 89, 93, 118, 119
Sperry Messenger airplane 155, 156
Sperry-Verville airplane *see* Verville-Sperry airplane
Spitfire 3, 189, 201, 206
Stace, Lt. D.F., USAS: 1921 Mitchell 63, *75*, 76, 117; 1924 Mitchell 154, 158
Stewart, Alexander Turney 18

Stewart, Mrs. Alexander Turney 19
Stoff, Joshua ix
Stolle, Brett E. ix
Stoner, Lt. Rex USAS: 1924 Pulitzer 154, 158, 159, 162, 197
Street, Lt. St. Claire USAS 32, 35, 76, 77, 99, 118
Sunday, Billy 9, *10*
Sunday, Mrs. Billy 9, *10*
Supermarine Company 202; Racers 206; S-4 airplane 189, 201; S-5 airplane 202; S-6 airplane 202; S-6B airplane 202, 206
Surface radiators *see* wing radiators

Taylor, Dr. Henry 71
Texas Wildcat **49**, *50*, 56, 76, 121, 143, 204; crash before 1920 Gordon Bennett 51, *52*, 53, 54, 86
Thomas, Douglas Benjamin 30, 46
Thomas Morse airplanes *see* MB-3; MB-3A; MB-6; MB-7; R-5
Thomas Morse Company 26, 28, 45–46, 54, 76–78, 80–81, 92, 292, 204; *see also* MB-3; MB-3A; MB-6; MB-7; R-5
Thurston, Arthur, "Mike" 48, 51, 78–80, 121
Tourtellot, Lt. George, USAS 117
Transcontinental Race, 1919 3, 6, 11, 8–*10*, *11*, *12*, 13, 30, 36, 192
Transcontinental Reliability and Endurance Test, 1919 8; *see also* Transcontinental Race, 1919
"Trixie" 9, *11*
Turner, Roscoe 152

Union Army 6
United States Air Force (USAF) 8, 92, 139, 197; Gen. Mitchell's failed effort to establish 72
United States Army 8, 9, 11, 161, 162, 193; evaluation of Pulitzer Races 198–207; 1919 "Transcontinental Race" 13–14; "Round the World Flight" (1924) 152; *see also* National Air Race, 1922; National Air Race, 1923; National Air Race, 1924; National Air Race, 1925; Pulitzer Race, 1920; Pulitzer Race, 1921; Schneider Race, 1924; Schneider Race, 1925
United States Army Air Force (AAF) 8, 72, 86, 197
United States Army Air Service (USAS) 5, 8, 11–12; evaluation of Pulitzer Races 198–207; *see also* National Air Race, 1922; National Air Race, 1923; National Air Race, 1924; National Air Race, 1925; Pulitzer Race, 1920; Pulitzer Race, 1921; Schneider Race, 1924; Schneider Race, 1925; United States Army Air Service Racers
United States Army Air Service Racers: R-2 76, 195, 209; *see also* Curtiss R2C; Loening R-4; Thomas Morse MB-6; Thomas Morse R-5; Verville R-1; Verville-Sperry R-3
United States Army Balloon School 40
United States Marine Corps 78, 197
United States Navy 9; evaluation of Pulitzer Races 198–207; *see also* National Air Race, 1922; National Air Race, 1923; National Air Race, 1924; National Air Race, 1925; Pulitzer Race, 1920; Pulitzer Race, 1921; Schneider Race, 1924; Schneider Race, 1925
USAS *see* United States Army Air Service
USMC *see* United States Marine Corps
USN *see* United States Navy

Veiled Prophet *see* St. Louis
Verville, Alfred "Fred" 21, 22, 38, 77, 80, 81, 90, 93, 99, 119, 120, 148, 152; criticial of Army's development of R-3 119
Verville (Army-built) airplanes: R-1 *34*, 36, 76, 77, 89, 93, 99, 119, 196, 210; VCP 21, *22*, *38*; VCP-R 21, *23*, 24–26, 31–*34*, *35*, 36, 38, 41, 45, 195; speed attempts 38; U.S. speed record 195
Verville-Sperry R-3 airplane 76, 80, 118, 156; comparative costs of VCP-R and R-3 84; development and tests 118, 189; 1922 Pulitzer 76, 77, 81, *82*, 89, 90, 93, 99, 102, 118, 196, 210; 1923 Pulitzer 118, 120, 129, 155, 156; 1924 Pulitzer 99, 138, 147, *148*, 151, 152, 154, 158, 159, *161*, 176, 196, 202, 210
Vin Fizz (airplane) 7
Vought airplanes: UO-1 16, 118; V.E. 7 20, 29, 35, 63, 70, 74

Wait, William 48, 81–82, 84, 91–92, 94, 97, 100, 172, 183, 186
Waldon, Sidney D. 40
War Department 41, 72, 142, 178
Washington Conference on the Limitation of Armaments 80
"Whistling Billy" 49
Whitehead, Lt. Ennis, USAS 90, 91; 1922 Pulitzer 77, 99, 196
Wilcox, Harold V. 76
Williams, Ens./Lt. Alford, "Al," USN 59; behavior 183, 187; development of airplanes for post–1925 Schneiders 201; end of Pulitzers 206; 1922 Pulitzer 76, 77, 85, 86, *87*, 94, 96, 97, 144, 145, 1925; 1923 Pulitzer 117, *124*, 128–131, *132*, *133*, 154, 189, 190, 194, 196, 198; 1925 Pulitzer 148, 166, 171, 172, 176, 181, 185–187, *188*, 189, 197, 198; world speed record 136–138, 147, 162

Williams-Brow competition for world speed record 136–138; *see also* Brow-Williams competition for world speed record
Wing radiators 78–81, 83, *85*, 119, 122, 124, 145, 148–149, *150*, 151, 158, 162, 173, 199
Wise, John 105, 108, 112; record-distance balloon flight, 1859 105, 108
Woodhouse, Henry 13, 14, 74
Wright, Katherine 132, *133*, 155
Wright, Orville 72, 73, 81, 108, *123*, *124*, 132, 141, 155, 156
Wright, Wilbur 81, 108, 139, 141
Wright Airplane Engines 21, 22, 80; E-3 153; E-4 153; H-2 195, 209, 210; H-3 78, 80, 81, 93, 117, 196, 210; Tornado T-2 86, 210; Tornado T-3 121, 196, 210; Whirlwind 70
Wright airplanes: F2W-1 *122*, *123*, 124, *128*, 129–131, 133, 136, 196, 210; 1910 biplane at 1923 National Air Race 155, 156; NW-1 77, 81, 86, *88*, 89, 93, 94, 99, 102, 115, 117, 123, 210
Wright Brothers (Orville and Wilbur) 7, 13, 19, 47, 108, 111, 119, 121–124, 131, 132, 139, 142, 206; rivalry and legal battles with Curtiss 47, 81, 132, 156
Wright Company (Wright Brothers Aeroplane Company) 24, 47, 48, 56, 80, 86, 88, 93, 97, 119, 124, 165, 192, 202, 204; *see also* Wright Airplane Engines; Wright airplanes
Wright Field, Dayton, Ohio 141, 142, 151, 152, 156, 158, 163, 165, 194, 196
Wright-Patterson Air Force Base 139, 142

XPW-8B *150*

Yackey Sport airplane 153

ZR-1 dirigible *126*, 127, 134, 194